CLIMATE CHANGE MITIGATION: AN INDIAN PERSPECTIVE

CLIMATE CHANGE MITIGATION: AN INDIAN PERSPECTIVE

Dr. Ashish Kumar Srivastava and Ms. Sneha Kumari

Delve PUBLISHING

www.delvepublishing.com

Climate Change Mitigation: An Indian Perspective

Dr. Ashish Kumar Srivastava and Ms. Sneha Kumari

Delve Publishing
224 Shoreacres Road
Burlington, ON L7L 2H2
Canada
www.delvepublishing.com
Email: orders@arclereducation.com

ISBN: 978-1-77469-924-9 (Hardcover)

Delve Publishing publishes wide variety of books and eBooks. For more information about Delve Publishing and its products, visit our website at www.delvepublishing.com.

ABOUT THE AUTHORS

Dr. Ashish Kumar Srivastava is an Assistant Professor at the Faculty of Law, University of Lucknow. An LL.M. gold medalist from Law School, Banaras Hindu University, he did his Ph.D. from National Law University, Jodhpur. Dr. Srivastava started his career in National Law University, Jodhpur, in 2008. He has several publications to his credit in reputed journals. He delivers lectures in Indian Railways Institute of Transport Management, Hindustan Aeronautics Limited, Human Resource Development Centre, University of Lucknow, Public Sector Undertakings, Allahabad Bank, Policy Research Institutes, Uttar Pradesh Academy of Administration & Management and various reputed organisations. Dr. Srivastava is the author Property Laws and Law of Partnership & Limited Liability Partnership, Corporate Governance: Emerging Dimensions.

Dr. Srivastava can be contacted at asrinlu@gmail.com, srivastava_ashisk@lkouniv.ac.in.

Ms. Sneha Kumari graduated in Law from Law School, Banaras Hindu University. She did her masters from Hidyatullah National Law University, Raipur, Chhattisgrah, India. She has been recently awarded degree of Ph.D. on Climate Change Mitigation. She has published many research papers in reputed Indian journals. She has keen interests in environmental law, climate change mitigation etc.

Ms. Sneha Kumari can be contacted at singh.sneharox@gmail.com.

Dedicated To

'Mother Earth'

and

"Bharat Ratna Mahamana Pandit Madan Mohan Malviya"

TABLE OF CONTENTS

LIST OF CASES

- A.R.C Cement Ltd v. State Of U.P (1993) supp. (1) SCC 57
- Abhimanyu Rathor v. State of H. P (2014) SCC OnLine HP 5202
- Bombay Dyeing and Manufacture Co Ltd v. Bombay Environmental Action-Group (2006) 3 SCC 434
- Bombay Dyeing and Mfg. Co. Ltd. v. Bombay Environmental Action Group and Ors (2006) 3 SCC 434
- Charanlal Sahu v. Union of India (1990) 1 SCC 613.
- Court of Its Own Motion v. State of H. P (2014) SCC On Line NGT
- Damodar Rao v. S.O. Municipal Corporation, Hyderabad
- G Sunderrajan v. Union of India (2013) 6 SCC 620
- Goa Foundation v. Goa State Coastal Zone Management MANU/MH/0964/2000
- Himachal Pradesh v. Ganesh Wood Products AIR 1996 SC 149
- Home Power Ltd vs. DCIT (151 TTJ 616)
- In Karnataka Industrial Areas Development Board v. Sri C. Kenchappa & Ors (2006) 6 SCC 371
- Indian Council for Enviro-Legal Action v. Union of India (1996) 3 SCC 212
- Karnataka Industrial Area Development Board v. C Kenchppa (2006) 6 SCC 371.
- Kinkry Devi v. State A.I.R. 1988 H.P. 4
- M.C Mehta V. Kamal Nath1997 1 SCC 38
- M.C Mehta v. Union of India (1996) 8 SCC 462
- M.C Mehta v. Union of India AIR 1987 SC 965
- M.C Mehta v. Union of India AIR1997 SC 734
- M.C. Mehta v. Union of India AIR 1987 SC 1086
- Mahendra Pandey v. UoI, Application No. 470/2016, (Jan. 2, 2019)
- Manu Anand v. State of Kerala MANU/KE/0461/2016
- Manushi Sangthan, Delhi v. Govt. Of Delhi & Ors (2010) DLC 168.
- Nar Bahadur Bhandari and Ors v. the State of Sikkim and Ors MANU/SI/0023/2010.
- Narmada Bachao Andolan v. Union of India AIR 2000 SC 3751

- Om Dutt Singh v. State of Uttar Pradesh, Application No. 521/2014
- Outdoors Communication v. PWD and Municipal Corporation of Delhi MANU/DE/2913/2007
- R.L. and E Kendra, Dehradun V. State of U.P.AIR 1985 SC 652
- Ratandeep Rangari v. State of Maharashtra & Others Application No. 19/2014 (WZ) (Oct. 15, 2015)
- Reliance Natural Resources Ltd. v. Reliance Industries Ltd, (2010) 7 SCC 1
- Ridhima Pandey v. Union of India & others. Application No. 187/2017 (Jan. 15, 2019)
- S.P. Gupta v. Union of India AIR 1982 SC 149
- S.P. Gupta v. Union of India, (1981) Supp SCC 87
- Society for Protection of Environment & Biodiversity v. UoI, Application No. 677/2016
- State of Maharashtra v. Suraj Pal (2015) 1 Bom CR (Cri) 576
- Sukhdev Vihar Residents' Welfare Ass'n v. Union of India [UoI], Application No. 22 (THC)/2013 (Feb. 2, 2017)
- Swami Parmanand Bhatta Company v. Union of India MANU/UP/2765/2010
- T.N Godaverman Thirumalkapad v. Union of India (1997) 2 SCC 267

LIST OF ABBREVIATION

AAI	Airport Authority of India
ACA	Airport Carbon Accreditation
ACI	Airports Council International
AERB	Atomic Energy Regulatory Board
AIA	Accredited Independent Agencies
AILA	Independence Association of Latin America and the Carib bean
AIR	All India Reporter
AJAY	Atal Jyoti Yojana
AMRUT	Atal Mission for Rejuvenation and Urban Transformation
AOSIS	Alliance of Small Island States
APGENCO	Andhra Pradesh Power Generation Corporation
ASP	Adaptive Social Protection
AWG-KP	Ad Hoc Working Group under Kyoto Protocol
AWG-LCA	Ad Hoc Working Group on Long-term Cooperative Action
BCC	Behavior Change Communication
BEE	Bureau of Energy Efficiency
BEVs	Battery Electric Vehicles
BG	Broad Gauge
BHAVINI	Bharatiya Nabhikiya Vidyut Nigam Limited
BHEL	Bharat Heavy Electrical Limited
BPL	Below Poverty Line
BRICS	Brazil, Russia, India, China, and South Africa
BS	Bharat Stage
BUR	Biennial Update Report
CAA	Climate Ambition Alliance
CAMPAs	Compensatory Afforestation Fund Management and Plan ning Authorities
CBDRRC	Common but Differentiated Responsibilities and Respective Capabilities
CCAP	Climate Change Action Program

CCI	Cricket Club of India
CCS	Carbon Capture Storage
CCU	Carbon Capture Unit
CDM	Clean Development Mechanism
CDRI	Coalition for Disaster Resilient Infrastructure
CFCs	Chlorofluorocarbons
CFL	Compact Fluorescent Lamps
CHCs	Custom Hiring Centers
CIAL	Cochin International Airport Limited
CII	Confederation of Indian Industry
Cl	Chlorine
CNG	Compressed Natural Gas
CO_2	Carbon Dioxide
COP	Conference of Parties
CSE	Centre for Science and Environment
CSR	Corporate Social Responsibility
DAE	Department of Atomic Energy
DCs	Designated Consumers
DDUGJY	Deen Dayal Upadhyaya Gram Jyoti Yojana
DFCCIL	Dedicated Freight Corridor Corporation of India Limited
DFCs	Dedicated Freight Corridors
DFP	Designated Focal Points
DGAD	Directorate General of Anti-Dumping and Allied Duties
DGCA	Directorate General of Civil Aviation
DIAL	Delhi International Airport Limited
DNA	Designated National Authorities
DOE	Designated Operating Entities
DRR	Disaster Risk Reduction
DSR	Direct Seeded Rice
ECBC	Energy Conservation Building Code
ECO	Energy Conservation and Commercialization
EESL	Energy Efficiency Services Limited
EPA	Environment Protection Act
ERU	Emission Reduction Units
ET	Emissions Trading

EU	European Union
EV	Electric Vehicle
EVI	Electric Vehicles Initiative
FCCC	Framework Convention on Climate Change
FFV	Flexible Fuel Vehicles
FRI	Forest Research Institute
FWCC	First World Climate Conference
GCF	Green Climate Fund
GDP	Gross Domestic Product
GEF	Global Environment Fund
GHG	Greenhouse Gas
GIM	Green India Mission
GOI	Government of India
GRIHA	Green Rating for Integrated Habitat Assessment
GW	Gigawatt
HCFCs	Hydro Chlorofluorocarbons
HFCs	Hydro Fluorocarbons
IBRD	International Bank for Reconstruction and Development
ICAP	India Cooling Action Plan
IDS	Institute of Development Studies
IGBC	Indian Green Building Council
IGBP	International Geosphere Biosphere Programs
IIEC	International Institute for Energy Conservation
IIT	Indian Institute of Technology
IMD	India Meteorological Department
IMO	International Maritime Organization
INC	Intergovernmental Negotiating Committee
INDC	Intended Nationally Determined Contribution
IPCC	Inter-Governmental Panel of Climate Change
IR	Indian Railways
IROAF	Indian Railways Organization for Alternate Fuels
ISA	International Solar Alliance
JI	Joint Implementation
JISC	Joint Implementation Supervisory Committee
JV	Joint Venture

LDCs	Least Developed Countries
LE	Life Extension
LMDCs	Like-Minded Developing Countries
LPG	Liquefied Petroleum Gas
LULUCF	Land Use, Land-Use Change, and Forestry
MARPOL	International Convention for the Prevention of Pollution
MESASEA	Middle East, South Asia and South East Asian
MIDC	Maharashtra Industrial Development Corporation
MIDH	Mission for Integrated Development of Horticulture
MNC	Multinational Company
MNRE	Ministry of New and Renewable Energy
MoEF	Ministry of Environment and Forests
MoEFCC	Ministry of Environment, Forest and Climate Change
MoES	Ministry of Earth Sciences
MOP	Meeting of the Parties
MoS	Ministry of Shipping
MSW	Municipal Solid Waste
MT	Million Tone
$MtCO_2$	Metric Tons of Carbon Dioxide
MW	Megawatt
NAF	National Adaptation Fund
NAFCC	National Adaptation Fund on Climate Change
NALCO	National Aluminum Company
NAM	National Agro-forestry Mission
NAPCC	National Action Plan for Climate Change
NAREGA	National Rural Employment Guarantee Act
NASA	National Aeronautics and Space Administration
NBM	National Bamboo Mission
NCEEF	National Clean Energy and Environment Fund
NCEF	National Clean Energy Fund
NCEPC	National Committee on Environmental Planning and Coor dination
NCR	National Capital Region
NDC	Nationally Determined Contribution
NEMMP	National Electric Mobility Mission Plan

NFSM	National Food Security Mission
NGHM	National Green Highways Mission
NGOs	Nongovernmental Organizations
NGT	National Green Tribunal
NISE	National Institute of Solar Energy
NMAE&T	National Mission on Agriculture Extension & Technologies
NMCG	National Mission for Clean Ganga
NMEEE	National Mission for Enhanced Energy Efficiency
NMGI	National Mission for Green India
NMSA	National Mission for Sustainable Agriculture
NMT	Non-Motorized Transport
NO	Nitric Oxide
NPCIL	Nuclear Power Corporation India Limited
NTPC	National Thermal Power Corporation
OEDCD	Organization of Economic Corporation and Development
OHCHR	Office of the High Commissioner for Human Rights
ONGC	Oil and Natural Gas Corporation Limited
OSOWOG	One Sun One World One Grid
PAT	Perform, Achieve and Trade
PBCMs	Plastic Bottle Crushing Machines
PFCs	Per fluorocarbons
PIL	Public Interest Litigation
PKVY	Paramparagat Krishi Vikas Yojana
PMKSY	Pradhan Mantri Krishi Sinchayee Yojana
PMUY	Pradhan Mantri Ujjwala Yojana
R&D	Research and Development
R&M	Renovation and Modernization
RECAI	Renewable Energy Country Attractiveness Index
RSBY	Swasthya Bima Yojana
RTS	Rooftop Solar
SAPCC	State Action Plan on Climate Change
SAUBHAGYA	Saubhagya Scheme Pradhan Mantri Sahaj Bijli Har Ghar Yojana
SBI	Subsidiary Body for Implementation
SBM	Swachh Bharat Mission

SBSTA	Subsidiary Body for Scientific and Technological Advice
SC	Supreme Court
SCC	Supreme Court Cases
SD	Sustainable Development
SDG	Sustainable Development Goal
SECI	Solar Energy Corporation of India
SERCs	State Electricity Regulatory Commissions
SHC	Soil Health Card
SHW	Severe Heat Wave
SLNP	Street Lighting National Program
SMAF	Sub-Mission on Agroforestry
SO_2	Sulfur Dioxide
SRI	System of Rice Intensification
SSLs	Solar LED Street Lights
SWM	Solid Waste Management
TERI	The Energy and Resources Institute
UDHR	Universal Declaration of Human Rights
UJALA	Unnat Jyoti by Affordable LED for All
ULBs	Urban Local Bodies
UN	United Nations
UNCED	UN Conference on Environment and Development
UNDRR	United Nations Office for Disaster Risk Reduction
UNEP	United Nation Environment Program
UNFCCC	United Nation Framework Convention on Climate Change
USAID	United States Agency for International Development
UTs	Union Territories
VCFEE	Venture Capital Fund for Energy Efficiency
VCLT	Vienna Convention on the Law of Treaties
Vol	Volume
WCED	World Commission on Environment and Development
WIM	Warsaw International Mechanism
WMO	World Meteorological Organization

ACKNOWLEDGMENT

Climate change mitigation is the focal point of discussion in the modern world. Conceiving the idea of adaptation of a PhD thesis into a book is not easy. The authors offer their floral tribute to the 'almighty' for enabling them to conceive and implement the idea of adapting the PhD thesis into a book. The authors are equally indebted to their 'Alma Mater' Law School, Banaras Hindu University, for inculcating the value of human-oriented legal research.

Gratitude cannot be expressed in words towards the Faculty of Law, University of Lucknow, for providing an ecosystem wherein work of such nature could be carried out. Authors equally record and register their deep sense of gratitude towards all those authors, scientists, politicians, industrialists, and international and national agencies which are working hard for climate change mitigation. We are equally grateful to UNFCC and IPCC for impeccable research and reports on the issues pertaining to climate change mitigation.

Authors are eternally indebted to all libraries for their open-hearted response in sorting, collecting, and making accessible the rich resources from their stacks and databases, especially to the library of Indian Law Institute, Banaras Hindu University, University of Delhi, Dr. Ram Manohar Lohiya National Law University, and the University of Lucknow. The authors are very much pleased to have access to all databases on a remote basis via Tagore Library of the University of Lucknow, through which we could access Jstor, Hein online, etc.

The authors are very much thankful to their students and colleagues for their valuable and constructive suggestions and guidance received during this work. We are thankful to Professor Amarpal Singh, Head & Dean, Guru Gobind Singh Indraprastha University, New Delhi, for his valuable continuous support and encouragement. We have left no stone unturned to record Indian initiatives in climate change mitigation, for which we are thankful to all government databases for a wonderful update of websites.

It shall be very impolite if we do not express our gratitude towards 'Cambridge Scholars Publishing' especially to 'Adam Rummens' who processed our proposal form in no time and encouraged us at every step for sumptuous and fruitful writing. Cambridge Scholars Publishing made us believe that a subject of this stature needs wider, deep, and more pervasive reach beyond India and the South Asian region. We are delighted to see it happen through this platform offered by Cambridge Scholars Publishing. We run out of words to convey our gratitude to Cambridge Scholars Publishing.

We are thankful to our friends, family, parents, spouses, and children (Adhish & Mishti), whose incessant and eternal support helped us at every stage of writing this work. We express our heartfelt thanks to each and every individual and institute contributing to the making of this book. We solicit the valuable advice of our avid readers and hope and believe that this book will receive a warm response from 'mass' and 'class' alike.

—Dr.AshishKumarSrivastava

Ms.SnehaKumari

PREFACE

"The world has enough for everyone's needs, but not everyone's greed."

—Mahatma Gandhi

The 'Blue Planet' is the most beautiful and the only habitable planet of the 'Milky Way.' Humans have been walking on this planet for 6 million years. The 'Earth' provides us with everything which makes human life sustainable, right from glass to granite. The environment on the earth makes life possible and livable. Human life, along with all its faculties, is in consonance with environmental forces. The earth has everything, e.g., beautiful and the tallest mountains, deepest and widest seas, deserts, plateaus, plains, hills, forests, and coastal lines.

The environment and ecology of the earth have been very conducive to the existence of all life processes in the form of flora and fauna. Roughly 1.5 million species of flora and fauna exist on this beautiful planet. This 'Biological Diversity' and the 'Bio-resources' are a treasure trove for mankind. Out of these species, the human species are blessed with the cognitive brain. Humans have made their lives more convenient and, in the process, have churned out all natural resources to their optimum level for their personal use. Humans have made airplanes, megastructures, satellites, nuclear bombs, and wonderful monuments. In the process of evolution of mankind after the invention of 'fire' and the 'wheel,' we witnessed the maximum scientific discoveries and inventions in the 19th, 20th, and 21st centuries. The industrialization and post-industrialization era supported by fossil fuel-based industries redefined human 'wants' and 'desires.' The inventions and discoveries were based on natural resources supported by MNCs, and with the growing population, the churning and exploitation of resources picked up the pace, and we started anthropogenic pollution as the whole idea of development was faulty and fallacious, being at the cost of nature. Humans have inherent vices like greed, vengeance, anger, lust, gluttony, and hoarding material things. Humans want to create perpetual properties for themselves and their successors. The idea of private property and private wealth created havoc.

The greed of humans paced up the churning of the earth. Human greed started damaging the ecology, environment, and envelopes of the earth which kept humans safe from extra-terrestrial and climatic forces. Robert Nozick and John Milton promoted the idea of the 'Capitalist State' and 'profit maximization,' under which human beings learned only the idea of 'Get rich quick,' and one is obliged to make money through business, trade, commerce, and manufacture. This started the eternal mass production and mass consumption cycle. The dividing line between human wants and desires got blurred, corporations became aggressive and frenzied shopping became the order of the day.

Human beings were in dire need of solutions to their existing problems. In search of remedies for human maladies, fossil fuels were discovered, and plastic was invented in the 19th and 20th centuries. With this, the idea of a wholesome environment came to an end. Now we can think of an environment wherein a permitted level of pollution exists. The five great elements (water, air, fire, ether, earth) of human life got polluted. This anthropogenic pollution was geared up by energy needs which were mostly non-renewable. The industries and anthropogenic activities started causing damage to the blue planet, and the first damage we witnessed in the early nineties was 'Ozone Layer Depletion.' The carbon emissions, GHGs, and CFCs became household names.

In 1972, United Nations held a conference in Stockholm on 'Human Environment' wherein the 'Stockholm Declaration & Action Plan' was chalked out for the protection and preservation of the earth and ecology. Stockholm conference declared:

"*Man is both creature and molder of his environment and has the fundamental right to freedom, equality and adequate conditions of life, in an environment of a quality that permits a life of dignity and well-being.*"

The conference resulted in the creation of the United Nations Environment Programme. In 1987, the Brundtland Commission published the report entitled 'Our Common Future,' which mooted the idea of 'Sustainable Development.'

The Indian constitution was environmentally blind, and law-making on environmental issues started around the seventies, e.g., Wildlife (Protection) Act, 1972; Water (Prevention & Control of Pollution) Act, 1974; Air (Prevention & Control of Pollution) Act 1981, Environment Protection Act (EPA), 1986, The Scheduled Tribes and Other Traditional Forest Dwellers (Recognition of Forest Rights) Act, 2006, National Green Tribunal Act, 2010. However, critics have always been apprehensive about its efficiency and say that "It has large fangs but no venom." The permitted pollution was regulated by the pollution control board, which could not perform efficiently. Environment Impact Assessment (EIA) based anthropogenic developmental activities was an idea mooted off lately in India. Amidst the intended law-making, the Indian Judiciary, under the leadership of 'Green Judge,' Hon'ble Justice Kuldip Singh, Advocate Mr. M.C. Mehta, and T.N. Godaverman Thirumulpad delivered the most iconic judgments like Oleum Gas Leak, Bhopal Tragedy Case, Bichhari Village Case, Silent Valley Project Case, Ratlam Municipality case were decided on multiple issues of deforestation, construction on the coastal line, preservation of the Ganges, quarrying of stones and limes, internally displaced people due to the construction of mega structures and dams. The Indian Supreme Court vigorously declared and included many principles in its legal framework, like the 'polluter pays principle,' sustainable development, inter-generational justice, precautionary principle, public trust doctrine, no-fault liability, and strict liability principle. The 'Public Interest Litigation' (PIL) became a wonderful instrument of justice, and mindless development was contained in India.

In the nineties, apart from ozone layer depletion, we started facing many problems related to anthropogenic pollution, and a need was felt that only soft policies and declarations were rhetoric. We need a concrete action plan. We started witnessing 'Climate Change' and 'Global Warming.' Naturally, we found in 1988 the Intergovernmental Panel on Climate Change (IPCC) to address the issue. In 1992, we signed the United Nations Framework Convention on Climate Change (UNFCCC). Since then, the Conference of Parties (COPs) has been incessantly working towards climate change mitigation. Till now, 26 CoPs have been held, and the recent Glasgow CoPs were very pivotal.

UNFCCC defines climate change in following words;

"Climate Change means a change of climate which is attributed directly or indirectly to human activity that alters the composition of the global atmosphere and which is in addition to natural climate variability observed over comparable time periods."[1]

Climate change plainly means change in climate and this is because of 'Anthropogenic Activity/pollution.' 'Radiative forcing' due to carbon emissions is making the world warmer. Since the ice age, Earth became warmer by 9°. It is estimated that at the current rate of GHG emissions, the Earth will be warmer by 2°F by 2025 and 5°F by 2100. The first-ever report on climate change assessment has been prepared by The Union Ministry of Earth Sciences (MoES, India) named "Assessment of Climate Change over Indian Region." The report states that India's average temperature has already increased by around 0.7°C during the period of 1901–2018 due to greenhouse gas emissions and by the end of 2100, it is expected to rise approximately by 4.4°C. The report states that rapid change in temperature means adverse impact on India's natural ecosystem, agriculture, and freshwater resources, simultaneously causing escalating damage to infrastructure. This ultimately leads to the adverse impact on the country's biodiversity, food, water, energy security, and public health. The Intergovernmental Panel on Climate Change (IPCC) has published its special report in 2018 on the implications of a 1.5°C rise in average global temperature. The report states climate change from human-induced causes is already taking place and is both accelerating and intensifying. Average global temperatures have risen by about 1°C since the start of the Industrial Revolution. The harmful effects are visibly evident like the extreme weather, rising sea levels, and diminishing Arctic Sea ice, among other changes. Temperature rise, even of 1.5°C, may result in disastrous and irreversible changes. 70–90% of coral reefs across the world would die at 1.5°C temperature rise. At 2°C rise, none would be left. According to the report, global warming so far may have led to the loss of over 40–50% of the coral reefs. The loss of fish populations due to the loss of coral reefs will have an impact on food security, which in turn affects human health. Climate change will result in the rising of sea level, loss of low-lying islands, immersion of low-lying coastal lines, tornadoes, cyclones, deforestation, wildfires in forests, etc. The poor and the most vulnerable will be severely hit.

IPCC Assessment Report 6, 2022 contemplates that Global warming is likely to reach 1.5°C between 2030 and 2052 if it continues to increase at the current rate. Climate-related risks for natural and human systems are higher for global warming of 1.5°C than at present, but lower than at 2°C. Most adaptation needs will be lower for global warming of 1.5°C compared to 2°C. Apart from global warming, anthropogenic activity is resulting in the loss of life processes on Earth. In 2021, according to IUCN, 142,500 species are on the IUCN Red List, with more than 40,000 species threatened with extinction, including 41% of amphibians, 37% of sharks and rays, 34% of conifers, 33% of reef building corals, 26% of mammals and 13% of birds[2].

Adaptation and mitigation are the only tools available against climate change. CoPs 1 to 26 left no stone unturned in setting out the tone and tune of climate change mitigation.

[1] Art. 1(2) of United Nations Framework Convention on Climate Change, (1992).

[2] https://www.iucnredlist.org/about/background-history (accessed on 07 July 2023).

CoP 21 resulted in the adoption of the 'Paris Agreement, 2015' on 12th December 2015. It was adopted by 196 countries and came into force on 4th November 2016. In June 2017, President Donald Trump (USA) surprised the world by his formal withdrawal from the Paris Agreement, which was a big blow to the whole movement. But in January 2021, President Joe Biden signed an executive order to rejoin the Paris Agreement.

But this is not the first time that the United States of America did so. In March 2001, George W. Bush, the USA President, withdrew the United States from the Kyoto Protocol on the plea of hurting the USA economy. The USA's dropout gave a big blow to the whole decarbonization movement.

The Paris Agreement, 2015, is the fruit of hard and untiring labor of the world. The Paris Agreement aims to contain the limit of global warming below 2°, preferably 1.5°C, compared to pre-industrial levels. The Agreement is entered into with an understanding that in developing nations, the requirement is global peaking of GHG emissions soon and thereafter rapid reduction to achieve long-term temperature goals set out in the Paris Agreement. The Agreement creates binding obligations on member nations. The agreement requires countries to submit their Nationally Determined Contributions (NDCs). Each country will submit the limit of reduction of GHG emissions in its NDCs. A financial, technical, and capacity-building framework has been outlined by the Paris Agreement to achieve its goals. The agreement aims that by 2030, zero-carbon solutions could be competitive in sectors representing over 70% of global emissions like power and transport.

The Paris Agreement delineates the energy needs and developmental goals of developing nations and makes many provisions suited to the national needs of developing nations. It allows them to reach the global peak of GHG emissions and asks developed nations for financial, technical, and capacity building of developing nations by developed nations. Developing nations often have been found fuming on the issue of climate justice and global warming. They complain that developed nations have achieved their developmental goals by reaching their global peak of GHG emissions much earlier, but they still are developing and have not fully attained their developmental goals. Consumption in developed nations is much higher than in developing nations, and it makes an indelible impact on climate and earth. But the whole argument falls flat as the developing nations have the largest populations, which outnumbers any country because of its population, and their population growth rate is overwhelming.

Climate Mitigation Litigation in the post-Paris Agreement era has surged. There have been cases across jurisdictions. Almost 1000 climate change mitigation petitions have been filed in 25 countries. In the highly debated case of The States of Netherlands V. Urgenda Foundation (2019), Dutch citizens filed a case against the Dutch Government to reduce carbon emissions. The initial ruling in 2015, requiring the government to meet an emissions goal of 25% reduction from 1990 levels by 2020, was upheld through the Supreme Court of appeals, affirming that a reduction in emissions was necessary for the Dutch government to protect human rights. In Juliana v. United States of America (2015), judgment delivered on 5th April 2018, the Supreme Court held that the Colombian government had not efficiently tackled the problem of deforestation in the Amazon, in breach of the fundamental rights to water, air, a dignified life, and health, among others, in connection with the environment. The Supreme Court recognized the Colombian Amazon as a 'subject of rights,' entitled to protection, conservation, maintenance, and restoration led by the government and its territorial agencies. The Colombian government was ordered to formulate a short-, medium-, and long-term plan within four months to counteract the rate of deforestation in the Colombian Amazon and the impact of climate change, with the participation of the plaintiffs and affected communities. In Armando Ferrão Carvalho and Others v. The European Parliament and the Council (2018), ten families from Portugal, Germany, France, Italy, Romania, Kenya, and Fiji, and Sáminuorra (a Swedish Youth Association), brought an action in the EU General Court. The action seeks to compel the EU to make more stringent greenhouse gas emissions reductions than the 40% target by 2030 (compared to 1990 levels) to avoid climate change impacts and threats to fundamental rights to life, health, occupation, and property. In the days to come, there will be more climate change mitigation litigation.

'Wall-E' movie, a sci-fi animated film, of 2008 shows us a plot where, in 29th century, earth has turned in a garbage bin non-habitable for human due to raised toxicity and pollutions and humans have been provided habitat on a Starliner 'Axiom'(Spaceshp) somewhere in space. Searching life back on earth humans find a small sapling in a planter and they find a hope to return back on earth and captain of the airship says, *"Oh! It's good to be home."* This film clearly represents the status of modern world which is running an adrenaline race of frenzied consumption turning the earth in to a dump yard and garbage bin in near future with the mind set of "I'll be gone, you'll be gone" (IBGYBG) making life of flora and fauna including humans impossible on earth. Exploration of extra-terrestrial planets and objects by private players with the mindset that this mess on earth can't be cleaned and there is much space in space is an idea which is going to burst sooner or later. *Human beings think that they are master of the planet but forget in the process that they are only a guest.*

In documentary film *An Inconvenient Truth* directed by Davis Guggenheim about former United States Vice President Al Gore's campaign to educate citizens about global warming calls upon his viewers to learn how they can help him in these efforts. Gore concludes the film by saying:

"Each one of us is a cause of global warming, but each one of us can make choices to change that with the things we buy, the electricity we use, the cars we drive; we can make choices to bring our individual carbon emissions to zero. The solutions are in our hands, we just have to have the determination to make it happen. We have everything that we need to reduce carbon emissions, everything but political will. But in America, the will to act is a renewable resource."

Covid-19, the pandemic, claimed more than 60 lakh lives. Though the exact reason for Covid-19 could not be established, one hypothesis is that it originated from wildlife and transmitted to humans. Consuming exotic wildlife meat is resulting in zoonotic diseases. This is also disrupting the food cycle and life cycle. Wildlife trade has surged due to the need for leather, meat, and medicines (aphrodisiac). Wildlife trade is transnational and organized. This is also causing an imbalance in nature as tigers keep forests intact. If there are no tigers, there will be no forests, and receding forests will make the earth hotter.

India is an important player in the world. The large population of 1.4 billion people resides in India, and it has the Himalayas, the Ganges, the Thar, Deccan plateau, Konkan-Malabar-Coromandel coastline, and the Gangetic plains. It has its own history. It has been severely invaded and looted and has been a colony of the British Empire. India is highly diverse and plural. The Indian Constitution was environmentally blind, but after a landmark judgment of Maneka Gandhi v Union of India AIR 1978 SC 597, when the Supreme Court of India read Article 21 of the Indian Constitution, it stated that the law must be just, fair, and reasonable. It introduced the idea of 'substantive and procedural justice' and 'due process of law' in the Indian legal order. This left an indelible imprint on environmental litigation. M.C. Mehta and T.N. Godaverman led the whole PIL litigation and created the path of climate justice in India. In 1976, Article 48A was added to the Indian Constitution, which provides that "The State shall endeavor to protect and improve the environment and safeguard the forests and wildlife of the country." The PILs filed in the Supreme Court on environmental issues made many doctrines of environmental law an integral part of Indian law, such as the public trust doctrine, polluter pays principles, precautionary principle, strict liability principle, etc. All kinds of issues like the construction of dams, internally displaced persons, the fundamental right to the environment, the fundamental right to drinking water, illegal quarrying, protection of mangroves and coastal lines, protection of biological diversity, gas leaks and strict liability, mega projects and rights of indigenous peoples, preservation of forests, etc.

India has a painful history of environmental litigation. 'Bhopal Gas Tragedy' or Union Carbide Corporation v. Union of India is a painful case where in the UCC plant of Bhopal, 'Methyl Isocyanate' gas leaked and claimed between 15,000 to 20,000 human lives and affected human generations. There was a series of litigation, and to date, Rs. 1500 crores are lying unclaimed in banks for disbursal. This shows that India was not prepared for such kind of litigation. The experience gained from the Union Carbide case came in handy in subsequent environmental litigation. The National Green Tribunal Act, 2010 established the National Green Tribunal in India for the resolution of environmental disputes.

India has always believed in the nature, earth and planet. In Vedic period 'Yajnya' were performed to purify the environment. Every life process however small was given its due importance with the idea of *'Vasudhaiv Kutumbkam'* (Earth is a family). Indians started worshipping nature, rivers, mountains, trees etc., so that its wholesomeness remain chaste and unpolluted. Anthropogenic activities remained low till colonial period so response to the issue of climate issues has been less. However, especially after independence, India started making its mega structures, e.g., dams, bridges, tunnels etc., and by seventies India became concerned about environmental issues. After nineties India also witnessed surge in GHGs emissions due to globalization, liberalization and privatization of economy. India joined UNFCCC in 1992 and has been a very active participant in all its action plan towards climate change mitigation. India vigorously participated in all COPs. Indian population accounts for 17.5% of the world population and the land surface of India is 2.5% of the world surface area. Most of the Indian populations are poor who has no access to electricity[3] and they rely on the solid biomass for cooking and 76 million population of India do not have access to safe drinking water.[4] India 1.77

[3] Kundan, P., (2015). *The Extent of India's Energy Poverty*. Down To Earth. accessed on September 2, 2021, https://www.downtoearth.org.in/news/energy/the-extent-of-india-s-energy-poverty-48966 (accessed on 07 July 2023).

[4] Saptarshi, D., (2017). *76 Million Don't Have Safe Drinking Water: India's Looming Water Crisis*. NDTV. https://swachhindia.ndtv.com/76-million-dont-have-safe-drinking-water-indias-looming-water-crisis-5606/ (accessed on 07 July 2023).

million people are without house.[5] India is the fourth- largest emitter of greenhouse gases (GHGs) in the world.[6] India being developing country has to meet its development goal along with its obligation under Paris Agreement. India only contributes about 7% to global carbon emission.[7] The quantification of goals in India's NDCs under Paris Agreement is threefold: first, reducing the emission intensity of its gross domestic product (GDP) by 33%–35% (vis-à-vis 2005) by 2030; second, achieving 40% cumulative electric power installed capacity from non-fossil fuel-based energy resources by 2030; and third, creating an additional carbon sink of 2.5–3 billion tons of CO_2 equivalent by 2030 through additional forest and tree cover.[8] Even though there would have not been obligation to mitigate but it was incumbent need to mitigate due to India's venerability to climate change.[9] The annual mean temperature during year 1901–2019 showed a significant increasing trend (0.61°C/100 years), with a higher increasing trend in maximum temperature (1.00°C/100 years) and relatively lower increasing trend (0.22°C/100 years) in minimum temperature. The year 2019 was the seventh warmest year on record since 1901.[10] March 2022 has been the hottest March in last 122 years. In the year 2019, eight cyclonic storms formed over the north Indian Ocean.[11]

India started its National Action Plan on Climate Change (NAPCC) and at CoP 21 India towards its obligations under the Paris Agreement, has submitted its Nationally Determined Contribution (NDC) comprising of eight contributions for 2021–2030. The three quantitative contributions are: (i) to reduce the emissions intensity of its GDP by 33–35% by 2030 from 2005 level, (ii) to achieve about 40% cumulative electric power installed capacity from non-fossil fuel based energy resources by 2030 with the help of transfer of technology and low cost international finance, (iii) to create an additional carbon sink of 2.5 to 3 billion tons of CO_2 equivalent through additional forest and tree cover by 2030. The other targets pertain to sustainable lifestyles; climate friendly growth path; climate change adaptation; mobilization of climate change finance; capacity building and technology. In this context, India's current share of non-fossil sources based installed capacity of electricity generation become more than 40% in November,

---

[5] Somesh, J., (2013). *1.77 Million People Live without Shelter*. Business Standard. https://www.ighhub.org/resource/177-million-people-live-without-shelter (accessed on 07 July 2023).

[6] The United Nations Environment Program, (2020). *Emissions Gap Report 2020*. https://www.unep.org/emissions-gap-report-2020 (accessed on 07 July 2023).

[7] Jacob, K., (2021). *India's Percentage CO_2 Emissions Rose Faster Than the World Average*. The Hindu. https://www.thehindu.com/news/national/indias-percentage-co2-emissions-rose-faster-than-the-world-average/article33965283.ece (accessed on 07 July 2023).

[8] *Supra* note 4.

9 See Lakshmanan, Pushpa, K., et al., (2017). Paris agreement on climate change and India. *Journal of Climate Change, 39*(1), 1–10.

[10] Jacob, K., (2020). *2019 Was The 7th Warmest Year in India Since 1901, Says IMD. The Hindu.* https://www.thehindu.com/sci-tech/energy-and-environment/2019-was-the-7th-warmest-year-since-1901-says-imd/article30515266.ece (accessed on 07 July 2023).

[11] Akshit, S., (2019). *8th Cyclone Formed in the North Indian Ocean region: Pawan joins Kyarr, Fani, Maha, Vayu, Hikaa, Bulbul and Pabuk; to cross Somalia on December 7*. Down To Earth. https://www.downtoearth.org.in/news/climate-change/8th-cyclone-formed-in-the-north-indian-ocean-region-68307 (accessed on 07 July 2023).

2021. Installed capacity of solar energy in India has increased by more than 18 times from 2.63 GW in March 2014 to 48.56 GW in November 2021.

Green Climate Fund (GCF) is the operating entity of the financial mechanism of the UNFCCC. It aims to deliver a 50:50 balance between mitigation and adaptation allocations in its portfolio. So far, four projects have been approved to India with USD 314.82 million of GCF funding. The total value of these four projects is 1491.07 million USD to address both adaptation measures and mitigation action.

India knows that power and transport are its sector which needs utmost attention. India made Vehicle Scrappage Policy in 2021 where under 15 years old commercial vehicle and 15 years old commercial vehicle shall be deregistered. To provide boost to electric vehicle India aims of switch 30% of private cars, 70% of commercial vehicles, and 80% of two and three-wheelers to EV by the year 2030. India under capacity building initiative of climate change mitigation started 'International Solar Alliance (ISA)' in 2015 with 105 members which are sunlit countries with a tagline of OSOWOG (One Sun, One World, One Grid).

Adaptation, mitigation and capacity building are the tools to curb the menace of climate change. In capacity building adapting to green technologies is required. Transfer of green technology from developed to developing nations has its own risk like infringement of intellectual properties via commercial exploitation. Cost of technology is also a cause of concern. Older fossil fuel based technologies are cheaper and has a proper infrastructure of support and service. Green technologies are still costly, unstable and evolving. Poverty plays a pivotal role in climate change mitigation. Ensuring food security is top priority of developing nations and environment protection obviously will be second priority but food security also gets affected in the process if we do not address the climate change issues. We know that climate change will affect, the poor, the hardest.

India has started schemes for almost all sectors vigorously to reduce its carbon emissions and to respond to the issues of climate change in a measured way. India, between 2015–2022, started many schemes responding to the issue of climate change mitigation, e.g., Clean Coal Policy, Partial Risk Guarantee Fund for Energy Efficiency (PRGFEE), The National Mission for Enhanced Energy Efficiency (NMEEE), National Smart Grid Mission, Deen Dayal Upadhyaya Gram Jyoti Yojana (DDUGJY), Pradhan Mantri Ujjwala Yojana (PMUY), Saubhagya Scheme Pradhan Mantri Sahaj Bijli Har Ghar Yojana (SAUBHAGYA), The India Cooling Action Plan (ICAP), Energy Conservation Building Code (ECBC), Green Rating for Integrated Habitat Assessment (GRIHA), The Green National Highways Corridor, The National Electric Mobility Mission Plan (NEMMP), Dedicated Freight Corridors (DFCs), Green Tax on Vehicles, Flexible Fuel Vehicles (FFV), Green Port project, National Mission for Sustainable Agriculture (NMSA), Green Aviation Policy, 2019, National Bamboo Mission (NBM), Paramparagat Krishi Vikas Yojana (PKVY), Pradhan Mantri Krishi Sinchayee Yojana (PMKSY), Soil Health Card (SHC), National Food Security Mission (NFSM), Mission for Integrated Development of Horticulture (MIDH), National Agro-forestry Policy–2014 (NAP), Climate Change Action Program (CCAP), National Adaptation Fund on Climate Change (NAFCC), Plastic Waste Management Rules, 2016, India Climate Change Knowledge Portal, Atal Mission for Rejuvenation and Urban Transformation

(AMRUT) (Smart Cities Development), Swachh Bharat Mission (SBM), Solid Waste Management (SWM), Green India Mission (GIM), National Mission for Clean Ganga (NMCG). Transport and power are core in all such schemes. India also drafted major policies and rules on climate change mitigation between 2015–2022, e.g., Hazardous Wastes (Management, Handling & Transboundary Movement) Rules, 2016, Fly Ash Notification 1999, amended in 2016, E-Waste (Management) Rules, 2016, Bio-medical Waste (Management & Handling) Rules, 2016, Construction and Demolition Waste Management Rules, 2016, Municipal Solid Waste Management Rules, 2016, Plastic Waste (Management and Handling) Amendment Rules, 2018, Draft Battery Waste Management Rules, 2020.

Apart from the litigation of environmental issues Supreme Court of India categorically observed about climate change in the case of *Karnataka Industrial Area Development Board v. C Kenchppa*,[12] the Hon'ble Supreme court directed that adverse impact on the environment must be properly comprehended by the Appellant before acquiring any lands for industrial purposes, in doing so, the Hon'ble court stressed on the importance of sustainable development while noting that:

"The entire world is facing a serious problem of environmental degradation due to indiscriminate development. Industrialization, burning of fossil fuel and massive deforestation are leading to degradation of environment. Today the atmospheric level of carbon dioxide, the principal source of global warming, is 26% higher than pre-industrialized concentration. The earth's surface reached its record level of warming in 1990. In fact, six of the seven warmest years on record have occurred since 1980, according to the World Watch Institute's 1992 report. The rise in global temperature has also been confirmed by the Intergovernmental Panel on Climate Change set up by the United Nations and its final report published in 1990. The global warming has led to unprecedented rise in sea level. Apart from melting of the polar ice it led to inundation of low-lying coastal region Global warming is expected to profoundly affect specie, and ecosystem. Melting polar ice, glaciers, thermal expansion of seas would cause worldwide flooding and unprecedented rise in sea level if gas emissions continue at the present rate. Enormous amount of gases and chemicals emitted by the industrial plants and automobiles have led to depletion of ozone layers winch serve as a shield to protect life on the earth, from the ultra-violet rays of the sun."

The National Green Tribunal in *Court of Its Own Motion v. State of H. P.*[13], cautioned against the dangers of global warming in the following manner:

"Global warming has a direct impact on environment and ecology of any zone. Global warming, a succession of hot summer through the 1980s seemed to bear out the conservationist's warming that the earth's climate was warming up. This global warming, if it is indeed the case, is a consequence of the large quantities of carbon dioxide released into the atmosphere by industries power stations and motor vehicle. The earth's tem-

[12] (2006) 6 SC C 371.

[13] (2014) SCC On Line NGT.

perature will rise, the polar cap will melt, raising sea levels and causing catastrophic flooding in low lying areas and large areas of productive land will become desert. This global warming-heat-has affected various countries in the world. The glaciers on the Mount Blanc were melting causing slush: the impact of what keeps on happening in the solar atmosphere and the sun directly affect our planet, hence, there is a need to tackle global warming. Environmentalists present a grim picture of the effect of global warming with the mean temperature of the earth increased by about 1.6°C. This is a declaration made at a conference of parties to the Framework Convention on Climatic Changes. Global warming represents the increase in the average temperature of earth's near-surface air and oceans (ref: "according and pollution laws in India" second edition).

According to a statistical report from the World Meteorological Organization, global temperature has registered an increase of more than 0.6°C in the past 100 years. In the near future contamination of underground and surface fresh water supplies is likely near coastal areas (Ref: Law relating to Environmental Pollution & Protection, 5th Edition. Vol 1, by Dr. N. Maheshwara Swamy).

Global warming has its impacts in other parts of the World, as in the Indian sub-continent. It is likely to affect the glaciers. There will be early and untimely melting of ice resulting in various environmental issues Rohtang Pass being one of the eco-sensitive and fragile areas of the glacier is likely to get affected more than other areas. Thus, there is a need for evolving schemes and mechanism to take greater care of the glacier in the interest of environmental and ecological balance."

Indian corporations are complying Environmental, Social & Governance (ESG) norms like never before. Boards of all prominent companies are working towards reduction of carbon footprints by adopting mitigation and adaptation measures. United States Security Exchange Commissions has mandated that listed public companies shall have to share their reduction plan publicly. Likewise the Security and Exchange Board of India must now enact a GHGs disclosure regulation. Indian challenges relating to climate change mitigation are very unique. India is geographically located in South Asia. Weather usually is hot. We are the second largest population on the earth. We are one of the biggest carbon emitters, e.g., in 2016 India's total GHG emissions, excluding Land Use Land-Use Change and Forestry (LULUCF) were 2,838.89 million ton CO_2 and 2,531.07 million ton CO_2 with the inclusion of LULUCF. Energy, Transport and construction are key emitting sectors. Energy sector contributes 75.01%, Agriculture – 14.37%; Industrial process – 7.98%; Waste – 2.65% in the total carbon emissions. Key climate risks in India are cold waves, hot waves, cyclones, floods and droughts. Vulnerable Sectors in India are agriculture and Food, Water, Coastal, Health, Forests and other natural ecosystems. India shall have to walk a tight rope as Indian energy needs are demanding and power supply remains in deficit. Out of installed generation capacity of 395,075 MW, fossil fuel energy is 235,929 which is 59.7%. Indian population growth rate is 1.04%. With the growing population energy demands and anthropogenic activities are going to increase on Indian subcontinent. Installing renewable energy sources of 450 GW by 2030 is an ambitious project. Switching over to Electric vehicles seems topsy-turvy as fire and blasts causing loss of human lives have been reported due to overheating of batteries. Switching over to cleaner technologies will have cost implications which India might not be ready to bear at the moment.

We know that climate change is happening due to anthropogenic reasons and we humans shall have to take lead in mitigating the climate change with full vigor. Individual and institute, rulers and ruled, manufacturer and consumers shall have to play an active role to save our plant and with this firm belief we pledge to save our planet.

Oṃ dyauḥ śāntir *antarikṣaṃ* śāntiḥ;

pṛthivī śāntir āpaḥ śāntir *oṣadhayaḥ* śāntiḥ

vanaspatayaḥ śāntir *viśvedevāḥ* śāntir *brahma* śāntiḥ;

sarvaṃ śāntiḥ śāntir *eva* śāntiḥ *sā* *mā* śāntir *edhi*

oṃ śāntiḥ śāntiḥ śāntiḥ[14]

— *Yajurveda 36:17*

14 "Om. May peace radiate there in the whole sky as well as in the vast ethereal space everywhere.
May peace reign all over this earth, in water and in all herbs, trees and creepers.
May peace flow over the whole universe.
May peace be in the Whole Universe.
And may there always exist in all peace and peace alone.
Om peace, peace and peace to us and all beings!"

Chapter

1 CLIMATE CHANGE MITIGATION

Contents

1.1. CLIMATE CHANGE

"What Is the Use of a House if you Haven't Got a Tolerable Planet to Put it on"?[1]

Our planet has everything like sea, flora, fauna, forest, plant, food, animal, glaciers, natural resources, marine life, balanced monsoon, agriculture, expected human life, for which we don't have to spend even a single penny. We have got it without any effort. Suppose one day they disappear. From where we will acquire these things again, can we buy it? Or can it be generated again through new technology? Can we produce it through industry? The answer to this question is no. Human cannot generate, produce and manufacture natural environment and natural resources. If human being continues to exploit environment, natural resources in unsustainable manner then one day their capacity to support human life will end. We, the human being, are known as the greatest creation of God, who has control over his life, can discover new things, has capability to make their life better then why can't human, preserve the climate and natural environment which they have got without any struggle and effort.

Climate change has become inevitable problem for our planet. Climate change has many effects like disturbance of monsoon, melting of glacier, rise in sea level, tsunami, earthquakes, cyclone, desertification, rise in global temperature, shrinking ice sheets, declining Arctic Sea ice, ocean acidification, decreased snow cover etc.

UNEP Report of 2012 discusses about certain glaring effect of climate change. Between 1906 and 2006, Africa's glacier lost 82% of land area; over last 25 years, 25% of global land area has seen productivity hit due to soil carbon loss and the Japan tsunami produced 6.15 million tons of debris in one city, equal to 103 years of normal waste.[2]

Other evidence of rapid climate change is Global Sea level rose about 17 centimeters in the last century[3]. The rate of rise in last decade has almost doubled to the last century. If it continues then country like Maldives and other island will be immersed in sea. The Greenland and Antarctic ice

[1]Henry, Da. T., US Transcendentalist Author (1817–1862).

[2] "*Annual Report FY 2012.*" United Nations Environment Program. https://www.unep.org/resources/annual-report/unep-2012-annual-report (accessed on 07 July 2023).

[3] Church, J. A., & White, N. J., (2006). 20th century acceleration in global sea level rise. *Geophysical Research Letters*, *33*, L01602. doi: 10.1029/2005GL024826. https://agupubs.onlinelibrary.wiley.com/doi/full/10.1029/2005GL024826 (accessed on 07 July 2023).

sheets have decreased in mass. Data from National Aeronautics and Space Administration (NASA's) Gravity Recovery and Climate Experiment show Greenland lost 150 to 250 cubic kilometers of ice per year between 2002 and 2006, while Antarctica lost about 152 cubic kilometers of ice between 2002 and 2005. The increased heat in atmosphere is absorbed by ocean has resulted into warming of 0.302° Fahrenheit since 1969.[4] The acidity of surface ocean waters has increased by about 30% as a result of emission of more carbon dioxide into the atmosphere.[5]

It has been cleared by IPCC 5th assessment report that climate change is anthropogenic. Human activities, especially in the industrialized countries, releases a number of gases mainly carbon dioxide, CFCs, methane and nitrous oxide that slow the escape of infrared radiation (heat) from the surface of the earth into space. It is estimated that due to the increased human emission of greenhouse gases, earth will warm up at an average of nearly 2°F by 2025 and 5° by 2100. This may not seem much. However, we cannot lose sight of fact that the earth has warmed up only by 9° since the coldest period of the last ice age, about 18,000 years ago. Moreover, the predicted range of increase during the next 100 years will be over 0.5 degree per decade which is much faster than any climate change recorded in the history.[6] The first ever report on climate change assessment has been prepared by Union Ministry of Earth Sciences (MoES) named as "Assessment of Climate Change over Indian Region." Report states that India's average temperature has already increased by around 0.7°C during the period of 1901–2018 due to greenhouse gas emission and by the end of 2100 it is expected to rise approximately by 4.4°C.[7] The report states that rapid change in the temperature means adverse impact on India's natural ecosystem, agriculture, fresh water resources, simultaneously causing escalating damage to infrastructure. This ultimately leads to the adverse impact on countries bio-diversity, food, water, energy security and public health.[8] The Intergovernmental Panel on Climate Change (IPCC) has published its special report in 2018 on the implications of a 1.5°C rise in average global temperature. The report states climate change

---

[4] Levitus, et al., (2009). "Global ocean heat content 1955–2008 in light of recently revealed instrumentation problems. *Geophysical Research Letters*, *36*, L07608.

[5] *What is Ocean Acidification?* PMEL Carbon Program. http://www.pmel.noaa.gov/co2/story/What+is+Ocean+Acidification%3F (accessed on 07 July 2023).

[6] *IPCC Assessment Report*. http://www.ipcc.ch/pdf/assessment -report /art5/syr/SYR_AR_FINAL_full.pdf (accessed on 07 July 2023).

[7] Government of India Report on "*Assessment of Climate Change over Indian Region*" (Ministry of Earth Science, 2020). https://pib.gov.in/PressReleaseIframePage.aspx?PRID=1706938 (accessed on 07 July 2023).

[8] Ibid.

from human-induced causes is already taking place and is both accelerating and intensifying. Average global temperatures have risen about 1°C since the start of the Industrial Revolution. The harmful effects are visibly evident like the extreme weather, rising sea levels and diminishing Arctic sea ice, among other changes. Temperature rise, even of 1.5°C, may result in disastrous and irreversible changes. 70–90% of coral reefs across the world would die at 1.5°C temperature rise. At 2°C rise, none would be left. According to report global warming so far may have led to the loss of over 40–50% of the coral reefs. The loss of fish populations due to loss of coral reefs will have an impact on food security which in turn affects the human health.[9]

We, human being, are actually responsible for climate change. Climate change is not seen but already happening. Disregarding the fact of climate change, does not make it less true and it is going to get much worse. Climate change will make the food system more unstable. It will also exacerbate health problems and will result into displacement of people. Due to unexpected climate change infrastructure may be weakened and may lead to fuel conflict. The entire ecosystem will be forced to move. It will be touching every area of life. The repercussion of climate change could be worse than we are predicting. To minimize the impact of climate change, the United Nation Conference on Human and Environment took place in Stockholm in 1972. This conference was first to initiate the world-wide participation from all over the world for the protection of the environment in order to operationalize right to life.[10]

The UNFCCC was adopted in 1992 as result of the report of the World Commission on Environment and Development (WCED) known as "Our Common Future" published in 1987 to solve the emerging environmental problem of climate change. It was supplemented by the Kyoto protocol. In December 1997, parties to the UNFCCC were scheduled to sign protocol to the convention which established a binding target and time table for greenhouse gas emission reduction. The Principle of 'common but differentiated responsibility' was evolved to fix the low carbon emission by developed nations. The Johannesburg summit 2002, Rio plus 20 was further step for protection of environment. The UNFCCC has been given a boost by several 'Conferences of Parties.' The recent one is COP 21 which is also known as Paris agreement. Till now COP 25 have taken place.

---

[9] *IPCC Special Report on "Global Warming of 1.5°C,"* (2018), https://www.ipcc.ch/site/assets/uploads/sites/2/2019/06/SR15_Full_Report_High_Res.pdf (accessed on 07 July 2023).

[10] Sarath, C., (2002). Human rights and environment protection. *Cochin University Law Review,* 175, 176.

1.2. PROBLEMS AND CHALLENGES

Climate change is bound to occur due to the many reasons and human beings are mainly responsible for climate change by their unsustainable activities. And it is the duty of all the individuals living on the earth to mitigate the climate change by their individual and collective efforts. The country like India which is developing has the dual responsibility, i.e., environment protection through low carbon emission and development with the use of fossil fuels. Both responsibilities need simultaneous focus. One cannot be ignored in order to fulfill another. India needs to be developed for providing better living standards to its residents. This can only be done through industrialization which is the main contributing factor for climate change. Then the question comes should we stop development? Which is the fundamental need of every person living in India. The answer is; India should not stop development but must ensure climate change mitigation through low carbon emission. But to reduce and prevent anthropogenic emission there is need of adoption of new technologies and practices in all sectors like energy, transport, industry, agriculture, forestry and waste management. Then again problem comes that India is developing country and there is lack of technology and financial resources. The main focus for mitigation in India is on large emitting sectors, i.e., energy, transport and waste management, coastline pollution, air traffic pollution are ignored for mitigation, which also contribute in climate change. Economic growth of India will slow as temperature will become warm. In India most of the people live below the poverty line, who are more vulnerable due to flimsy housing, poor health care which results into their victimization of natural disaster. Due to unanticipated climate change new type of poverty will be created and it is found that poverty cannot be eliminated from India without first tackling climate change.

Many communities have faced the consequences of climate change in the form of disturbance of monsoon, melting of glacier, rise in see level, tsunami, cyclone, earth quacks, global warming etc. Reasons of climate change are mostly anthropogenic and results are disastrous. Greenhouse gas emission could reach 58 gigatons annually in 8 years if no action is taken. In Asia and India in the Indo-Gangetic plains of South Asia, there could be a decrease of about 50% in the most favorable and high-yielding wheat area as a result of heat stress at two times CO_2. Sea level rise will submerge low-lying areas and will specially affect rice growing region. India has reason to be concerned about climate change. A massive population in India depends on climate sensitive sector like agriculture, forestry, and fishery for livelihood. The adverse impact of climate change in any form will increase severity of drought and flooding. This will intimidate food securities and livelihood of the person residing in India. Due to poor infrastructure, deficiency of financial resources, vast population, lack of technology and weak institutional mechanism India cannot go for adaptation of renewal technologies. The only option remains to India is mitigation of climate change. Societies can respond to climate change by reducing greenhouse gas emission and enhancing sink and reservoirs. Climate change is a collective problem which needs collective focus. India is a developing country and development is fundamental for economic growth at same time environment protection measures should not deprive the country from economic growth. India being developing country need to take action for climate change mitigation along with development and cannot escape from the liability under international conference on environment protection.

1.3. SCOPE OF CLIMATE CHANGE

The work aims to investigate the areas, e.g., to study the climate change, its impact and need for climate change mitigation; To find out the interrelation between climate change, poverty and population; To look into international environmental law relating to climate change mitigation; To look into the Indian legislative regime relating to climate change mitigation; To analyze the focused and neglected areas of climate change mitigation in India; To search for India's contributing factors for climate change and its regulation; To search for the best method for focusing on neglected areas of climate change mitigation; To address the economic and social consequences of response measures; To critically apprise and compare into the steps to be taken by India for mitigation under Paris convention 2015; To look for the future mitigation opportunity; and to examine the problem of implementation environment laws and suggest the best method of implementation.

The work aims to investigate questions like; Whether climate change mitigation is sufficient or should India look for adaptation also?; Is there any link between climate change mitigation, adaptation, and sustainable development?; Whether Climate change mitigation is an onerous duty only on developing nations due to their energy needs?; Whether developing nations need to balance between right to environment and right to development?; Whether polluter pays principal, sustainable doctrine, public trust doctrine are part and parcel of climate change mitigation?

1.4. EDIFICE OF WORK

The authors have referred *Handbook of Climate Change Mitigation* edited by Wei-Yin Chen, Seiner, J., Toshio Suzuki, Maximillian Lackner, which helped the authors to understand climate change and its impact international efforts to combat global warming. The book, *Handbook of Climate Change Mitigation and Adaptation* edited by Wei-Yin Chen, Toshio Suzuki, Maximillian Lackner helped the authors to understand the rational for the reality of climate change mitigation and adaptation.

The book entitled as *Environmental Economics and Development* written by Jiwitesh Kumar Singh and Debendra Kumar Das helped the authors to understand relation between Sustainable Development and poverty. The book on *Environment Protection, Land and Energy Law* written by Sanjay Upadhaya and Videh Upadhaya, helped the authors to know about international obligation and environment protection. The book on *Planning and Environment Protection: A Review of Law and Politics* Edited by Chris Millor, helped the authors to understand environmental impact assessment.

The Article entitled as 'Mitigating climate change: Indian perspective' written by Dr Ashish Kumar and Anurag Kumar Srivastava helped the authors to build the concept about climate change and its impact, climate change mitigation, UNFCCC and Indian perspective of climate change mitigation. The Article on 'Climate Change and Mitigation' written by Kerstin Nibleus and Rickard Lundin helped the authors to understand contributing factors for climate change.

The Article on 'Climate Change Mitigation and Sustainable Development' written by Laura H. Kosloff helped the authors to understand the link between climate change mitigation and sustainable development. The objective of achieving 'sustainable' development is premised on the assumption that human economic needs and environmental protection can be accommodated – that environmental and economic goals do not have to be mutually exclusive.

The Article on 'Climate Change and Domestic Mitigation Efforts' written by Vijaya Gupta helped the authors to understand Indian vulnerability to climate change and domestic efforts to mitigate greenhouse gases. The Reports of IPCC and UNEP has been used by the authors for better and efficient presentation of work extensively for validation and authenticity.

1.5. WHY THIS BOOK?

Climate change is a prominent issue. It is assumed and acknowledged that climate change mitigation is the only solution to this human-caused problem. It is believed that the UNFCCC and the Paris Convention of 2015 have made a strong effort to encourage the global community to take direct and collective action to mitigate climate change. It is understood and believed that India, as a concerned state and a champion of developing nations, has taken steps to address climate change and promote a sustainable environment without losing sight of its development goals and energy needs. The authors of this work aim to examine and investigate the legal aspects of climate change mitigation, with a specific focus on the Indian perspective, in order to propose a more comprehensive and holistic approach to climate change mitigation that does not compromise development.

This research will help society understand the concept of climate change and the need for climate change mitigation. Through this research, society will also become aware of India's vulnerability to climate change, as well as its willingness and efforts to mitigate it. The research also explores how individuals can contribute to climate change mitigation.

Climate change mitigation is a broad topic with many aspects. It is a global issue, not just a national one. This research is limited to studying climate change and its impact, climate change mitigation, and Indian efforts and obligations for climate change mitigation. The study focuses on areas that need attention for climate change mitigation in India, as well as the response measures taken by the judiciary and legislature in India and India's obligations under the Paris Agreement of 2015. The authors have not studied the response measures taken by other developing and developed countries for climate change mitigation. The study is limited to legal aspects only.

The book is divided into seven chapters for a clear and organized presentation. Chapter 1, titled "Climate Change Mitigation," introduces the work and its plan. Chapter 2, titled "Climate Change and Its Impact: Meaning and Concept," provides a detailed introduction to the meaning, nature, and scope of the term "climate change." The authors describe what climate change is and the serious concerns associated with it. They discuss the causes and adverse impacts of climate change, as well as the interrelation between climate change and other concepts such as poverty, population, and migration.

Chapter 3, titled "The Climate Change Mitigation: The Legal Framework," attempts to define mitigation and adaptation. The authors explore the various options available to humanity to curb and mitigate climate change. They provide various methods of mitigation and adaptation, with a special focus on renewable energy. They also discuss the interrelation between mitigation and adaptation and the challenges associated with implementing these strategies.

Chapter 4, titled "Climate Change Mitigation: International Legal Order," highlights the need for an international climate change regime. It discusses the evolution of the international climate change regime and the development of what is now known as the International Climate Change Regime. The authors emphasize the scope and gradual development of this international regime to address the crisis of climate change, with a specific focus on the Kyoto Protocol and the Paris Agreement of 2015, as well as post-Paris negotiations.

Chapter 5, titled "Climate Change Mitigation: Indian Legal Order," examines various laws and policies related to environmental protection and climate change from a national perspective. This chapter explores domestic legislation and institutions that aim to address the crisis of climate change. The authors focus on the role of the judiciary in protecting the environment and mitigating climate change, particularly highlighting the principle of sustainable development.

Chapter 6, titled "Climate Change Mitigation and Paris Agreement, 2015: Indian Initiatives," focuses on India's obligations under the Paris Agreement and its efforts in various sectors to mitigate climate change and fulfill its commitments outlined in its Nationally Determined Contributions (NDC). The authors discuss the challenges faced by India in meeting its promises and highlight recent initiatives taken by the government. They also emphasize climate-friendly lifestyle practices in India for mitigating climate change.

Chapter 7, titled "Climate Change Mitigation Prospects," explores the future of climate change mitigation.

Chapter

2 CLIMATE CHANGE AND ITS IMPACT: MEANING AND CONCEPTS

Contents

2.1. INTRODUCTION

Since its beginning, the development of human civilization is majorly dependent on the favorable geographical location and conditions which include climatic conditions, fertile land, clean air, and water. Many ancient civilizations like the Indus Valley Civilization and the Mesopotamian Civilization developed on the river banks because of the availability of arable land, clean water, and air. A climate that has influenced the evolution of life on earth since its inception to present, man has driven their life due to the changing seasons, atmosphere, and climate.[1] The gradual change in climate has been supportive of the life of men, animals, pets, plants, and ecosystems. But today the change in climate has become a matter of concern; the reason is the drastic change in climate due to explicit use of resources and technological and industrial revolution. The concentration of greenhouse gas like Carbon Dioxide, Methane, and Sulphur Dioxide has adversely affected the climate cycle, life, and livelihood of the people.[2] The increase in temperature, depletion of ozone layer, raising of sea level, change in the rainfall pattern, scarcity of water, cyclone etc., are significant concerns. Due to all these reasons, climate change has become a very significant issue at the political, economic and ecological platforms from the last two decades.[3] Climate change represent a major threat to the world and it would be affecting each country differently.

2.2. CLIMATE CHANGE: MEANING AND DEFINITION

Climate is changing since the evolution of the Earth, but recently it was observed that it has taken up speed which has become threat to mankind and very existence of the planet. Oxford dictionary define climate change as.

"A change in global or regional climate patterns, in particular a change apparent from the mid to late 20th century onwards and attributed largely to the increased levels of atmospheric carbon dioxide produced by the use of fossil fuels"[4].

---

[1] Ramkrishna, P. S., (2010). *Primer on Characterizing Biodiversity: Trans-Disciplinary Dimension* (p. 12). New Delhi: National Book Trust.

[2] Srinivasan, A., (2008). *Climate Regime Beyond 2012: Reconciling Asian Development Principle and Global Climate Interest* (p. 130). Hayama, Japan: Institute for Global Environmental Strategies.

[3] Anderson, S., Steinar et al., (2002). *Leaders, Pushers and Laggards in the Working of the Climate Regime* (Vol. 12, pp. 41–45). Global Environmental Change.

[4] https://en.oxforddictionaries.com/definition/climate_change (accessed on 07 July 2023).

Another definition is *"The climate change phenomenon refers to seasonal changes over a long period with respect to the growing accumulation of greenhouse gases in the atmosphere."*[5] Dictionary.com defines climate change as "*a long-term change in the earth's climate, especially a change due to an increase in the average atmospheric temperature*"[6] Collins's Dictionary defines climate change *as "change occurring in the Earth's overall climate and in particular climates, now regarded as a result of human activity and resulting generally in global warming."*[7]

The climate change has been defined by United Nation Framework Convention on Climate Change and Intergovernmental Panel on Climate Change.

"*A change of climate which is attributed directly or indirectly to human activity that alters the composition of the global atmosphere and which is in addition to natural climate variability observed over comparable time periods.*"[8]

"*Climate change as referred to in the observational record of climate occurs because of internal changes within the climate system or in interaction between its components, or because of change in external forcing either for natural reasons or because of human activity. It is generally not possible clearly to make attribution between these causes. Projections of future climate change reported by IPCC generally consider only the influence on climate of anthropogenic increases in greenhouse gases and other human-related factors.*"[9]

Climate change has gone beyond alarming level. In the reports found that in November 2006, Arctic temperatures observed to be 20°C above normal. If we consider this was one of occurrence, in February 2018 the Arctic temperature were again found to be over 20°C above normal. This alarming rise in temperature draw attention of many climatologists and environmentalists.

Our planet's average surface temperature has risen by more than 1°C since the late 19th century, Indian cities is also getting warm with as year passes. The Earth's climate can be thought of as long-term weather as many

---

5 https://beeatna.ae/en/definition-of-climate-change (accessed on 07 July 2023).

6 http://www.dictionary.com/browse/climate-change (accessed on 07 July 2023).

7 https://www.collinsdictionary.com/dictionary/english/climate-change (accessed on 07 July 2023).

8 Definition given by UNFCCC.

9 Definition given by IPCC.

of us will tell, it's always changing. So why all bother this time because this time changing of climate is different?[10]

The giveaway is unnatural steadiness of climb. Natural climate variation during the past century has been cyclic. 'Sunspots have an 11-year cycle while El Nino has a four to seven years cycle. Doctor J. Srinivasan[11], says Human induced changes, however, because a monotonic increase in temperature. The world agrees that this warming has been brought about by humans.[12] The burning of fossil fuels such as coal, petroleum and natural gas along with our efforts at deforestation have increased the concentration of carbon dioxide (CO_2) and other greenhouse gases that have warmed our planet. But the warming is found to be speedy as the normal temperature has been increased to 1°C which is really alarming and a great matter of concern. Why did the Paris Agreement[13] in 2015 have world leaders clamoring to restrain temperature increased to less than 2°Centigrade? To a layman this may seem to be a bit or when temperature within a single month change more than that, why are the world scientist and leaders so concerned about what seems to be a small increase in temperature.

A warming of earth by couple of degrees for a few days or months may lead to higher cooling costs and slightly worse tempers. However, a sustained warming of the climate even by as little as a degree is a very big deal indeed the key word here being sustained[14] – because a sustained increase in a temperature affects the world in three broad ways: it changes rainfall pattern, it melts major ice caps and it displaces animal and plants.

In 2014, deadly flood ravaged to Jammu and Kashmir and left over 500 dead. Omar Abdullah, the Chief Minister of the State, bemoaned 'I had no government, my secretariat, the police headquarters, the control room, fire

---

[10] Mridula, R., (2018). *The Climate Solution: India's Climate-Change Crisis and What We Can Do About It* (p. 15). New Delhi: Hachette India.

[11] A Distinguished Scientist, Divecha Centre for Climate Change, Indian Institute of Sciences.

[12] The 2018 US National Climate Assessment—whose authors include 300 leading scientists and thirteen federal government agencies—concludes that "it is extremely likely that human activities, especially emissions of greenhouse gases, are the dominant cause of the observed warming since the mid-20th century. For the warming over the last century, there is no convincing alternative explanation supported by the extent of the observational evidence."

[13] The Paris Agreement is a legally binding International Treaty on Climate Change. It was adopted by 196 Parties at COP 21 in Paris, on 12 December 2015 and entered into force on 4 November 2016. Its goal is to limit global warming to well below 2, preferably to 1.5°C , compared to pre-industrial levels.

[14] Randers, J., & Goluke, U., (2020). An earth system model shows self-sustained thawing of permafrost even if all man-made GHG emissions stop in 2020. *Scientific Reports, 10.* https://www.nature.com/articles/s41598-020-75481-z (accessed on 07 July 2023).

services hospital all the infrastructure was underwater I have no cell phone and no connectivity.'

The Kashmir flood was caused by the combination of unusual heavy rains and invoice urban development which involves encroaching or building over water bodies, riverbank and marshlands. Warmer air holds more water vapor this is a basic Law of physics. Which means as the climate has warmed, rainfall has become more episodic and more intense. Scientists have found that over many parts of India the number of 'rain days' (or days when it rains) has been decreasing while the incidence of extreme rainfall have been increasing which depicts that a warmer world may have more storms.[15] Recent example is rainfall in Henan, China[16], it is also being witness in Karnataka, Maharashtra, and Uttarakhand[17] Our existing infrastructure-cannot handle intense rainfall, even if they were maintained perfectly. The drainage system is built which can sustain certain maximum intensity of rainfall beyond which the rainfall can overwhelm the system and will cause flooding. The occurrences of flood have become frequent in the past 100 years and damaged have skyrocketed graphic, which presents an evidence that our climate has indeed changed. In a very short span of time in June 2013, Uttarakhand received 847% more rainfall than 'normal' in a multiday cloudburst. Large scale deforestation and developmental work had already destabilized the Himalayan slope and made landslides more likely. This unprecedented cloudburst together with the melt water from the Chorabari glacier caused multiple landslides. Debris from these landslides dammed the Mandakini River which flooded Kedarnath.[18] The disaster was supplemented by the rusting of Lake above the town. The disaster left over 4,000 people dead, officially, and many more deaths occurred which are undocumented. As per the World Bank report, the damage and reconstruction costs exceeded $600 million for this disaster.

The impact of the Uttarakhand flood became worst by combination of reasons; the flood is struck pilgrimage center during the peak season

[15] Mridula, R., (2018). *The Climate Solution: India's Climate-Change Crisis and What We Can Do About It* (p. 16). New Delhi: Hachette India.

[16] Sonal, G., (2021). *Explained: China's Heaviest Rainfall in 1,000 Years, Which Resulted in Devastating Floods*. The Indian Express. https://indianexpress.com/article/explained/explained-chinas-heaviest-rain-in-1000-years-floods-7415657/ (accessed on 07 July 2023).

[17] *Ganga Water Level Rises Due to Heavy Rainfall in Uttarakhand,* (2021). Asian News International. https://www.ndtv.com/india-news/ganga-water-level-rises-due-to-heavy-rainfall-in-uttarakhand-2496911 (accessed on 07 July 2023).

[18] Nagraj, A., (2019). Impacts of global warming in India: narratives from below In: Navroz, K. D., (ed.), *India in a Warming World: Integrating Climate Change and Development* (p. 78). New Delhi: Oxford University Press,) 78.

of Chardham Yatra, which meant that more people were at risk. India's economic development has also ensured that more pilgrims than ever are making the trip. For instance, the number of pilgrims to Kedarnath exploded from 90,000 in 1987 to 5,75,000 in 2012. Catering to this inflow, hundreds of buildings were constructed on the floodplains of the rivers and the rubble thrown into the rivers from construction reduced the river channel's size thus reduced river carrying capacity. [19]

Climate change-induced sea-level rise take place for two key reasons: first, over 90% of the excess heat trapped by GHGs since 1971 has gone into oceans worldwide, causing the warmer waters to expand and second, melting ice from glaciers and the fact that the great ice sheets on Greenland and West Antarctica 'are losing mass at an accelerating rate.'[20]

Scientist estimates that if all the ice of the glaciers were to melt, the sea level rise by 17 meters this is a lot but it is not a catastrophic and will not occur suddenly. However, if the Greenland ice sheet were to melt entirely, it would increase the sea levels by 6 meters. If the Antarctic ice sheet were to completely melt the sea would rise by about 60 meters. That would change (or destroy) human civilization.

Many of us remain relatively sanguine because scientists have said sea-level rise would be a gradual and manageable. But in 2015 Dr. Hansen[21], wrote the research paper which found that the polar ice sheet could be melt rapidly for which there is an evidence of ice melting and sea level rise +5–9 meters and extreme storms in the prior interglacial period was less than 1°C warmer than today. If the giant ice sheet were to melt suddenly, many of the world's biggest cities would be submerged under water. His paper brought rapid sea level rise from a distant future problem to a problem that would occur well within our lifetimes. The thought of losing Mumbai and Chennai to see in the next hundred years should make us pause. But even much milder sea level rise would increase a destructive power of storm surges.[22] In 2016 in Rajasthan, the city of Phalodi all-time record high temperature was

---

[19] Mridula, R., (2018). *The Climate Solution: India's Climate-Change Crisis and What We Can Do About It* (p. 19). New Delhi: Hachette India.

[20] Nagraj, A., (2019). Impacts of global warming in India: narratives from below. in Navroz, K D., (ed.), *India in a Warming World: Integrating Climate Change and Development* (p. 67). New Delhi: Oxford University Press.

[21] The retired head of the NASA and scientist credited with the bringing global warming to the public knowledge.

[22] Mridula, R., (2018). *The Climate Solution: India's Climate-Change Crisis and What We Can Do About It* (p. 19). New Delhi: Hachette India.

found in India as it hit 51°C on 19 May 2016.[23] The Japan Meteorological Agency recently announced that the Cherry Blossom season of Japan is in full bloom. Usually, the Cherry trees of Japan reach their full bloom in April. This time the Cherry Blossom has occurred ten days earlier. This is the first time the Cherry trees have bloomed this earlier in Japan. Previously, the earliest Cherry Blossoms in Japan occurred before 1,200 years. According to Osaka University of Japan, the flowering of the Cherry trees was this earlier in 812 AD. The Climate Change and global warming has been identified as main reason behind this early arrival of cherry Blossom season. The average temperature of Japan increased from 47.5°F in 1953 to 51.1°F in 2020. The leaves of the Cherry trees secrete a hormone that stops the flowers from blooming further. The secretion of this hormone stops when the temperature increases. After the weather becomes warm, the flowers bloom completely. This year, due to climate change, the flowers reached this stage earlier than usual as the hormones stopped secreting.[24]

2.3. CLIMATE CHANGE AND ECOLOGY

A change in the climate is profound change, a modification in some of the interacting forces that, in combination with its distance from the sun, give this planet its distinctive environment. What may be harmed by human choices to bring about avoidable change in the climate depends upon what is morally considerable, the range of things whose interest deserve to be considered. However, since the climate is planetary level phenomenon, virtually anything that is indeed morally considerable may well be affected by the extent or speed of climate change. For example, the species initially flourished on earth now are species that found a hospitable spot in the climate we have now. Species adapt-some migrate-and specifications are occurring, but natural adaptation and natural evolution occur on a timescale far slower than the timescale that appear likely if increases in the atmospheric concentration of anthropogenic greenhouse gases continue. Few species can adapt nearly as rapidly because the climate may change. The primary cause of species extinction is loss of habitat; rapid climate change could be the ultimate destroyer of their habitats. Insofar as existing species, or

---

[23] *India Records its Hottest day Ever as Temperature Hits 51C (that's 123.8F),* (2016). Agency France-Presse. https://www.theguardian.com/world/2016/may/20/india-records-its-hottest-day-ever-as-temperature-hits-51c-thats-1238f (accessed on 07 July 2023).

[24] *Photos: Cherry Blossoms Bloom Early in Japan as Climate Warms*, (2021). Hindustan Times. https://www.hindustantimes.com/photos/news/photos-cherry-blossoms-bloom-early-in-japan-as-climate-warms-101617442427392.html (accessed on 07 July 2023).

the ecosystem in which they live, or the natural processes through which they would otherwise change, have some value and are worthy of protection from avoidable human destruction or distortion-insofar as anthropogenic extinction in an ethical issue-anthropogenic rapid climate change is morally suspect for the harm it will inflict beyond human species.[25]

Here, the focus will be on the harm to the human, which is at least of two kinds, the second somewhat less evident than the first. First is straightforward if indirectly caused, physical harm of many kinds climate change being centrally involve in relatively simple changes in weather, some of which will be dangerous. Consider water and insects, some location that now have abundant rain fall will have inadequate rain fall; drinking water may become inadequate. Other location that suffers drought may receive heavy rain fall leads to flood and causing people to drown. Changes such as deficit and surplus and drinking water and uncomplicated utterly ordinary, but people and crops and livestock, and thus indirectly more people perish from drought and floods. And war, in which of course more people (and crops and livestock) die, are fought over access to water for drinking, irrigation, and manufacture[26].

2.3.1. Climate Change and Global Warming

Climate change is caused by the emission of heat trapping gases mostly carbon dioxide from vehicle, industry, power plant and deforestation. As these gases build up, they act like a thick blanket over hitting the planet changing our climate and threatening our health economy and natural environment. The term global warming and climate change are often used interchangeably, but the two phenomena are different. Global warming is rise in the global temperature due to an increased heat trapping carbon emission in the atmosphere. Climate change on the other hand is a more general term that refers to change in many climatic factors such as temperature precipitation around The World. These changes are happening at different rates and in a different ways for example United States become wetter over the 20th century while the Sahel region in Central Africa became drier.[27]

[25] Henry, S., (2014). *Climate Justice Vulnerability and Protection* (p. 196). New York: Oxford University Press.

[26] *Ibid.*

[27] Sarkar, A. N., (2010). *Global Climate Change: Beyond Copenhagen* (p. 33). New Delhi: Pentagon Earth.

2.4. CHANGING OF CLIMATE

When more CO_2 and other heat trapping gases added to the atmosphere, the world will become warmer. Historical records of temperature and precipitation have been extensively analyzed in many scientific studies. These studies show that the global average surface temperature has increased by over 1 degree Fahrenheit (0.6°C) during the 20th century. About half of this rise has occurred till the late 1970s. The warmest years in the 20th century occurred since 1980. In 1998, global temperature set a new record by a wide margin, that of the 1997, by about 0.3-degree Fahrenheit. Higher latitude is warmer than equatorial region, and night time temperatures have risen more than daytime temperature.[28]

As the earth warms, more water evaporates from the ocean, rivers, lakes and other water bodies eventually to fall as rain and snow. During the 20th century, annual precipitation has increased about 10% in the mid and high latitudes. The warming is also causing permafrost to thaw, and melting sea ice, snow cover, and mountain glaciers.

Global sea level rose to 4 to 8 inches during 20th century because ocean water expands as it warms and because melting glacier are adding water to the oceans. According to the intergovernmental panel on climate change scientific evidence confirms that human activities are a discernible cause of substantial part of the warming experienced over the 20th century. Temperatures in recent decades are higher than at any point of time in at least the past 1000 years. It is very unlikely that these unusually high temperatures can be explained solely by natural climatic variations. The intensity and pattern of temperature changes within the atmosphere implicates human activities as a cause.

The relevant question is not whether the increase in greenhouse gases is contributing to global warming, but rather, what will be the amount and rate of future warming and associated climate changes, and what impacts those changes have on human and natural systems.

The first ever report on climate change assessment prepared by Union Ministry of Earth Sciences (MoES) named as "Assessment of Climate Change over Indian Region." Report states that India's average temperature has already increased by around 0.7°C during the period of 1901–2018 due to greenhouse gas emission and by the end of 2100 it is expected to

[28] Gurkirat, K., (2006). *Environmental Crime* (p. 53). New Delhi: Shree Publishers and Distributors.

rise approximately by 4.4°C.[29] The report states that rapid change in the temperature means adverse impact on India's natural ecosystem, agriculture, fresh water resources, simultaneously causing escalating damage to infrastructure. This ultimately leads to the adverse impact on countries bio-diversity, food, water, energy security and public health.[30]

2.5. CLIMATE CHANGE—AN ENVIRONMENTAL PROBLEM

The most popular perspective on climate change is to see it as an environmental issue. Majorly, the international climate policy is aimed to prevent menacing anthropogenic climate change by lowering down the net GHG emissions. The 2015 Paris agreement recognizes the severity of presence of carbon dioxide in the atmosphere and sets its goal for complete elimination of net emissions to prevent dangerous climate change. For reducing this emission at any particular point of time is basically a function of three factors first, the level of increase in temperature considered safe; second, the concentration level required to check warming from exceeding that temperature limit; and third the selection of the emission pathway to attain the required concentration level. In the Paris Agreement, the member states approved the goal to limit the temperature increase fairly below 2°C compared to pre-industrial emission levels and also worked towards the limiting of global warming to 1.5 degree centigrade. The reduction in the emission to the 2°C limit would require stabilization of GHGs concentration in the atmosphere at not more than 450 ppm and global emissions of GHGs to peak and then fall by 40% to 70% by 2050.[31]

2.6. CLIMATE CHANGE—AN ECONOMIC PROBLEM

Climate change can also be seen as an economic problem. From this perspective the goal of climate change policy is to achieve the effective outcome, i.e., the outcome with the highest net profit accordingly we should reduce emission to the extent so that the benefit of the reduced emission outweighs the cost of the reduction. When the station is cheaper than

[29] Government of India, (2020). *Report on Assessment of Climate Change over Indian Region*. Ministry of Earth Science, PIB, , https://pib.gov.in/PressReleaseIframePage.aspx?PRID=1706938 (accessed on 07 July 2023).

[30] *Ibid.*

[31] Daniel, B., Jutta, B., et al., (2017). *International Climate Change Law* (p. 6). London: Oxford University Press.

mitigation then the proposed policy for calculating cost and benefit will be extremely difficult. Many of the advantages of reducing emissions are non-market boots that are difficult to value immediately but will be realized in future. However, some rough weighing of cost and benefits is the basis of most decision making.

With defining the goal of efficiency, the economic perspective emphasizes on the pathways of achieving the goal of reducing emissions as much cost effective as possible. In general, a policy can, be cost effective if it equalizes the marginal cost of adherence over time and place. If GHGs emissions can be reduced reasonably at lower cost in the future then now or by one country more inexpensively than another, then it may be possible to reach the same climate benefit at a lower price by relocating some of the population reductions into the future or two nations with considerably lower mitigation costs.[32]

It is universally agreed that the rising level of pollution in cities and towns have been resulting in deleterious effects on human health. Air pollution alone is estimated to cause around seven million deaths in one year globally. But many of us fail to understand is that asthma, impaired lungs, high risk of cardiovascular diseases and cancer, and shortened life spans are a few of the multitude of effects of a degraded environment and these translate into enormous costs for the economy as well. A diseased populace produces an inefficient workforce, which works below its optimal level, and significantly lowers the margin between the output and input. In an environment where development is measured by the growth in Gross Domestic Product (GDP), one may not even see the correlation between the impact of environmental degradation on human health and economy, but research findings reveal a grim picture. According to the 2016 data of the World Bank, the health toll of air pollution costs India around 8.5% of its GDP, or more than $560 billion[33]. A major problem with mainstream economic indicators like the GDP is that they do not factor in ecosystem services, which refer to the broad spectrum of benefits derived from the ecological processes. These include efforts to recharge groundwater table, keep the nutrient cycles running, among others. However, such aspects do not feature in the mainstream narratives around "growth" and "development," simply because our economic models have

[32] Daniel, B., Jutta, B., et al., (2017). *International Climate Change Law*. London: Oxford University Press.

[33] Subhayan, C., (2016). *Economic Toll of Air Pollution May Have Cost India 8.5% of GDP: World Bank Report*. Business Standard. https://www.business-standard.com/article/economy-policy/economic-toll-of-air-pollution-may-have-cost-india-8-5-of-gdp-world-bank-report-116090900235_1.html (accessed on 07 July 2023).

failed to quantify the transactions in the ecosystem. Our biodiversity has always been perceived as an absolute and altruistic entity that would bless us with more, without demanding anything from us in return. The link between the environment and economy always existed and is going to be strengthen in the future.[34]

A study found that a single pollutant l lead towards a net loss in primary productivity, because of the reduced plant and biomass production, and increased plant susceptibility to stress[35]. On the other hand, environmental degradation has already begun to show its adverse impacts on the economy, for example, the real estate sector in the National Capital Region (NCR), New Delhi, has been incurring severe costs to transport water for construction purposes from distant places. This happened because of the dip in the groundwater reserves, which was resulted from excessive extraction, concretization, and deforestation. The increase in construction cost would result in decline in the profit margin for the employers in the sector. Another sector that has been reeling under the impact of environmental degradation is tourism which ultimately results in economic loss. [36]

2.7. CLIMATE CHANGE—AN ETHICAL PROBLEM

Equity and climate justice are also important aspects of climate change. Simple Cost-benefit analysis aims to maximize total economic value and does not get down to the ethical issues augmented by climate change. If one country acquires the economic benefits by polluting activity and another cost of pollution, the policy is still effective as long as, in total, the benefits exceed the cost. In contrast, the ethical perspective mainly focuses on the issues related to distributive and corrective justice which includes equitable distribution of pathways to mitigate and adapt climate change contrary to environmental impact and economic efficiency, for which there are relatively well accepted metrics, there is less unanimity about what equity and climate justice necessitate. Several accounts focus on historical responsibility, others on duty to next generations, other on an equitable division of burdens on the

[34] Ankita, N., (2020). *The Promise of Environmental Action for Economic Growth* (Vol. 55, No. 1). Economic and Political Weekly Engage.

[35] Persson, L., et. al., (2010). *Impacts of Pollution on Ecosystem Services for the Millennium Development Goals.* Stockholm Environment Institute, Project Report.

[36] Ankita, N., (2020). *The Promise of Environmental Action For Economic Growth* (Vol. 55, No. 1). Economic and Political Weekly Engage.

ground of present capacity, and yet other on the egalitarian principle that people have an equal right to the atmospheric space.[37]

2.8. HISTORY OF CO_2 AND CLIMATE CHANGE

The prediction of future climate by scientists is done through models. These models help in capturing the scientist's understanding of the magnitude and behavior of different pieces-CO_2 levels, ocean currents, clouds, and Ice sheets and so on and how these pieces affect the climate and interact with each other. This helps in predicting how the temperature of the globe along with its handmaidens, drought, flood and the rest will be changed as the CO_2 levels rises but the that rising understanding CO_2 levels have an impact on the rising temperature is relatively recent.[38]

In 1986, Svante Arrhenius, a Swedish scientist a childhood prodigy key who could read at the early age of three became the first person to check fossil-fuels directed global warming. He explained how radiation leaving Earth is trapped by CO_2 which is heating the Earth-which is called as greenhouse effect. Ironically, he pointed out the advantages of the greenhouse effect by saying: "we often hear lamentations that the coal stored up in the earth is wasted by the present generation without any thought of the future by influence of the increasing percentage of the carbonic acid in the atmosphere, we may hope to enjoy ages with more equitable and better climates, especially as regards the colder regions of the earth he even predict that arctic temperature would rise by 8–9 degree centigrade if atmospheric CO_2 levels rose two to three times from the level around 1900"

In these years the research gets stalled, perhaps because the globe wasn't warming or because any warming was contemplated advantageous. It wasn't until the advent of the cold war when weather was realized to be of utmost importance to provide a military advantage, that studies into climate re-emerged and with this thought came funding from various organizations.

Charles David Keeling, the son of Lumberjack gave deep insight which changed the overall understanding of climate change's greatest player-CO_2. Armed with a PhD from North-western University in chemistry, in the year 1950, he started measuring CO_2 levels in California with his specific device he designed himself. He took frequent readings of CO_2 level throughout the

---

[37] *Ibid.*

[38] Mridula, R., (2018). *The Climate Solution: India's Climate-Change Crisis and What We Can Do About It* (p. 30). New Delhi: Hachette India.

day and night because, in his words, "I was having fun... the prospects of spending more time at a Big Sur state park to take suites of air and water samples instead of just a few didn't seem objectionable, even if I had to get out of sleeping back several times in the night."

On his colleague's request, Keeling measured the carbon isotopic ratios in the given samples of air. This fortuitous decision would come up with tremendous insight into how world functioned – a reminder if we needed one, the role played by the sheer luck in great discoveries

Keeling finding surprised his colleague. He first observed that there was clearly a systematic pattern of the level of CO_2 in the air. The levels found to be lowest in the afternoon and highest in the night. This finding reveals that the night air held much less of the heavier carbon-13 Isotope. This concluded that this was because of the plants breathing-in the day time photosynthesis removed CO_2 from the air, but at night, breathing does not involve the process of photosynthesis so they released CO_2 which was found to be richer in the carbon-12 they preferred.[39]

Keeling next measured CO_2 levels in air samples across the world. What he found amazed him: "The CO_2 concentration in the air was nearly the same everywhere-around 310 parts per million but popular scientific consensus at that time was that tropical air held 350 PPM of CO_2 while Arctic air held as little as 150 PPM of CO_2 he also found that CO_2 levels varied by season: they globally pic in spring and then fell, as plant life begin to sprout leaves and photosynthesize over the summer and autumn. This was particular to Northern hemisphere as in the southern hemisphere due to the existence of less plant matter, photosynthesis matters less."

Apart from this, keeling discovered a troubling and the most concerning pattern: global CO_2 levels were rising everywhere on the earth. Keeling measurement of CO_2 over next 40 years have become definitive symbol of global warming. Keeling Curve shows that CO_2 concentration in the atmosphere is at their highest-level past 800000 years- by a large margin.[40] This is how the main actor in the climate change scenario had been identified.

2.9. ADVERSE IMPACT OF CLIMATE CHANGE

Why we need to mitigate climate change. For this first we need to understand what can be the adverse impact of climate change on human life, animal,

[39] Mridula, R., (2018). *The Climate Solution: India's Climate-Change Crisis and What We Can Do About It* (p. 31). New Delhi: Hachette India.

[40] *Ibid.*

environment, health, agriculture, weather, business, industry, water, tourism, transport, forest, river, sea. The climate change is anthropogenic and it has many adverse effects. It is not going to leave any one whether it is developed country and developing country. But it will hit more the developing country and poorest country. Some impacts are visible and some are invisible. Due to climate change, there will be extreme weather conditions such as drought, heat waves under which it will be very difficult for man to survive.[41] Water scarcity is another impact due to which man, agriculture and natural environment is going to suffer. Business and industry are also going to be suffered because due to warm weather employee will not be able to work comfortably, power cut due to extreme weather condition like flood and storm will impair industry. Poor will be the easily victim of extreme weather condition as it will cause many health problems such as food poisoning, Chikungunya, Malaria and Jaundice etc. It is not the end; these are the anticipated impacts and it can't be imagined unanticipated impact. There may be some impact which may be the beyond the human imagination.[42]

2.10. IMPACTS OF CLIMATE CHANGE

As this climate change is majorly human induced so, the people have to limit their activities which are contributing in emission of GHGs. According to the IPCC, fourth assessment report 2007 even if we stabilize greenhouse gas concentration in the atmosphere the warming will continue for the decades, and its effect will be like rising in sea level will continue for a Century. While some direct effect of climate change is already visible and some are invisible. Most visible and noticeable effect includes frequent extreme whether events such as heat waves, flood droughts, rising in sea levels etc., key effect of climate change is summarized as follow:

2.10.1. Extreme Weather Event

Frequent long-term change in climate has been noticed in recent year including extreme weather such as droughts, heavy precipitation, heat waves and the intensity of tropical cyclones. IPCC 4th assessment report has observed more powerful storm and hotter, longer dry period. Drying has also been observed over large region, i.e., the Sahel, the Mediterranean southern

---

[41] Clarissefrass-Ehrfeld, (2009). *Renewable Energy Sources A Chance to Combat Climate Change* (p. 113). The Netherlands: Wolters Kluwer Law & Business, Kluwer Law International.

[42] Sarkar, A. N., (2010). *Global Climate Change: Beyond Copenhagen* (p. 61). New Delhi: Pentagon Earth.

Africa and part of southern Asia. In Africa large catchment of basins of Niger, Lake Chad and Senegal; total available water has decreased by 40 to 60% and desertification has been worsened by lower average rainfall especially in southern, northern and western Africa.[43]

In 2015 from May to Jun severe heat wave (SHW) incidences over the south peninsula and eastern parts of the country including the States of Andhra Pradesh, Telangana and Odisha was witnessed. In 2016, from March to May SHW conditions was prevailed over north-eastern, central and peninsular parts of the country. In 2017, from May to October lightning and heavy rainfall resulted floods, lightning and rainfall events caused loss of life in various parts of Odisha. Flood and heavy rains caused loss of life in Gujarat. A massive landslide caused deaths at Kotrupi, Himachal Pradesh on 13th August, 2017. Floods in Ghaghara, Gomati and Rapti rivers also claimed lives during 4th to10th September in 2017.[44]

Extreme Weather Event in India in the Years (2018–2021)

Year	**Weather Event**	**Affected Areas**
2018	Heavy rainfall in June-September resulting in floods	Heavy rainfall resulting in floods caused loss of life in Kerala, Karnataka and Uttar Pradesh.
2018	Lightning and thunderstorm	Lightning and thunderstorm events caused loss of life in various parts of Uttar Pradesh from April to September, in Jharkhand from 2nd May to 14th September, in West Bengal from 30th March to 14th November, in Bihar from 30th March to 8th June
2018	Severe cold wave in January and December	Intense cold wave related incidents in northern parts of the country including Uttar Pradesh, Bihar and Jharkhand.

[43] Clarissefrass-Ehrfeld, (2009). *Renewable Energy Sources A Chance to Combat Climate Change* (p. 14). The Netherlands: Wolters Kluwer Law & Business, Kluwer Law International.

[44] Krishna, A. R., & Friederike, O., (2019). Changing climate and weather: Evidence from attribution science. In: Navroz, K D., (ed.), *India in a Warming World: Integrating Climate Change and Development* (pp. 45–58). New Delhi: Oxford University Press.

2018	Cyclonic storm 'DAYE' over east central Bay of Bengal and adjoining Myanmar in September	The storm caused extremely heavy rainfall at isolated places over Odisha, north Andhra Pradesh, Chhattisgarh, Vidarbha and heavy to very heavy rainfall at a few places over Telangana
2018	Very severe cyclonic storm 'TITLI' over east central Bay of Bengal	Loss of life due to Very Severe Cyclonic storm 'TITLI' in Odisha and Andhra Pradesh.
2019	Severe heat wave (SHW)	Intense heat wave conditions prevailed over eastern and central parts of the country including Bihar and Maharashtra.
2019	Lightning and thunderstorm	Lightning and thunderstorm events caused loss of life in various parts of Jharkhand from March to October, in Bihar from May to October and Maharashtra from April to October.
20119	Extremely severe cyclonic storm (ESCS) 'FANI' over east central equatorial Indian Ocean and adjoining southeast Bay of Bengal	Loss of life due to ESCS 'FANI' in Odisha.
2019	Very severe cyclonic storm 'VAYU' over southeast and adjoining east central Arabian Sea	No adverse weather over west coast of India was reported due to this system
2020	Thunderstorm and lightning	Thunderstorm & lightning caused loss of life in various parts of country including Bihar, Jharkhand and Madhya Pradesh
2020	Heavy rainfall resulting in floods, Landslide	Heavy rainfall resulting floods caused loss of life and properties in Assam, Bihar, Rajasthan, Kerala, Karnataka, Maharashtra, Odisha, South Chhattisgarh and Telangana.

2020	Severe cyclonic storm 'NISARGA' over the Arabian Sea	Claimed 4 lives in Maharashtra
2020	Super cyclone 'AMPHAN' over the Bay of Bengal.	Claimed 86 lives in West Bengal.
2021	Cyclone Tauktae	It affected southern Gujarat on 17 May 2021 and was classified as a Very Severe Cyclonic Storm (VSCS). As many as 24 people were killed across three Indian states. Twelve people died in Maharashtra, eight in Karnataka, and four people in Gujarat.
2021	The cyclonic storm, Cyclone Yaas, formed in the Bay of Bengal	Hit the coast of West Bengal and adjoining Odisha coasts in May 2021. The name of the cyclone is given by Oman.

2.10.2. Melting of Glaciers

There is strong evidence that extreme weather events, such as Hurricanes, floods, droughts, and heat waves are increasing and becoming more severe and frequent because of climate change. According to WHO 1,50,000 people are already dying every year as a result of climate change. It is for this reason that the world needs to act now before our planet becomes damaged beyond repair.

The melting of snow in the Arctic and Antarctic due to global warming and climate change is reported frequently. However, the melting of the Himalayas glaciers goes larger. At present, 12% of the Earth's landmass is covered permanently with snow, the majority of which found in polar region.[45] In Ladakh, the northern most region of India, all life depends on snow. Ladakh is a high-altitude desert/cold desert with only 5mm of rainfall. Ladakh water comes from the snow melt, the snow that fall on the land and provides the moisture for farming and pastures. Glacier runoff in the Himalayas is the largest source of the fresh water for Northern Indian states and provides more than half of the water to the Ganga.

[45] Samuel Markings, *What Percent of the Earth Is Permanently Covered with Snow & Ice?* https://education.seattlepi.com/percent-earth-permanently-covered-snow-ice-4666.html (accessed on 07 July 2023).

According to the IPCC glaciers in the Himalayas are receding faster than in any other part of the world and if the present rate continues the likelihood of them disappearing by the year 2035 and perhaps sooner is very high if the earth keeps getting warmer at the current rate. According to the IPCC report the total area of glaciers in the Himalaya will shrink from 193051 squre miles to 35,000 square miles by 2035. The lives of billions of people are at stake. That is why we have started a participatory process for Himalayan communities to engage in the discussion on climate change including issues of the climate justice, adaptation, and disaster.[46]

The government of India has set up a national climate action plan[47] which has eight missions. One of the missions is for sustaining the Himalayan Ecosystems[48]. However, the Himalayan communities are missing in the mission.

The Intergovernmental Panel on Climate Change pointed out in its comprehensive fourth Assessment Report on Climate change that if the Himalayas glaciers continued to recede at the rate existing at the time, there is a very high risk of their disappearing by the year 2035, perhaps sooner to. Accordingly strict measures have been taken to curb warming of the earth at the prevailing rate. It warned with a view to making people aware of the fact that water scarcity could affect more than a billion people in Asia itself. Water shortages, it is a common knowledge now, mean a big threat for 500 million inhabitants of the Indus, Ganga, and Brahmaputra River basins, who rely on permanent supply of melted water from the Himalayas. The problems of the water stress are already prevalent in the region due to increasing demands of domestic consumption, agriculture, industry, and an ever-increasing population. Any future reduction in the availability of freshwater would have serious consequences for the economy, environment and daily life of people living in the affected areas.[49]

---

[46] Tumpa, M., (2010). Climate change and sustainable development. In: Sanjay, K. S., (ed.), *Environment Law and Climate Change* (p. 129). New Delhi: SBS Publishers & Distributors Pvt. Ltd.

[47] National Action Plan for Climate Change (NAPCC) is a Government of India's program launched in 2008 to mitigate and adapt to the adverse impact of climate change. The action plan was launched on 30th June 2008 with 8 sub-missions and currently had 9 submissions.

[48] The Department of Science & Technology, Ministry of Science & Technology was entrusted with the responsibility of coordinating National Mission for Sustaining Himalayan Ecosystem (NMSHE).

[49] Tumpa, M., (2010). Climate change and sustainable development. In: Sanjay, K. S., (ed.), *Environment Law and Climate Change* (p. 129). New Delhi: SBS Publishers & Distributors Pvt. Ltd.

2.10.3. Shift of the Natural World

Scientists have observed the disappearance of many species and plants. In the Alps some of the plant species have been migrating and some plants earlier found at the top of the mountain have been disappeared[50].

2.10.4. Water Supply

Due to flood and drought the demand of water may be increased and it won't be possible for government to meet the water supply to its citizen[51]. Drought can cause million to leave their homeland. Drought also lowers the available nutrition for people, hitting women and children particularly hard.[52]

2.10.5. Health

Climate change will affect the health of the population. The poor are more vulnerable to climate change. Summer death and winter death would be common. Victim may have mental stress due to extra weather event[53]. Any change in climatic conditions is expected to affect human health. The direct impact of climate change on human health is mortality due to increased temperature, disasters resulting in flood, loss of life and infrastructure due to cyclones etc., impact on water and Vector-borne diseases, malnutrition and respiratory diseases.[54] Relationship between climate and health is an old one as it is evidenced by increase in the visit of patients to clinics after severe heat, rain and cold. However, India with limited resources is expected to face a host of health effects due to climate change, including vector-borne and water-borne diseases such as malaria, cholera, and dengue.[55]

---

[50] Clarissefrass-Ehrfeld, (2009). *Renewable Energy Sources A Chance to Combat Climate Change* (p. 14). The Netherlands: Wolters Kluwer Law & Business, Kluwer Law International.

[51] *Ibid.*

[52] Sarkar, A. N., (2010). *Global Climate Change: Beyond Copenhagen* (p. 58). New Delhi: Pentagon Earth.

[53] Clarissefrass-Ehrfeld, (2009). *Renewable Energy Sources A Chance to Combat Climate Change* (p. 15). The Netherlands: Wolters Kluwer Law & Business, Kluwer Law International.

[54] Poonam, K. S., & Ramesh, C. D., (2012). Climate change and human health: Indian context. *Journal of Vector Borne Diseases, 49*(2) 55. link.gale.com/apps/doc/A299885344/HRCA?u=anon~77d962c6&sid=googleScholar&xid=ce501a65 (accessed on 07 July 2023).

[55] Ramana, D. V., Paul, J. S., & George, L., (2013). Climate Change and infectious diseases in India: Implications for health care providers. *Indian J. Med. Res., 138*, 847–852.

2.10.6. Agriculture

Agriculture will be affected by flood and drought situation.[56] Due to this irrigation and the feeding of livestock would be expensive. Intense rainfall and drought would lead to soil erosion and damage[57].

2.10.7. Business and Industry

Intermittent power cut in the industry due to flood, strong heat waves, tsunami and many other weather conditions may restrict the potential growth of the business and industries. Non-supply of water in industry due to extreme weather condition will stop the business in industry.[58] For example in summer season due to non-supply of water to Maharashtra Industrial Development Corporation (MIDC) in Nagpur and Chandrapur stop the functioning of power plants in that area.

2.10.8. Transport

Due to extreme weather condition damage may be caused to road and rail infrastructure and restriction may be created in navigation and transportation. Flooding may restrict transportation in coastal area. Passenger traveling in hot weather also feel discomfort.

2.10.9. Tourism and Leisure

Climate change will also affect tourism. Due to extreme whether condition it will be very difficult to maintain heritage site. Poor quality of water and low flow of water may restrict recreational activity in river and lake.

2.11. IMPACTS OF CLIMATE CHANGE ON DIFFERENT COUNTRIES OF THE WORLD

2.11.1. Impacts of Climate Change in North America

Many cities in America are situated in the coastal areas. If sea level would raise due to climate change these cities could lose their land. Due to heat

[56] Kavi Kumar, K. S., (2007). *Climate Change Studies in Indian Agriculture* (Vol. 42, No. 45, 46, pp. 13, 14). Economic and Political Weekly.

[57] Sarkar, A. N., (2010). *Global Climate Change: Beyond Copenhagen* (pp. 60, 61). New Delhi: Pentagon Earth; see generally Aggarwal, P. K., (2003). Impact of climate change on Indian agriculture. *Journal of Plant Biology, 30*, 189–198.

[58]

waves the human health of the people of America would be deadly affected. It will have impact on agriculture and ecosystem too.[59]

2.11.2. Impacts of Climate Change in Latin America and Caribbean Country

Climate change would have impact on whole world but it would directly destroy the rainforest of Amazon. Increased temperature in mid-century associated with decreased soil moisture would lead to gradual replacement of tropical forest by Savanna in eastern Amazonia. El Niño-Southern Oscillation directly affects Latin America by bringing rain to western South America and vast reduction in fish stock of Peru and Chile. If due to climate change El Niño-southern oscillation occurs frequently it would lead to increased flood and decline in fishery output in western South America. Extreme observed effect as one of the observed effects of climate change would adversely affect the commercial crops of country like banana and coffee. Production of some other important crops would also be harmed and this will pose major threat to human beings and livestock living in that country. Small island of Caribbean Sea is vulnerable to rise in sea level. And very existence of them is under threat of climate change. Climate change intensify the occurrence of hurricanes which would adversely affect the Caribbean countries.[60]

2.11.3. Impacts of Climate Change in Middle East

The hottest desert is in the Middle East country which is inhabitable to many life forms but some regions are habitable but due to climate change they would also be inhabitable. Climate change would increase the scarcity of water due to heat and evaporation in that region which would lead to conflict between nations. Egypt is one of the fertile lands' countries in this region and due to rise in sea level there would be coastal erosion and this will pose major threat to nation.[61]

2.11.4. Impacts of Climate Change in Africa

Africa being a continent of poor nation is vulnerable to climate change. Africa is not even capable to adopt and implement mitigation and adaptation

---

[59] Clarissefrass-Ehrfeld, (2009). *Renewable Energy Sources A Chance to Combat Climate Change* (p. 16). The Netherlands: Wolters Kluwer Law & Business, Kluwer Law International.

[60] *Ibid.*

[61] *Ibid.*

strategies. It is well known that economy of many countries is based on agriculture and in case of drought food security is under threat. In current time many countries of Africa are facing the problem of water as a consequence of climate change. Desertification and loss of land is expected to accelerate due to climate change. The grassland of Sahel is already shrinking. Reduced rain fall in southern Africa could also grow the Kalahari and Namib Desert in size. Water stress, food insecurity, health issues may lead to the conflict between stressed nations in Africa unleashing migration of environmental refugee on the African continent. Small island state like Comoros and Seychelles are under threat due to rising of sea level.[62]

2.11.5. Impacts of Climate Change in Asia

In Asia, due to the climate change there is expectation of migration of ecosystem leading to temperate biomes such as steppes and temperate forest replacing much of the area currently covered by the Siberian Taiga. In central Asia, the semi-arid grassland is at risk of desertification as precipitation may decrease in those areas. Global climate change would bring more droughts and wild fires to south Asia. Fresh water availability in Central South and South-East Asia, which are large river basin is projected to decrease by 2050. Climate change may disturb the soil moisture which are necessary for some sensitive crops this may reduce the crop yielding in Asia. Whole Asia would be facing challenges due to rise in sea level, however, for some Asian nations like Bangladesh and Vietnam which are totally flat and a sea level rise would be hazardous for these countries. Sea level rise would directly affect the small island country like Maldives. Major cities of Japan which are at the coast of sea are under threat. Asian countries like Japan, South Korea, and Taiwan could afford mitigation and adaptation strategy but the nation which are situated in Bay of Bengal are very poor and it would be very difficult for them to cope up with climate change.[63]

2.11.6. Impacts of Climate Change in Polar Region

Even though there are no permanent inhabitants reside in this area but the glacier which covers the Antarctica and Greenland is expected to melt; and this has drawn the attention because melting of ice which become the

---

[62] Clarissefrass-Ehrfeld, (2009). *Renewable Energy Sources A Chance to Combat Climate Change* (p. 20). The Netherlands: Wolters Kluwer Law & Business, Kluwer Law International.

[63] *Ibid* 21.

primary source of rise in sea level.[64] Antarctica heading towards Climate Tipping Point by 2060. Climate Tipping Point is the threshold that leads to large changes in the state of the system. The fifth Assessment Report of the Intergovernmental Panel on Climate Change (IPCC) defined Climate Tipping Point "as the irreversible change in the climate system. It is often referred to as runaway Climate Change." However, it is not same as that of the Runaway Greenhouse Effect. There are several ice shelves in Antarctica, if as these ice shelves will break up, the towering ice cliffs would not be able to stand on their own. The study found that these ice shelves are breaking at a faster rate.[65]

2.11.7. Impacts of Climate Change in India

Climate change is a phenomenon that is now considered strongly associated with human activities and it has also direct impact on human in many ways. It has many forms and all the forms has potential to major threat in India. India is vulnerable to climate change due to economic and geographical reason. Climate change would lead to destruction, loss of infrastructure, loss of human life, and harm to natural environment in India. Climate change has the potential to influence the earth's biological systems. Climate affects water, food, air quality, diseases, physical comforts and human health. India has reason to be concerned about climate change. A vast population depends on climate sensitive sectors like agriculture, forestry, and fishery for livelihood in the India.[66] The monsoon in India supports the agriculture but due to climate change pattern agriculture sector is also under threat and it has become active area for research[67]. In India, agriculture is very important as 58% of population are engaged in agriculture. Water availability resources of Himalaya region which are sources of Major River like Ganga and ground water recharge is decreasing which will create more problems for people dependant on that river for drinking water and agriculture in India[68]. Most of

---

[64] *Ibid* 22.

[65] *Antarctica is Headed for a Climate Tipping Point by 2060, with Catastrophic Melting if Carbon Emissions Aren't cut Quickly,* (2021). The Times of India. http://timesofindia.indiatimes.com/articleshow/82731637.cms?utm_source=contentofinterest&utm_medium=text&utm_campaign=cppst

[66] Srinivasan, J., (2019). Impact of climate change on India. In: Navroz, K. D., (ed.), *India in a Warming World: Integrating Climate Change and Development* (pp. 32–41). New Delhi: Oxford University Press.

[67] Clarissefrass-Ehrfeld, (2009). *Renewable Energy Sources A Chance to Combat Climate Change* (p. 21). The Netherlands: Wolters Kluwer Law & Business, Kluwer Law International.

[68] India, Ministry of Finance, Economic Survey, 2012-13, (pp. 256, 257).

the population in India are poor who are more vulnerable to natural disasters. As we know the flood create the health problem in country like India where no sanitation machineries are available after flood and water contaminated due to human and livestock[69]; give rise to disease like diarrheal and other bacterial and viral illnesses. The adverse impact of climate change would increase the severity of flood, drought and will ultimately lead to instability of economy and food and exacerbated health problem in India.[70]

The Intergovernmental Panel on Climate Change (IPCC) report has shown that the impacts at 2°C are "catastrophic" for the poor and developing nations like India, news reports are also providing evidences of how change in the climate has been affecting the nation and its plans to cope up. Many reports and studies released at the end of the year 2018 too provided facts and figures on the state of India's climate. It has widespread impacts. The Economic Survey 2018 envisaged that climate change is not productive. Climate change is taking a toll on India's agricultural productivity and farmers' income. Overall 600 million Indians will be moderately or severely affected by changes in temperature and rainfall patterns, warned the new World Bank report which is released in June, 2018. Farming of 30 traditional crops has been abandoned in Pithoragarh district of Uttarakhand in the last one decade due to climate change and migration which has been said in a study by Wildlife Institute of India, Dehradun. Most of the existing habitats of the Nilgiris Tahr in the Western Ghats will become unsuitable as global warming intensifies, found in the study by Ashoka Trust for Research in Ecology and the Environment. Yak the lifeline of pastoral nomads in high altitudes of the Indian Himalayan region is facing the threat of gradually rising temperatures in the region. A World Bank report released in June, 2018 said climate change could cost India 2.8% of GDP, and lower living standards of nearly half of its population by 2050. It added that average annual temperatures are expected to rise by 1–2% over three decades.[71]

The Global Climate Risk Index which is released on 25 January 2021 by Germanwatch-a NGO based in Bonn, Germany. In this index, India has been ranked at seventh position among the countries that are most effected

[69] Bhattacharya, S., Sharma, C., et al., (2006). Climate Change and Malaria in India. *Current Science, 90*(3), 369–375.

[70] Ramana, D. V., Paul, J. S., & George, L., (2013). Climate Change and infectious diseases in India: Implications for health care providers. *Indian J. Med. Res., 138*, 847–852.

[71] Kiran, P., (2018). *Rewind 2018: How Much India Changed with Climate Change*. Down To Earth. https://www.downtoearth.org.in/news/climate-change/rewind-2018-how-much-india-changed-with-climate-change-62609 (accessed on 07 July 2023).

by climate change in the year 2019. According to the report, Monsoon continued for a month longer than normal in the year 2019 in India. In the period from June to September 2019, 110% of the long-period average was recorded. Due to Heavy rain severe flooding was caused in India that resulted into death of 1,800 people across14 states. The flood also resulted the displacement of 1.8 million people. As per the report, eight tropical cyclones came in India in which six of them intensified to be very severe. Report further highlighted that, 'Extremely severe' cyclone "Fani" affected 28 million people and killed 90 people in India and Bangladesh.[72]

2.12. CLIMATE CHANGE CHALLENGES FOR DEVELOPING COUNTRIES

Developing countries like India and China has been developing through industrialization and this rapid growth results in emission of greenhouse gases and due to this, these countries are becoming major source of greenhouse gases in the atmosphere at international level. The major reason for emission in developing country are use of fossil fuels in energy production and transportation. And for the development of any country these are basic thing without which a country cannot be developed. Development is the right of every country but it can't be at the cost of environmental degradation. The main challenge which is being faced by developing country regarding climate change mitigation is the low potential for financing projects focused on a low carbon emission. The other challenge is the developing country have contributed least towards global climate change so it is inequitable for them to make effort and funding into mitigation technologies at the first place.[73]

2.13. CAUSES OF CLIMATE CHANGE

Past 10,000 years of the earth's climate have been unusually calm; the climate been calm enough to allow agriculture to develop and to support civilization, now we are tipping the world out of this comfort zone by releasing too

[72] *Global Climate Risk Index, 2021*, https://germanwatch.org/en/cri (accessed on 07 July 2023).

[73] Abel, N., & Alexander, K., (2013). Climate change mitigation in developing countries using ICT as an enabling tool. in Walter, L. F., et al., (eds.), *Climate-Smart Technologies: Integrating Renewable Energy and Energy Efficiency in Mitigation and Adaptation Responses* (pp. 20–24). Verlag Berlin Heidelberg: Springer; see generally Ting, W., Shili, Y., John, C. M., & Peijun, S., (2012). Developed and developing world responsibilities for historical climate change and CO_2 mitigation. *Proceedings of the National Academy of Sciences of the United States of America*, *109*(32), 12911–12915.

large quantity of greenhouse gas, such as CO_2, into the atmosphere at high rate. Greenhouse gases have an important quality to them-they allow sunlight to pass through them freely but block certain types of energy in the infrared spectrum emanating from the earth. Thus, they trap the earth's heat increasing its surface temperature. In that sense they work like a greenhouse (hence the name) that restrict the sunlight to heat the air inside but prevents the heated air from escaping.[74]

Earth's climate is dynamic and it has always been changing but today's concern is the speed of changing. The reason of climate change is natural as well as manmade. Natural cause includes volcanic eruption, continental drift, ocean currents, the earth's tilt, and comets and meteorites.[75]

In 19th century, industrial revolutions took place and for this fossil fuel were used. People got job opportunities in industries for which they started moving towards urban area which led to consumption of natural resources for industry, transportation etc., population agglomeration also took place in the urban area which led to the higher concentration of greenhouse gases in atmosphere which is mainly responsible for climate change.

Human are the main reason for increasing earth's temperature by burning fossil fuels, cutting down rainforests and farming livestock[76]. Burning of fossil fuel like coal and oil in large quantity has increased the concentration of greenhouse gases in the atmosphere due to this temperature is increasing. These fossil fuels are burned for meeting the energy demand which is very important for development of any country and living a comfortable life. It was observed by The *Royal Society 2010* "*There is strong evidence that the warming of the Earth over the last half-century has been caused largely by human activity, such as the burning of fossil fuels and changes in land use, including agriculture and deforestation.*"

2.14. NATURAL CAUSES OF CLIMATE CHANGE

The climate of the earth has been changing since the beginning due to the natural reasons like *the earth's orbital changes, ocean current, volcanic eruptions and solar variations.*

74 Mridula, R., (2018). *The Climate Solution: India's Climate-Change Crisis and What We Can Do About It* (p. 28). New Delhi: Hachette India.

75 *Causes* of *Climate Change,* http://edugreen.teri.res.in/explore/climate/causes.htm

76 https://climate.nasa.gov/causes/ (accessed on 07 July 2023).

2.14.1. Volcano

Any volcanic eruption throws a large amount of sulfur dioxide (SO_2), water vapor, dust, and ash into the atmosphere, the large volumes of gases and ash influences climatic patterns for years by increasing planetary reflectivity causing atmospheric cooling. Tiny particles called aerosols are produced by volcanoes. Because they reflect solar energy back into space, they have a cooling effect on the world.[77]

2.14.2. Ocean Current

The oceans are also considered as the main component of the climate system. Relations between the ocean and atmosphere can also produce phenomena such as El Niño which occur every 2 to 6 years. The oceans play a significant role in determining the atmospheric concentration of CO_2.

2.14.3. Earth Orbital Changes

The earth makes one full orbit round the sun each year. Earth is tilted at an angle of 23.5° to the perpendicular plane of its orbital path. Any Changes in the tilt of the earth can lead to small but climatically important changes in the strength of the seasons, more tilt means warmer summers and colder winters; less tilt means cooler summers and milder winters. Slow changes in the Earth's orbit led to small but climatically significant changes in the strength of the seasons over tens of thousands of years. Climate feedbacks intensify these small changes, thereby producing ice ages.[78]

2.14.4. Solar Variations

The Sun is the leading source of energy for the Earth's climate system. Even though the Sun's energy output appears constant from an everyday point of view, small changes over an extended period of time can lead to climate changes. Some scientists suspect that a portion of the warming in the first half of the 20th century was due to an increase in the output of solar energy. As the sun is the essential source of energy that is instrumental in our climate system, it would be rational to assume that changes in the sun's energy output would cause the climate to change. Scientific studies demonstrate that solar

---

[77] *Causes of Climate Change,* https://www.metoffice.gov.uk/weather/climate-change/causes-of-climate-change (accessed on 07 July 2023).

[78] *How Earth's Climate Changes naturally (and Why Things Are Different Now)*, https://www.quantamagazine.org/how-earths-climate-changes-naturally-and-why-things-are-different-now-20200721/ (accessed on 07 July 2023).

variations have performed a role in past climatic changes. For example, a decrease in solar activity was thought to have triggered the Little Ice Age between approximately 1650 and 1850, when Greenland was largely cut off by ice from the year 1410 to the 1720 and glaciers advanced in the Alps.[79]

2.15. HUMAN CAUSES OF CLIMATE CHANGE

It has been proved that climate change is majorly due to the anthropogenic emissions. It includes emissions of greenhouse gases, greenhouse-gas precursors, and aerosols associated with human activities. These include the burning of fossil fuels, deforestation, land-use changes, livestock, fertilization, etc., that result in a net increase in emissions[80].

2.15.1. Human's Contribution to the Greenhouse Effect

While the greenhouse effect is an essential environmental perquisite for life on Earth, there really can be too much of good thing. The problems begin when human activities distort and quicken the natural process by creating more greenhouse gases within the atmosphere. For Example, burning of natural gas, coal and oil including gasoline for automobile engines raises the level of carbon dioxide in the atmosphere. Some of the farming practices and land use changes increases the levels of methane and nitrous oxide.[81] Many factories produced long lasting industrial gases that don't occur naturally, yet contribute significantly to the rise the concentration of atmospheric phenomenon and "global warming" that is currently under way. Deforestation also contributes to global warming. Trees take carbon dioxide and give-off oxygen balance of gases in the atmosphere.[82] Population' growth is another cause in global warming, because as more people use fossil fuels for heat, transportation and manufacturing, the level of greenhouse gases continues to increase in atmosphere. As more farming occurs to feed many new people, more greenhouse gases enter the atmosphere. Ultimately, more greenhouse

---

[79] *Causes of Climate Change*, http://www.climatechangechallenge.org/Resource%20Centre/Climate-Change/3-what_causes_climate_change.htm (accessed on 07 July 2023).

[80] *Glossary of Terms and Definitions on Climate Change and Adaptation,* http://www.mrcmekong.org/assets/Publications/glossaries/Glossary-of-Terms-n-Definitions-on-CCA-Eng-04072013.pdf (accessed on 07 July 2023).

[81] See *IPCC Special Report on Climate Change and Land, 2019*; Chapter 2 of the report provides data on livestock methane emissions, about 66% is agricultural methane and about 33% of global methane emissions come from livestock, https://www.ipcc.ch/srccl/ (accessed on 07 July 2023).

[82] Cornelis, V. K. G., Sabina, L. S., et al., (2002). Mitigating climate change by planting trees: The transaction costs trap. *Land Economics*, *78*(4), 559–572.

gases mean infrared radiation trapped and held, which gradually increase the temperature of the Earth's surface and the air in the lower atmosphere. India, today the 2nd most populous country in the world, is likely to become the most populous country in world by 2026, surpassing China. India is one of the few countries of the world that enshrined in its Constitution a commitment to environmental protection and improvement. It has enacted various laws for the protection and improvement of environment. However, the problem of environmental deterioration continues to worsen. [83]

2.15.2. Industrial Processes

Industrialization is one of the main reasons of climate change.[84] Industrialization led to emission of greenhouse gases, destroying natural habitat, urbanization, exploitation of natural resources. Industrialization is not only the reason of climate change but also responsible for environmental degradation.[85]

2.15.3. Agriculture

Due to increased population the land for agriculture is decreasing day by day. And due to this fertilizer are used to grow more and these fertilizers harm the land and manufacturing of these fertilizers releases Sulphur Dioxide (SO_2) in the atmosphere. Apart from fertilizer, the process of clearing land for next season of farming is also responsible.[86] Recently, Delhi witnessed smog in the month of October and it was observed that the burning of agricultural field for clearance was the main reason.[87]

---

[83] Tumpa, M., (2010). Climate change and sustainable development. In: Sanjay, K. S., (ed.), *Environment Law and Climate Change* (p. 136). New Delhi: SBS Publishers & Distributors Pvt. Ltd.

[84] Sarah, S., (2012). *Global Warming: The Predicament, Contributions and Initiatives*. Seminar Paper (April 19, 2012), http://www.philau.edu/collegestudies/Documents/SarahSchaub.pdf (accessed on 07 July 2023).

[85] *Causes of Climate Change*, http://www.climatechangechallenge.org/Resource%20Centre/Climate-Change/3-what_causes_climate_change.htm (accessed on 07 July 2023).

[86] Kavi, K. K.S., (2007). *Climate Change Studies in Indian Agriculture* (Vol. 42, No. 45, 46, pp. 13–18). Economic and Political Weekly.

[87] Pratik, P., (2019). *Toxic Air in Delhi Fuelled by Rice Fields That India Doesn't Need*. The Economic Times. https://economictimes.indiatimes.com/news/economy/agriculture/toxic-air-in-delhi-fueled-by-rice-fields-that-india-doesnt-need/articleshow/72016990.cms?from=mdr (accessed on 07 July 2023).

2.15.4. Transport

Transportation has an important role in the development of nation and ease for human being. But to run the existing transportation system oil and natural gas is needed which is fossil fuel. Burning of fossil fuel always results in emission of greenhouse gases.[88] For transportation low efficient vehicle is used in country like India and which emits more.[89]

2.15.5. Individuals

Each individual is responsible for climate change because energy is the need of individual, agriculture is done for individuals, and vehicles are also used by individuals need. Human beings cut trees for land, paper, fuel and furniture which play role of carbon sink if there will not be tree then who will absorb carbon dioxide. Each and every human being is responsible for climate change.[90]

2.15.6. Production of Energy

Since energy is billed as the 'growth engine' index of economic growth and human development of any nation in this 21st century, it needs no emphasis that the access to energy has direct impact on energy sources in the world. India derives its energy from coal, oil, gas, water which is conventional sources.[91] Today, energy has become the basic need of the human. For most of the things in our life we are dependent on energy such as charging mobile and laptop, cooking food, watching television, listening music, running industry and business, etc. The process of generation of energy includes the use of coal which is also a fossil fuel. In present time, most of the country is generating energy through thermal power plant and generating energy through thermal power plant is the main reason for climate change.[92] Due to this world is thinking to move towards green and renewable energy and efforts are being made.[93]

---

[88] IPCC 5th Assessment Report 2014: Chapter 8; Transport.

[89] Lee Schipper, Herbert, F., et al., (2009). Transport and carbon dioxide emissions: Forecasts, options analysis, and evaluation. 9 *ADB Sustainable Development Working Paper Series* (No. 9). http://www.indiaenvironmentportal.org.in/files/Transport-CO2-Emissions.pdf (accessed on 07 July 2023).

[90] Avram, H., (2011). Climate change and individual responsibility. *The Monist, 94*(3), 349–368.

[91] Varn, R., & David, G. V., (2009). *Climate Change and the Energy Challenge: A Pragmatic Approach for India* (Vol. 44, No. 31, pp. 78–85). Economic and Political Weekly.

[92] *See Chapter 7: Energy System of IPCC 5th Assessment Report, 2014*, https://www.ipcc.ch/site/assets/uploads/2018/02/ipcc_wg3_ar5_chapter7.pdf (accessed on 07 July 2023).

[93] *Causes of Climate Change,* http://www.climatechangechallenge.org/Resource%20Centre/Climate-Change/3-what_causes_climate_change.htm (accessed on 07 July 2023).

2.16. ENVIRONMENTAL IMPACT OF DIFFERENT SOURCES OF ENERGY

Environmental concerns are associated with all forms of energy sources including fossil fuels, nuclear energy and even renewable energy throughout the energy chain from exploration/mining, transportation and generation/ manufacture to the end-use/ decommissioning.

2.16.1. Coal

Coal emerges as the most important energy source for country like India. Since oil will not be able to meet world energy demand, the option left for meeting energy demand is coal. Coal in every form whether in mining or use for power generation contributes to environmental pollution Approximately, 75% of Indian electricity is derived from thermal power plant for which coal is the foremost need power generation in power plant which contributes to atmospheric pollution due to the release of particulate matter, carbon dioxide, sulfur and nitrous oxides.[94]

Coal mining means survey, explore and extract minerals, materials and coal lying wrapped up in the bowels of the earth. Coal mining affects the environment in the form of air pollution. Water table goes down due to mining. Large scale excavation, removal of top soil, dumping of solid wastes cutting of roads are the effect of mining on land.

2.16.2. Oil

Oil pollution is inevitable fact in 21st century when the millions of human beings are depended on the oil-based technology for traveling with speed and comfort. But this oil along with speed and comfort causes environmental pollution in extraction of oil and transportation and storage of oil.[95]

2.16.3. Nuclear Energy

Nuclear energy is conventional source of energy. It is used for both purpose military and civilian. There are two views regarding nuclear energy one is that nuclear energy can meet the energy demand and another is that nuclear energy is by product and the main product is deadly. The main problem with nuclear energy is the disposal of radioactive waste. Until and unless

---

[94] See Mustafa, B., (2005). Usage of energy sources and environmental problems. *Energy Exploration & Exploitation, 23*(2), 141–168.

[95] *Ibid.*

the problem of disposal is solved, we can't depend on nuclear energy for fulfilling our energy requirement.[96]

2.17. COMPOSITION OF GREEN HOUSE GASES

Human contributes to climate change in many ways such as requirement of energy leads to emission of greenhouse gas. Fossil fuel-based transportation is also an important reason. Large population size leads to more stress on earth, land, food industry which indirectly contributes to climate change. Students need paper for writing which is manufactured from cutting trees. Trees play a vital role in mitigating climate change but due to increased population the forest are being cleared unsustainably for their needs of land, woods for furniture, paper for writing, etc., these all leads to human contribution towards climate change.[97]

2.18. GASES THAT CONTRIBUTE TO THE GREENHOUSE EFFECTS INCLUDES

The question comes what are greenhouse gases due to which climate is changing. Greenhouse gases are those gases which block heat from escaping from earth's atmosphere. Greenhouse gas that remains semi-permanent in the atmosphere and do not respond physically or chemically to changes in the temperature are described as 'forcing' climate change. These greenhouse gases are carbon dioxide (CO_2), nitrous oxide, water vapor, chlorofluorocarbons (CFCs).[98]

- **Water Vapor**: Water vapor is the most abundant greenhouse gas and it acts as a feedback to the climate. Water vapor increases as the Earth's atmosphere warms, so does the possibility of clouds formation and precipitation, making these some of the most important feedback mechanisms to the greenhouse effect.
- **Carbon dioxide (CO_2)**: A minor but very important component of the atmosphere, carbon dioxide is released through natural processes such as respiration and volcanic eruptions and through human activities such as deforestation, land use changes, and

[96] See Nico, B., Robert, J. B., (2012). Economics of nuclear power and climate change mitigation policies. *Proceedings of the National Academy of Sciences of the United States of America, 109*(42), 16805–16810.

[97] *Causes of Climate Change*, http://edugreen.teri.res.in/explore/climate/causes.htm (accessed on 07 July 2023).

[98] *Causes of Climate Change*, https://climate.nasa.gov/causes/ (accessed on 07 July 2023).

burning fossil fuels. Humans have increased atmospheric CO_2 concentration by more than a third since the Industrial Revolution began. This is the most important long-lived "forcing" of climate change.

- **Methane**: A hydrocarbon gas which is produced both either through natural sources or human activities, including the decomposition of wastes in landfills, agriculture, and especially rice cultivation, as well as ruminant digestion and manure management associated with domestic livestock. On a molecule-for-molecule basis, methane is a far more active greenhouse gas than carbon dioxide, but also it is one of the gases which is much less abundant in the atmosphere.
- **Nitrous Oxide**: A powerful greenhouse gas which is produced by land cultivation practices, especially the use of commercial and organic fertilizers, fossil fuel combustion, nitric acid production, and biomass burning.
- **Chlorofluorocarbons (CFCs)**: Synthetic compounds which is completely of industrial origin used in a number of applications, but now largely regulated in production and release to the atmosphere by international agreement for their ability to contribute to destruction of the ozone layer. They are also greenhouse gases.[99]

2.19. GREENHOUSE GASES AND ATMOSPHERIC CHEMISTRY

Among all the Greenhouse gases present in the atmosphere, CO_2, CH_4 and N_2O are the most important. Among all three gases, CO_2 play a crucial role which is produced by burning of fossil fuels such as oil, natural gas and coal, followed by deforestation. Till 18th century, proportion of CO_2 was stable at 80 ppm which increased to 368 ppm in the year 2000. Livestock is the main source of methane, it is derived from decomposition of animal manure, rice cultivation and by treatment of human waste. Some amount of methane is also released during the production and transport of coal, natural gas and oil. N_2O is emitted by the combustion of solid waste and fossil fuels as well as emitted during agriculture and industrial activities. Some other Greenhouse Gases are Chlorofluorocarbon (CFCs), Hydro chlorofluorocarbons (HCFCs), Halons, Hydro fluorocarbons (HFCs), Per

[99] https://climate.nasa.gov/causes/ (accessed on 07 July 2023).

fluorocarbons (PFCs) and Sulphur hexafluoride (SF_6). Tropospheric ozone and stratospheric water vapors are also important GHGs. It might be noticed that tropospheric water vapor is essential for the hydrological cycle. The reactive gases such as CO, VOC and Nitrogen Oxide ($NOx=NO+NO_2$) are considered as indirect GHGs. They persist in very small amount in the atmosphere. However, through atmospheric chemistry, they control the occurrence of direct GHGs.

In the atmosphere by chemical processes all the GHGs can be removed except CO_2 and water vapor. GHGs with one or more H atoms, such as CH_4, HFC and HCFCs react with the hydroxyl radicals, OH. In troposphere these reaction takes place. However, N_2O, PFCs, SF_6, CFCs, and Halons do not react with radicals in the troposphere. These gases are destroyed in upper layer of troposphere, i.e., stratosphere in the atmosphere by UV radiation at short wavelength- less than 240 mm and live for a long time. They persist in environment for minimum 20 years because they need time to be transported to the region where they are lost in the atmosphere. CO_2 which is major GHGs does not directly influence the chemistry of the atmosphere. It is contributed in environment by the oxidation of CH_4, CO and VOC. The amount of CO_2 present in the troposphere depends on the quantity of NOx, CH_4, CO, VOC, O_3, water vapors and solar UV (>300nm) that reaches the troposphere (IPCC 2001). Hence it has a wide spatial and temporal variation. Consequently, the rate of loss of the gaseous reaction with OH also varies with place and time. Due to its dependence on CH_4 whose emissions have increased in recent years, OH level in the troposphere is expected to change since the pre-industrial era and are also expected to change in the future. The abundance of OH is strongly linked with abundance of O_3. The amount of O_3 produced due to stratospheric changes in humidity and temperature, and the interactions between tropospheric aerosol and trace gases control the chemical reactions in the troposphere, to some extent.

Such a coupling between the troposphere and the stratosphere is a climate chemistry feedback process between the concentration of GHGs themselves and the GHG induced climate change. Likewise, a chemistry feedback process led to the loss of OH, due to its interaction with CH_4 search feedback processes have a bearing on the abundance of O_3 and OH in the troposphere. In the stratosphere, the amount of O_3 and the atmospheric circulation pattern control the amount of UV penetrations and the movement of long life GHGs to regions of photochemical loss. There is another feedback process in the atmosphere that some of GHGs such as N_2O and CFC supply nitric oxide (NO) and chlorine (Cl) which cause depletion of O_3. In stratosphere depletion of ozone leads to an increase in chemical activity of proton in the troposphere thereby changing tropospheric OH and O_3.[100]

2.20. GREENHOUSE EFFECT AND CLIMATE CHANGE

When certain gases trap in the Earth's atmosphere, it warms the whole atmosphere which is termed as 'Greenhouse effect.' These gases let in the light but keep heat from escaping, like the glass walls of a greenhouse. Sunlight shines onto the Earth's surface, where it is absorbed and then radiates back into the atmosphere as terrestrial heat. In the atmosphere heat is trapped by some of the greenhouse gases and the rest escapes into the space. The more the greenhouse gases present in the atmosphere, the greater quantity of heat being trapped. In 1895, a Swedish chemist Avante Arrhenius discovered that humans could intensify the greenhouse effect by generating carbon dioxide, a greenhouse gas.

The amounts of greenhouse gases vary throughout earth's history, but they have been moderately constant over the time as well. Humans are accelerating the greenhouse effect and warming of Earth by increasing the GHGs in the atmosphere by burning of fossil fuels.

Scientist often use the term "climate change" instead warming. It is due to increase in the mean earth temperature, winds and ocean currents move heat around the globe in ways that can cool some area, warm others, and change the amount of rain and snow falling. Impact of the climate change is different in different areas.

The risk lies in the expeditious increase of carbon dioxide and other greenhouse gases that escalate this natural greenhouse effect. From thousands of years, the carbon chain in the environment is stable, whatever carbon is supplied in the atmosphere as much as they released from the atmosphere. Modern human activities like deforestation, burning of fossil fuels, and intensive agriculture-has amplify the huge quantities of carbon dioxide and other greenhouse gases in the atmosphere.

The amount of carbon dioxide increases by 148% than pre-industrial era.[101] There are various sources of GHGs emission like increasing power generations, agriculture, transportation, manufacturing, and other activities. Fossil fuels like coal, oil, and natural gas-release CO_2 during extraction and

[100] Dash, S. K., (2007). *Climate Change an Indian Perspective* (p. 63). Ahmedabad, India: Centre for Environment Education.

[101] *The State of the Global Climate* (2020), https://public.wmo.int/en/our-mandate/climate/wmo-statement-state-of-global-climate (accessed on 07 July 2023).

consumption. World's consumption of energy is expected to rise by 55% by 2030 based on current projection of population and economic growth.[102] The large share of that energy is coming from fossil fuels or thermal power plants.[103]

Anthropogenic activities like substantial uses of fossil fuel, deforestation, and certain agricultural and industrial activities lead to climate change. The increasing concentration of GHGs in the atmosphere has the potential over the next few generations to significantly alter global climate. This would result in large change in ecosystems, leading to possibly catastrophic disruption of livelihood, economic activity, living conditions, and human health. On the other hand, abatement of GHGs, would involve remarkable economic costs. While climate change is a global issue, countries bear different levels of responsibility for reducing GHGs concentrations in the atmosphere. Further, the adverse impacts of climate change will fall disproportionately on least contributing countries (developing nations like India) in the global emissions.

GHG emissions of India in the year 1994 was 1228 metric ton (Mt) CO_2 equipment, which is below 3% of global GHG emissions. In per-capita terms, it is 23% of the global average and 10% to Japan, per capita emissions in 1994. According to the IEA, in 2018 India's emission level was 2307.78 $MtCO_2e$, an escalation of 335.33% from 1990.[104] In terms of GHG, escalation of the economy has taken place in the purchasing power parity terms.

Anthropogenic climate change, significant responsibility for which clearly does not lie with India or other developing countries, may on the other hand have likely adverse impacts on India's precipitation patterns, ecosystems, agricultural potential, forests, and water resources, coastal and marine resources besides increase in range of several disease vectors. Large amount of resource is required for adaptation measures for climate change impacts; human misery is to be avoided. Accordingly, the following will comprise essential elements of India's response to Climate change.

1. Adherence to the principle of common but differentiated responsibilities and respective capabilities (CBDRRC) of

[102] *World Energy Outlook 2019*, https://www.iea.org/reports/world-energy-outlook-2019

[103] *Where Does Our Electricity Come From?* https://world-nuclear.org/nuclear-essentials/where-does-our-electricity-come-from.aspx

[104] Pooran, C. P., (2020). *India Has Seen Greenhouse Gas Emissions Increase by a Staggering 335% Since 1990.* India News Brief and Action Alert.

different countries in respect to both mitigation of GHGs and adaptation measures.

2. Reliance on multilateral pathways, as opposed to bilateral or plurilateral or unilateral measures.
3. Equal per capita entitlements of environmental resources to all over the countries.
4. Identify key vulnerability of India to climate change, in particulars impacts on water resources, forests, coastal areas, agriculture and health.
5. Assess the need for adaptation to future climate change and scope for incorporating these in relevant programs, including coastal zone planning and regulations, watershed management, forestry management, agricultural technologies and practices and health programs.
6. Encourage Indian industry to participate in the clean development mechanism (CDM) through capacity building for identifying and preparing CDM projects, including in the financial sector.
7. Participate in voluntary partnerships with other countries both developed and developing countries to address the challenges of the UN framework Convention on climate change.

The Kyoto protocol covers six greenhouse gases which are as follows: or unilateral measures. Carbon dioxide, Methane, Nitrous oxide, Hydro fluorocarbons, CFCs, Sulphur dioxide[105]

2.21. CLIMATE JUSTICE

This term encompasses a starting though complex reality some people are more responsible than others for the warming of the earth's atmosphere that is triggering catastrophic climate change. So, when the world started discussing what to do in response to climate Justice. This said they wanted to continue being able to develop without being bound by actions needed to curb greenhouse gases, while industrialized countries were asked to immediately start such action. Thus emerged the concept of "common but differentiated responsibility" acknowledging that all nations had the duty to act, but some could act later or be assisted through funding and technologies if the world wanted them to act faster.

[105] Tumpa, M., (2010). Climate change and sustainable development. In: Sanjay, K. S., (ed.), *Environment Law and Climate Change* (p. 133). New Delhi: SBS Publishers & Distributors Pvt. Ltd.

At the Rio conference in 1992[106], the United Nations Frame Work Convention on Climate Change accepted differentiated responsibilities[107] as the underlying principle for addressing the problem of climate change but the U.S. and Australia refused to cut their emissions citing high levels of emissions by China and India. A report by Green Peace India revealed that climate injustice occurs not only between nations, but within nation also.

In average emissions Indian citizen are way below the global average. Some Indian (the richest) are already near to this average. They are already well above levels considered sustainable. But is camouflaged by the fact that the vast majority of Indian-the poor are way below the average. In effect poor Indians are subsidizing the rich, following them a much greater share of the atmosphere than should be rightfully theirs.

Even more important, the government and civil society should work together to create greater degree of awareness among the people of this country on many issues involved in this complex problem. Greenpeace[108] provides partial answers to this. It assures the rich that their life styles need not be sacrificed. The solution rather is in 'decarburizing' the economy, moving towards replacing fossil fuels by renewable sources like solar, wind and biomass and towards greater efficiency in energy production and use. It advocates greater focus on public transport systems, mandatory fuel efficiency standards in cars. High speed trains, to check the increasing use of air travel. It also proposes a carbon tax on use of fossil fuels, proceeds from which could be used to help the poor to get access to cleaner forms of development and to mechanisms cope with the impact of climate change.

Thus, seen every which way, India's responsibility for global warming remains at rock bottom. The concept of basic allowance of harmless emission and subsistence allowance should prove handy during tough negotiations that lie ahead provided they can find a wide acceptance. It does not however mean that India needs to do nothing except negotiation smart. India ought to

---

[106] The United Nations Conference on Environment and Development (UNCED), also known as the Rio de Janeiro Earth Summit, the Rio Summit, the Rio Conference, and the Earth Summit (Portuguese: ECO92), was a major United Nations conference held in Rio de Janeiro from June 3rd to June 14th in 1992.

[107] Common but differentiated responsibilities (CBDR), principle of international environmental law establishing that all states are responsible for addressing global environmental destruction yet not equally responsible.

[108] *Greenpeace is* an organization founded in 1971 that stresses the need to maintain a balance between human progress and environmental conservation. Members take active but nonviolent measures against what are regarded as threats to environmental safety, such as the dumping of nuclear waste in the sea.

choose the lowest carbon path for development consistent with the need to bring 330 million people out of poverty loop. Above all India has to act on population front just as China has done.

Higher emissions are caused not only by higher consumptions but also by the higher number of people who consume. Between now and 2050, the world's population is likely to increase by about 3.50 million which would be entirely contributed by the developing world. For all its sins of emissions on the pollution front China is in an altogether different league. Its rate population growth in 2002 was 0.87% against India's 1.51 even though both were similarly placed in 1950. But India is very silent on this population issue. The world countries especially India has to realize that the post-Kyoto climate regime has been much more nuanced due to only two categories of the developed and the developing countries in the climate change treaty of today. The climate change could have a huge impact on our lives. So, to face the challenge of increasing human interference on climate change, all countries in the world must adopt strategies to decrease pollution and the governments should make laws relating to control and prevention of climate injustice and they should not be signatories to the international conventions on climate change but also enact national laws. India can make amendments to the Air (prevention and control) of pollution act 1981 with stringent provisions to curb carbon emissions.[109]

2.22. INTER RELATION OF CLIMATE CHANGE WITH OTHER THINGS AND CONCEPTS

Climate change has become a very big concern today. Climate change alone is not a topic it is interrelated to each and every aspect of human life in the form of cause and effect. The population growth is responsible for climate change in one way and this population has to face the consequences of climate change.

2.22.1. Population and Climate Change

The question comes in mind how increasing population and climate change are interrelated. The growth of population results in more pressure on earth's natural resources such as water, fossil fuels, land, and forest. There is various

[109] Sita, K. T., & Sunitha, K., (2010). Climate change and international law. In: Sanjay, K. S. (ed.), *Environment Law and Climate Change* (pp. 47–49). New Delhi: SBS Publishers & Distributors Pvt. Ltd.

linkage between population and climate change.[110] The growing population needs energy and in order to fulfill the energy need fossil fuels are burnt which results in concentration of CO_2 in atmosphere and mainly responsible for climate change. It has been seen in rise of energy demand in past decades and this is important for economic development of any country. In India 65% of energy is generated through power plant which uses coal for power generation and due to this pattern of energy production climate change is occurring.[111]

With the passage of time each and every person wants to live in urban area because of good education to children, better life style, transportation, etc. The growth of population is associated more with the industrial and urban development process rather than with agriculture process. The process of urbanization and industrialization which is leading to higher and higher concentration in larger town and megacities which in turn depict potential threat of greenhouse gases.[112] If urbanization trend would continue then it would result in concentration in metros which in turn would have serious urban environmental degradation implication. Such pattern of population growth and concentration in metropolitan cities has led to proliferation of illegal squatter settlements and urban sums with inadequate basic amenities comprising health care, sanitation, drainage, sewage, and potable water, etc.[113]

Rapid population growth leads to demand of more food for more numbers of mouth and more fertilizers and extensive cleaning of forest and agricultural land which all ultimately lead to climate change. Human behavior with regard to consumption of food over time and space depicts that urban people with generally higher incomes consume less of cereal than people in the countryside.[114]

One of the consequences of population growth is poverty. Poverty is said to be both cause and effect of environmental degradation. Poor people are

---

[110] Usha, T., (2016). Population growth climate change and law with special reference to India. In: Usha, T., (ed.), *Climate Change Law Policy and Governance* (p. 301). Lucknow, India: Eastern Book Company.

[111] *Ibid.*

[112] See Suzanne, P., (2009). Policy review: Thoughts on addressing population and climate change in a just and ethical manner. *Population and Environment, 30*(6), 275–289.

[113] Usha, T., (2016). Population growth climate change and law with special reference to India. Usha, T., (ed.), *Climate Change Law Policy and Governance* (p. 302). Lucknow, India: Eastern Book Company.

[114] Bansal, P. C., (1999). *Demand for Food Grains by 2020 A. D: A Monograph* (p. 11). observed research foundation, n.d.

vulnerable to climate change and become the easily victim of the impacts of the climate change. Due to poverty, they are not even able to meet their livelihood so it becomes impossible for them to fight with the adverse effect of climate change. Due to their poor condition of living, they are mainly dependent on the natural resources which they use in unsustainable and unregulated manner. As said by our earlier Prime Minister Mrs. Indira Gandhi that "the most pollutant we face is poverty" without eradication of it nothing significant can be done for environment protection. The reason is that the economically backward people use natural resources like, water and land in unsustainable way which play a vital role in polluting our environment. They use wood as cooking fuel which notably contribute in polluting the air around us. The extensive use of forest wood for different purposes ultimately reduces the carbon sink. It can be said that apart from energy production plants and fuel-based industries the poverty is also to the larger extent is the cause and effect of climate change.[115]

2.22.2. Climate Change and Migration

Earlier people used to migrate due to war or for the purpose of livelihood earning. But in recent times climate change has been emerged as a push factor of migration and the population which is migrating due to environmental challenges are termed as "climate induced migrants" or "climate refugees." Climate refugees are those people who are forced to migrate from their place due to climate change. In coming years there will be rise in sea level which is one of the observed effects of the climate change. The rise in the sea level will lead to submergence of many islands into water. The inhabitants of such islands will need a new place to abode and they will be shifted to another country or another place which would be already occupied by its original inhabitants. This will create extra pressure on earth, natural resources, land, forest, food, etc., where these people will be shifted. The original inhabitants living at that place could find difficulties in sharing all the resources with these refugees and this would create conflict among the people. Even there is no any provision in international law for these "climate induced migrants." If we look at the international law for the protection of the right of climate migrants then we find that in practice international law does not encourage mobility across international boundaries.[116]

---

[115] Usha, T., (2016). Population growth climate change and law with special reference to India. *Climate Change Law Policy and Governance* (p. 304). Lucknow India: Eastern Book Company, Lucknow.

[116] Aam, S., (2016). Climate change, migration and environmental justice. In: Usha, T., (ed.), *Climate Change Law Policy and Governance* (p. 222). Lucknow India: Eastern Book Company, Lucknow.

Article 13 of Universal Declaration of Human Rights (UDHR) provides "for freedom of movement" but it is also incomplete right because it is only providing right to leave your own country and right to enter in your country of which you are citizen. But international law does not provide for corresponding right to enter into any other country[117]. It is the matter of sovereignty of a country. In specific circumstances, international law compromises with sovereignty of state by requiring them to admit foreigners as part of their duty to uphold human right. The only kind of cross border mobility thus secured by international law is that of refugee, which is specifically defined by the refugee convention. Another problem in this regard is how person can be identified as climate refugees. In near future along with mitigation and adaptation of climate change the climate induced migrant also needs to be focused internationally.[118]

2.22.3. Climate Change and Energy

In present era, human civilization has become accustomed to luxurious life which they are getting at the cost of our environment. In the process of making their life more luxurious the people are continuously degrading their environment. It has been exploited in such a way that has gone beyond the control of human being and it is showing its dark side in the form of climate change. Energy has become the need of present human civilization. All kinds of developments including technological, economic and social is not possible without energy. And in the production and manufacturing of energy by using fossil fuel greenhouse gas is emitted which is mainly responsible for climate change. It has been rightly said that *'man should never bite the hands that feeds him.'* National Green Tribunal (NGT) in case of Court on Its Own Motion v. State of HP 2014 SCC Online NGT observed;

"Global warming has a direct impact on environment and ecology of any zone. Global warming, a succession of hot summer through the 1980 seemed to bear out the conservationists warming that the earth's climate was warming up. This global warming, if it is needed the case, is a consequence of the large quantities of carbon dioxide released into atmosphere by industry, power station and motor vehicles. The earth temperature will rise the polar caps will melt, raising sea level and catastrophic flooding in low lying areas of productive land will become desert this global warming heat has affected various countries in the world hence there is need to tackle global warming."

117 *Ibid,* 223.

118 *Ibid,* 224; Susan, M., (2010). Climate change, migration, and governance. *Global Governance, 16*(3), 397–414.

[119] Electricity is the main source of power in cities for each and every thing. People are dependent on the electricity which is mostly generated from thermal power plants by using fossil fuel which results into emission of greenhouse gas in the atmosphere. For transportation purposes oil and gas is used which is also fossil fuel and contributes to climate change.[120]

2.22.4. Climate Change and Human Rights

Human rights which are internationally protected are right to life, health, right to food, water, shelter, property, livelihood and culture with migration and settlement.[121] Safe and healthy environment is precondition for the enjoyment of fundamental human rights. Both regimes related to environmental protection and human rights are concerned with the development of human wellbeing. To live in a healthy and quality environment is fundamental and basic human right. The human rights are key to the overall development of any group of population at any place.[122] Even though the UNFCCC seeks to "protect the climate system for the benefit of present and future generations of humankind," but it is not designed to provide human rights protections, humanitarian aid or redress to individuals or communities facing the environmental harms. Instead the UNFCCC is an agreement between states to "anticipate, prevent or minimize the causes of climate change and mitigate its adverse effects." Much of the recent interest in the human rights dimensions of climate change has been due to the plight of the Inuit[123] and the Small Island States.

In their 2005 petition before the Inter-American Commission on Human Rights the Inuit were claiming that the impacts of climate change could be attributed to acts and omissions of the U.S., and violating their fundamental human rights, specifically the rights to the benefits of culture, to property, to the preservation of health, life, physical integrity, security, and a means of subsistence, and to residence, movement, and inviolability of the home.[124]

---

[119] Usha, T., (2016). *Climate Change Law Policy and Governance* (p. XIII). Lucknow, India: Eastern Book Company.

[120] *Causes of Climate Change*, http://edugreen.teri.res.in/explore/climate/causes.htm (accessed on 07 July 2023).

[121] Puneet, P., (2016). Human right and climate change. In: Usha, T., (ed.), *Climate Change Law Policy and Governance* (p. 207). Lucknow India: Eastern Book Company.

[122] *Ibid.*

[123] A member of an indigenous people of northern Canada and parts of Greenland and Alaska.

[124] Organization of American States, Inter-American Commission on Human Rights, Petition Seeking Relief from Violations Resulting from Global Warming Caused by Acts and Omissions of the United States (2005)

IT was argued that these rights were protected under several international human rights instruments, including the American Declaration of the Rights and Duties of Man.[125] The Inter-American Commission on Human Rights declined to review the merits of the petition, stating that the "information provided does not enable us to determine whether the alleged facts would tend to characterize a violation of rights protected by the American Declaration."[126] Even though the Inuit Petition did not fare well before the Commission, it drew attention to the links between climate change and human rights and led to a "Hearing of a General Nature" on human rights and global warming.[127] The Small Island States, and the Maldives in particular, launched a campaign to link climate change and human rights.[128] The Maldives' 2008 submission to the Office of the High Commissioner for Human Rights (OHCHR) as part of its preparation of its study on climate change and human rights described how climate change threatens its right to life. Like many other small island states, the Maldives is particularly at risk from the World Bank Study effects of rising sea levels as a result of warming water and melting ice sheets. Rising sea levels exacerbate inundation, storm surges, and erosion.[129]

2.22.5. Climate Change and Women

Climate change poses high risk to female population. The first risk which women may face due to climate change is risk of unemployment. Almost three quarters of working women in India work as agricultural laborers. They have very limited opportunities for employment. They have neither alternative rural employment nor ability to migrate alone. How they would continue to the strenuous manual work when there will be extreme hot climate.[130] Another indirect risk of climate change will be reduction in

---

[125] American Declaration on the Rights and Duties of Man, Apr. 1948, OAS Resolution XXX, OEA/Ser. L.V/II.82 doc.6 rev.1.

[126] Quoted in George, Jane, ICC Climate Change Petition Rejected, NUNATSIAQ NEWS (2006), http://www.nunatsiaq.com/archives/61215/news/nunavut/61215_02.html (accessed on 07 July 2023).

[127] Press Release, Inter-American Commission on Human Rights, IACHR Announces Webcast of Public Hearings of the 127th Regular Period of Sessions, No 8/07 (February 26, 2007), http://www.cidh.org/Comunicados/English/2007/8.07eng.htm (accessed on 07 July 2023).

[128] McInerney-Lankford, S., Mac, D., et al., (2011). *Human Rights and Climate Change: A Review of the International Legal Dimensions* (pp. 8–10). The World Bank.

[129] *Ibid,* 13–14.

[130] See Irene, D., (2002). Climate change: Learning from gender analysis and women's experiences of organising for sustainable development. *Gender and Development, 10*(2), 21–29.

agricultural yields which will result into the reduction of surplus pay for labor and reduction of need of labor which brings a question that what this vulnerable group will do for their livelihood. Another challenge arises due to climate change is of nutritional status of women. As per the patriarchal setting of our society, the women have already a lesser share of food into their plate and this will become adverse when there will be low income in the household. Due to the low income, women will compromise with their health and nutritional status as they meant to do such sacrifices and they will prefer to feed their family with their share of food. During drought situation women travel miles from their home in order to collect water for household purposes.

As a consequence, to this they get less time to rest, study even compromises as they dropout early from their regular schooling. Even at the time of mensuration and pregnancy they don't get rest due to water scarcity. The cyclone "Aila" which came in 2009 in the Indian Bengal Delta caused heavy destruction of life and property. It was observed that in the wake of cyclone Aila in 2009, approximately half of the men from the most affected blocks of the Indian Sundarbans ended up in migrating to other parts of the country in pursuit of alternative livelihoods. And the women were left behind to shoulder the sole burden of running the household and dealing with the aftermath of the cyclone. Cultivable lands were most affected due to the saltwater intrusion in the area. Nearly, 70% of total women were engaged in farming but cyclone made them helpless and jobless because women were not left with any other option of livelihood earning which also causes food scarcity. The repercussion led to wage discrimination, sexual exploitation and harassment in the workplace. As per the reports, many of these women from the Indian Sundarbans migrated to the red-light districts of Kolkata due to the climate change-induced distress. Many of these sex workers who migrated from Sundarbans identified themselves as bhasha (environmental refugees). Women struggle to get even basic necessities such as sanitary napkins and food for their new born in the consequences of natural disaster. Women who were forced to enter into prostitution in order to look after their family and children faced social ostracism and the threat of sexual exploitation at the hands of their clients, and were also under threat of sexually transmitted diseases like HIV/AIDS.[131]

---

[131] *Climate Change and Gendered Vulnerabilities: Accounting for Women and Patriarchal Systems in Climate Governance Policy,* (2020). Economic and Political Weekly Engage. https://www.epw.in/engage/article/climate-change-and-gendered-vulnerabilities (accessed on 07 July 2023).

2.22.6. Climate Change and Rural Poverty Levels in India

Poor and marginalized people are being affected and will be most affected by the threat posed by climate change and extreme weather events such as droughts, floods, cyclones and other natural calamities.[132] This has a direct or indirect impact on their lives and livelihoods through loss of crop yields, incomes, assets, employment, etc. In addition to this, a rise in food prices followed by crop loss hurts the poor who are net buyers of food. The Fifth Assessment Report of the Intergovernmental Panel on Climate Change (IPCC) notes that climate change and climate variability will "worsen existing poverty, exacerbate inequalities and trigger both new vulnerabilities and some opportunities for individuals and communities it will create new poor between now and 2100 in developed and developing countries, and jeopardize sustainable development."[133] "*Climate change driven impacts ... act as a threat multiplier meaning that the impacts of climate change compound other drivers of poverty ... Climate change intersects with many causes and aspects of poverty to worsen not only income poverty but also undermine well-being, agency, and a sense of belonging*"[134]. It will hinder economic growth and efforts to achieve the United Nations' (UN) Sustainable Development Goal (SDG) 1 that seeks to reduce the number of poor people by half and eradicate extreme poverty in all its forms everywhere by 2030. Funds that can otherwise be used for development purposes have to be diverted to invest in climate infrastructure and build climate resilience as well as provide safety nets for the poor.[135]

2.22.7. Climate Change, Air Pollution, and Human Health

As per the United Nations Industrial Development Organization, climate change is likely to have a greater impact on India as compared to other nations because of its unique combination of the geography, diverse

---

[132] Somanathan, E., & Rohini, S., (2009). *Climate Change: Challenges Facing India's poor* (Vol. 44, No. 31, pp. 51–58). Economic and Political Weekly.

[133] "IPCC, (2014a). "Livelihoods and poverty; chapter 13. In: Olsson, E., Opondo, M., Tschakert, P., Agrawal, A., Erickson, S H., Ma, S., Perch, L. N., & Zakielden, S. A., (eds.), *Climate Change 2014: Impacts, Adaptation and Vulnerability, Part A: Global and Sectoral Aspects, Contribution of the Working Group II to the Fifth Assessment Report of the Intergovernmental Panel on Climate Change* (pp. 793–832). B C Fields, and others (eds.), Cambridge, UK and New York, NY, US: Cambridge University Press.

[134] *Ibid.*

[135] Ninan, K. N., (2019). *Climate Change and Rural Poverty Levels in India* (Vol. 54, No. 2). Economic and Political Weekly. https://www.epw.in/journal/2019/2/special-articles/climate-change-and-rural-poverty-levels.html (accessed on 07 July 2023).

population characteristics and larger dependence on carbon-related energy. Climate change will definitely impact the basic requirements for maintaining health-clean air and water, adequate shelter, sufficient food and freedom from disease given the already high level of poverty, low nutritional status and poor health infrastructure in the country.

The question arises that how will this changing climate affect our health condition? Climate has direct influence on many of the key determinants of health: it leads to extremes and violent weather events; upturn of disease, organisms and vectors; affects the quality as well as quantity of air, food and water; and the stability of the ecosystem in which we are living. Climate change can directly and indirectly impact human health. Mostly, Indirect impacts arise from changes in the usual temperature patterns, which may cause disturbance in the natural ecosystems, change in the ecology of infectious diseases, distress agriculture and fresh water supplies, exacerbate pollutants level in the air, and cause large-scale reorganization of animal and plant communities.

When heat from the sunrays strikes on the surface of the seas, lakes and rivers induces evaporation then the phenomenon of rain is caused. The water vapor formed in this process rises upward to a height where it condenses into moisture. If required atmospheric conditions prevail it comes down as rain. But in the case of acid rain, water vapors reach to the atmosphere, condense, and react with atmospheric pollutants which are deposited on the soil, vegetation, surface water and reservoirs. The deposition ultimately results in damages because of the acidity of the pollutants. Acidic rainwater liberates mercury from the soil which can cause hinder in the brain development of the fetus. Fish-eating birds and humans consume mercury by eating fish of water bodies containing high levels of the metal. Acid rain also releases cadmium and aluminum. Cadmium can cause severe kidney disorders, besides accumulating in the outer layer of the kidney, causing wounds. On the other hand, Aluminum, causes problems for kidney patients. In process of dialysis during kidneys malfunction-it enters the blood stream directly without having passed the body's normal protective filters. This may lead to skeletal and brain damage. It may also cause Alzheimer's and Parkinson's diseases. Increasing industrial emissions, traffic and exhaust as well as increasing concentrations of SO_2, NOx, O_3 and suspended particulate matter, continuously damaging human health. High temperature increases the levels of ozone and other air pollutants at ground level, and hasten the onset of pollen season. Presence of pollen and other allergens in the air stimulate and aggravate Asthma and cardiovascular respiratory diseases.[136] The impact of climate change on water availability is one of the most significant for the human health. Higher temperature on earth's surface due to change in climate are hastening the rates of evaporation of surface water thereby reducing the availability of the fresh water. Due to lack of fresh water, people compromise with basic hygiene thus increasing incidence of diarrheal diseases. On the other hand, too much water due to floods causes contamination of freshwater supplies. Extreme events like rise in the sea level coupled with strong storm surges and flooding in the coastal belt can be followed by outbreak of diseases such as Cholera.

Perhaps the long-term impact to human health from climate change will be the disruption of natural ecosystems in which we are living, which provide an array of services that ultimately supporting human health. Biotic systems-whether in forests, aquatic environments, rangelands, or elsewhere-provide food, materials, and medicines; store and release freshwater; absorb and detoxify wastes; and satisfy human needs for recreation and wilderness. They are also intimately involved in sustaining the genetic basis of agriculture.[137]

2.22.8. Climate and Ecosystems

Climatic conditions of a region determine whether particular species of plant and animal can live, grow, and reproduce. Thus, the group of species that we are familiar with – the desert ecosystems of the arid southwest, the southern mixed deciduous forest or the productive grasslands of the great plains- are majorly influenced by climatic condition as well as other factors such as the land-use. The species in some ecosystems are so strongly influenced by the climate to which they are adapted that they are vulnerable even to modest climate change.

Alpine meadows at high elevations in the west exist because the plants that comprise them are adapted to the cold condition that would be too harsh for other species in the region. The desert vegetation of the southwest is adapted to the high summer temperature and aridity of the region. Forest in the east is adapted to relatively high rainfall. And soil moisture; if drought conditions were to persist, grasses and shrubs could begin to out-compete tree seedlings, leading to completely different ecosystems.

---

[136] Sangeeta, M., (2010). Effect of climate change on air pollution and human health in India: An analysis. In: Sanjay, K. S., (ed.), *Environment Law and Climate Change* (pp. 51–57). New Delhi: SBS Publishers & Distributors Pvt. Ltd.

[137] *Ibid.*

There are also many fresh water and marine examples of sensitivities to climate variability and change. In aquatic ecosystem, for example, many fish can breed only in water that falls within a narrow range of temperatures. Thus, species of fish that are adapted to cool waters can quickly become unable to breed successfully if water temperature rises. Wetland plant species can adjust to rising sea levels by dispersing to new locations, within limits. Too rapid sea level rise can surpass the ability of carbon in respiration than they gain in photosynthesis.[138]

2.22.9. Climate Change and COVID-19

At present whole world is facing the challenge of COVID-19 pandemic, simultaneously we can't afford to lose sight of the larger crisis that lies in wait: climate change.[139] The COVID-19 pandemic is causing havoc in countries across the whole world by causing a global health crisis and forcing economies to slow down due to the strict quarantine measures. COVID-19 pandemic is affecting the lives of millions of people and, also, the environment. The CO_2 emissions and human mobility have been reduced due to which air quality has improved and encouraged wild animals to come out and explore the cities.

Scientists have confirmed that air quality in certain regions have improved. As industries, aviation, and other means of transportation was totally stopped due to lockdown which resulted in reduction in air pollution in the countries which were severely affected by the virus, such as China, Italy, and Spain. A reduction in commuting due to work from home policies has also played its part in reducing carbon emissions.

In the long term, the COVID-19 pandemic will offer lessons and opportunities which will lead to environmental action. For example, we will have a new baseline of what can be achieved digitally like remote work, education, shopping, and more. In addition, as our governments, private institutions, and even social media succeed in partnering, we will possibly feel more capable of tackling other pressing issues of environment such as climate change. Our response to this health crisis will prepare us to deal with a climate crisis in the next decades. With all its negative impact on human civilization the COVID 19 have also given us an opportunity and insight to

---

[138] Gurkirat, K., (2006). *Environmental Crime* (p. 63). New Delhi: Shree Publishers and Distributors.

[139] of *COVID-19 and Climate Change, One Is the Tip and the Other Is the Iceberg,* https://science.thewire.in/environment/of-covid-19-and-climate-change-one-is-the-tip-and-the-other-is-the-iceberg/ (accessed on 07 July 2023).

develop some habits that are incidentally beneficial to the environment which may last since people have experienced scarcity. For example, the reduction of travel, personal consumption, and food waste, marriage celebration with minimum people.[140]

As the Secretary-General of the United Nations, António Guterres, indicated the threat of COVID-19 is temporary, meanwhile, the threat of droughts, floods, and extreme storms linked to climate change will remain for decades and will always require constant action. Human beings are part of nature and all activity that impacts the environment will ultimately impact us. Under UNFCCC we are committed to reduce carbon emission but no one wanted carbon emissions to be reduced this way. COVID-19 has a dark cost which resulted in loss of several life, healthcare systems, and mental health of people around the world. Nevertheless, it has also shown that communities can make a great difference when they take care of each other, and this could be an invaluable lesson when we face climate change.[141]

2.23. CONCLUSION

"It has been demonstrated beyond reasonable doubt that the climate is changing due to man-made greenhouse gases. We are already committed to future substantial change over the next 30 years and change is likely to accelerate over the rest of the 21stcentury."

Climate change is now not regional or national issue but it has become a global issue and its impact will not exempt anyone in this world. It will affect each and every country differently depending upon their geological condition, geographical location and economic development. In Asian region India is most vulnerable to climate change because of its physical condition, geographical location and emerging economy. If we will not reduce the emission to nearly zero then we will face the consequence beyond imagination. India needs to balance both economic development at one hand and mitigation of climate change on the other hand.

Climate change has been significantly described as a 'wicked' problem– it has no unambiguous definition, is riven with scientific uncertainty, and proposed solutions are complicated by their embedding in social, political,

---

[140] See Yamini, A., Vijay, C., et al., (2021). "India's resurgence of COVID-19: Urgent actions needed. 397 *The Lancet, 397*, 2232, 2233. https://www.thelancet.com/action/showPdf?pii =S0140-6736%2821%2901202-2 (accessed on 07 July 2023).

[141] Jason, P. *What is the Effect of COVID-19 on Climate Change?* Jauregui https://www.oneyoungworld.com/blog/what-effect-covid-19-climate-change (accessed on 07 July 2023).

and economic systems. If climate change is considered as an environmental issue for some people, for others it is predominantly about justice and equity, and for yet others, it is largely an economic and technical challenge. If we see climate change from an Indian perspective, India as a country faces enormous, and immediate, challenges of lifting its citizens out of poverty; providing access to basic needs such as health, education, energy, and water and sanitation; and addressing governance challenges of corruption and communal tensions. Due to these all reasons climate change appears less immediate, less certain, and therefore, less of a priority. There is a long-standing political perspective that India has contributed relatively little to causing the climate problem and should not be asked to be in the forefront of solving it. Climate change can interrupt agricultural systems, water availability, forest health, and coastal ecology, thereby affecting the lives and livelihoods of millions of people. Given India's vulnerability to climate change, India's interests lie in promoting effective global cooperation in addressing climate change. With low per capita and historical emissions, India may not have a responsibility for climate, but as the fourth largest emitter in it may have a responsibility to vulnerable populations to engage climate change.

In documentary film *An Inconvenient Truth* directed by Davis Guggenheim about former United States Vice President Al Gore's campaign to educate citizens about global warming calls upon his viewers to learn how they can help him in these efforts. Gore concludes the film by saying:

"Each one of us is a cause of global warming, but each one of us can make choices to change that with the things we buy, the electricity we use, the cars we drive; we can make choices to bring our individual carbon emissions to zero. The solutions are in our hands, we just have to have the determination to make it happen. We have everything that we need to reduce carbon emissions, everything but political will. But in America, the will to act is a renewable resource."

Chapter

3 THE CLIMATE CHANGE MITIGATION: THE LEGAL FRAMEWORK

Contents

3.1. INTRODUCTION

Now a days the term "climate change" has become the point of discussion whether it is scientific or political. The climate has been changing since the beginning of mankind. But why today has it has gain so much importance the reason is the speed of climate change. Today everyone knows that climate is changing too fast and it may be the greatest threats for mankind and existence of planet. Climate change has become very big concern for the existence of mankind and it is anthropogenic. Climate change has many effects like disturbance of monsoon, melting of glacier, rise in sea level, tsunami, earthquakes, cyclone, desertification, rise in global temperature, shrinking ice sheets, declining Arctic Sea ice, ocean acidification, decreased snow cover etc. It has been cleared by IPCC 5th assessment report that climate change is anthropogenic. Human activities, especially in the industrialized countries, releases a number of gases mainly carbon dioxide, CFCs, methane and nitrous oxide that slow the escape of infrared radiation (heat) from the surface of the earth into space. Both the countries have wide range of threat from climate change. The question comes how we can fight with climate change. There is two possible suggestions 'mitigation' and 'adaptation' for minimizing the impact of the climate change. Mitigation and adaptation both are complementary and supplementary of each other.[1] Climate change means change in climate cycle due to concentration of greenhouse gas in atmosphere; which lead to heavy rainfall, flood, and drought, melting, of glacier etc. The solution to this climate change is mitigation and adaptation through mitigation future effects of climate change can be reduced. So, there is need of adaption also; adaption alone is not sufficient because it is not a solution but it ensures in survival ship on climate change and it requires huge amount of money. So, there is incumbent need of mitigation policy and laws and rules. The options available for climate change mitigation is removing carbon sequestration in agriculture soil, reduction in carbon emission, de carbonization of energy input, carbon offset through bio-organic farming, switching to renewable energy, bio efficiency etc.

3.2. MEANING AND DEFINITION OF MITIGATION AND ADAPTATION

Climate change mitigation refers to the actions which are taken to reduce or limit the magnitude of long-term climate change. Climate change mitigation

[1] See Kerstin, N., & Rickard, L., (2010). Climate change and mitigation. Special report: Energy 2050. *Ambio, 39*(Supplement 1), 11–17.

generally comprises of reductions in anthropogenic emissions of greenhouse gases (GHGs) by increasing the capacity of carbon sinks, e.g., through reforestation[2]. The option through which climate change can be mitigated is removing carbon sequestration in agriculture soil, reduction in carbon emission decarbonization of energy input, carbon offset through bio-organic farming, switching to renewable energy, bio electricity, etc.

Climate change Mitigation is an anthropogenic intervention to reduce the anthropogenic forcing of the climate system; it includes strategies to reduce greenhouse gas sources and emissions and enhancing greenhouse gas sinks[3].

Climate Change Mitigation refers to efforts to reduce or prevent emission of greenhouse gases. Mitigation can mean using new technologies and renewable energies, making older equipment more energy efficient, or changing management practices or consumer behavior. It can be as complex as a plan for a new city, or as a simple as improvements to a cook stove design. Efforts underway around the world range from high-tech subway systems to bicycling paths and walkways. UN Environment takes a multifaceted approach towards climate change mitigation in its efforts to help countries move towards climate-resilient and low emissions strategies.[4]

3.3. ADAPTATION AND MITIGATION ACTIONS AS DEFINED BY THE FOURTH IPCC CLIMATE REPORT 2007

If we Refer to the fourth IPCC climate change report 2007, the single sector\ actions can contribute to the reduction of GHG emissions are as follows:

1. Investing in New energy infrastructure in developing countries, up gradation of energy infrastructure in industrialized countries, and policies that promote energy security.
   i. Investing new energy infrastructure will have long-term impacts on GHG emissions, due to reason of the long life-times of energy plants and infrastructure capital stock.

[2] *Climate Change Mitigation*. https://www.sciencedaily.com/terms/climate_change_mitigation.htm (accessed on 07 July 2023).

[3] *Definition Given by IPCC*. http://www.ipcc.ch/ipccreports/tar/vol4/index.php?idp=204 (accessed on 07 July 2023).

[4] *Definition Given by The United Nations Environment Program*. https://www.unenvironment.org/explore-topics/climate-change/what-we-do/mitigation (accessed on 07 July 2023).

The widespread diffusion of low-carbon technologies may take many decades, even if early investments in these technologies are made attractive.

ii. Investment in end-use energy efficiency improvement: efficiency improvement has a positive effect on energy security, local and regional air pollution abatement, and employment.

iii. Investment in renewable energy: this generally has a positive effect on energy security, employment and on air quality. The higher the market prices of fossil fuels, the more low-carbon alternatives will be competitive, although price volatility will be a disincentive for investors.[5]

2. **Transport sectors:** There are multiple mitigation options in this sector, but their effect may be counteracted by growth in the sector, and these mitigation options in this sector are confronts with many barriers, for example consumer preferences and lack of policy frameworks. Market forces alone, including rising fuel costs, are not expected to lead to significant emission reduction:

i. Light-duty vehicles: improved vehicle efficiency measures, leading to fuel saving, in many cases have net benefits, but the market potential is much lower than the economic potential due to the influence of other consumer considerations, such as performance and size.

ii. There is not enough information to assess the mitigation potential for heavy-duty vehicles.

iii. Biofuels are expected to play an important role in addressing GHG emissions in the transport sector, developing on their production pathway.

iv. Modal shifts from roads to rail and inland waterway shipping and from low-occupancy to high-occupancy passenger transportation as well as land-use, urban planning and non-motorized transport (NMT) offer opportunities for GHG mitigation, depending on local conditions and policies.

---

[5] *IPCC Fourth Assessment Report: Climate Change 2007: Impacts, Adaptation and Vulnerability*. https://www.ipcc.ch/site/assets/uploads/2018/03/ar4_wg2_full_report.pdf (accessed on 07 July 2023).

v. The aviation sector could achieve medium-term mitigation potential for CO_2 emissions from improved fuel efficiency, which can be achieved, e.g., through technology, operations and air traffic management. Nevertheless, such improvements are expected to only partially offset the growth of aviation emissions.[6]

vi. Realizing emissions reductions in the transport sector is often a co-benefit of addressing traffic congestion, air quality and energy security.

3. **Building industry:** Energy-efficient building (new and existing ones), besides limiting the growth of CO_2 emissions, can also improve indoor and outdoor air quality, improve social welfare and enhance energy security:

i. In general, opportunities for realizing GHG reductions in the building sector exist worldwide – by 2030, about 30% of the projected GHG emissions in this sector can be avoided with net economic the magnitude of the benefit.

ii. Multiple barriers make it difficult to realize this potential; these barriers include availability of technology, financing, poverty, higher costs of reliable information, limitations inherent in building designs and an appropriate portfolio of policies and programs.

iii. The magnitude of the above barriers is higher in developing countries, making it more difficult for them to achieve the GHG reduction potential of the building sector.

4. **Energy-intensive industries:** Even though many industrial facilities in developing countries are new and include the lasted technology with the lowest specific emissions, many older, inefficient facilities remain in both industrialized and developing countries. Upgrading these facilities can deliver significant emission reductions. But, the slow rate of capital stock turnover, lack of financial and technical resources, and limitations in the ability of firms, particularly small and medium-sizes enterprises,

[6] *IPCC Fourth Assessment Report: Climate Change 2007: Impacts, Adaptation and Vulnerability*. https://www.ipcc.ch/site/assets/uploads/2018/03/ar4_wg2_full_report.pdf (accessed on 07 July 2023).

to access and absorb technological information are key barriers to full use of available mitigation options.[7]

5. **Agriculture industry\practices:** can make a significant contribution at low cost to increasing biomass feedstock for energy use:

a. A large proportion of the mitigation potential of agriculture (excluding bioenergy) arises from:

   i. Soil carbon sequestrations, which has strong synergies with sustainable agriculture and generally reduces vulnerability to climate change (stored soil carbon may be vulnerable to loss through both land management change and climate change);

   ii. Reductions in methane and nitrous oxide emissions in some agricultural systems.

   iii. There is no universally applicable list of mitigation practices, as practices need to be evaluated for individual agriculture systems and setting.

   iv. The contribution of biomass to mitigation depends on demand for bioenergy from transport and energy supply, on water availability, and on requirements of land for food and fiber production. Widespread use to agriculture land for biomass production for energy may compete with other land uses and can have positive and negative environmental impacts and implications for food security.[8]

6. **Forest-related mitigation activities:** can considerably reduce emissions from sources and increase CO_2 removals by sinks at low costs, and cab be designed to create synergies with adaptation and sustainable development:

   i. About 65% of the total mitigation potential is located in the tropics and about 50% of the total could be achieved by reducing emissions from deforestation.

---

[7] *IPCC Fourth Assessment Report: Climate Change 2007: Impacts, Adaptation and Vulnerability.* https://www.ipcc.ch/site/assets/uploads/2018/03/ar4_wg2_full_report.pdf (accessed on 07 July 2023).

[8] *IPCC Fourth Assessment Report: Climate Change 2007: Impacts, Adaptation and Vulnerability.* https://www.ipcc.ch/site/assets/uploads/2018/03/ar4_wg2_full_report.pdf (accessed on 07 July 2023).

ii. Climate change can affect the mitigation potential of the forestry sector (i.e., native and planted forests) and is expected to be different for different regions and sub-regions, both in magnitude and direction.

iii. Forest-related mitigation options can be designed and implemented to be compatible with adaptation, and can have substantial co-benefits in terms of employment, income generation, and biodiversity and watershed conservation, RE supply and poverty alleviation.

7. **Waste Sector:** Even though post-consumer waste is only a small contributor to global GHG emissions the waste sector can positively contribute to GHG mitigation at low costs and promote sustainable development:

i. A wide range of mature, environmentally effective technologies are commercially available to mitigation emissions and provide co-benefits for improved public health and safely, soil protection and pollution prevention, and local energy supply.

ii. Waste minimization and recycling provide important indirect mitigation benefits through the conservation of energy and materials.

iii. The most important barriers in this sector are: lack of local capital for waste and wastewater management in developing countries and countries with economics in transition and lack of expertise on sustainable technology.[9]

3.4. PROBLEM IN MITIGATING CLIMATE CHANGE

If we know the solutions to the problem of climate change then what is the issues due to which this problem is not being solved. The reason is the world know the problem and solutions both but no one is ready to take responsibility to mitigate it. The world is divided in two parts developing and developed developing nations blames developed nations for climate change so they should take responsibility to mitigate it.[10] Developed nations holds

[9] *IPCC Fourth Assessment Report: Climate Change 2007: Impacts, Adaptation and Vulnerability.* https://www.ipcc.ch/site/assets/uploads/2018/03/ar4_wg2_full_report.pdf (accessed on 07 July 2023).

[10] Ting, W., Shili, Y., et al., (2012). Developed and developing world responsibilities for historical climate change and CO_2 mitigation. *Proceedings of the National Academy of Sciences of the United States of America, 109*(32), 12911–12915.

the opinion that each one is contributing to climate change so each country should be responsible for mitigation these contentions are the problem for the countries like India which is developing.[11]

Much of international climate change law focuses on mitigation which encompasses both major to limit greenhouse gases emissions and measures to preserve or enhance sinks. Policies to reduce emissions include Energy Efficiency standards subsidies for renewable energy, a carbon tax, emission trading system, funding of urban mass transit system, and Technology research and development (R&D). Sinks policies generally relate to Land use, land use change, and forestry, and include measures to reduce emission from deforestation and forest degradation and to encourage afforestation.

3.5. CLIMATE CHANGE MITIGATION INTERNATIONAL AND NATIONAL ISSUES

The first issue is whether to address greenhouse gas emissions on an economy wide basis or at a sectoral level? Generally, the UN climate regime has pursued to address aggregate National emission and has not separated out particular sector emission such as electricity generation or buildings. But a few sectors in international law receive specific attention, including emissions from international Maritime and air transport, which are addressed through The International Maritime Organization (IMO) and International Civil Aviation organization respectively, and forestry, which is the subject of REDD+.[12]

The second issue is whether to regulate greenhouse gases emission comprehensively or gas- by- gas? Even though CO_2 accounts for more than two third of total greenhouse gases emissions and, except for emissions from land use can be accurately accounted, the UN climate regime has not focused is specifically on CO_2. UNFCCC seeks to promote cost effectiveness by addressing greenhouse gases comprehensively, which allows States to focus on whichever gas can be reduced at the least cost. In contrast, efforts

[11] Navroz K. D., (2019). An introduction to India's evolving climate change debate: From diplomatic insulation to policy integration. In: Navroz, K. D., (ed.), *India in a Warming World: Integrating Climate Change and Development* (p. 3). New Delhi: Oxford University Press.

[12] REDD+ (or REDD-plus) refers to "reducing emissions from deforestation and forest degradation in developing countries, and the role of conservation, sustainable management of forests, and enhancement of forest carbon stocks in developing countries" (emphasis added); the most recent, elaborated terminology used by the COP.

to address climate change through the Montreal Protocol[13] have focused on particular gases- hydro chlorofluorocarbons (HCFCs) initially and, currently hydro fluorocarbons (HFCs).[14]

The third issue is whether to prescribe particular measures internationally give States flexibility? In general, International climate change law has not right to prescribe specific mitigation measures. International climate change law has adopted either a bottom-up approach that permits States to develop and report on their own policies or when it has prescribed rules internationally, the rules have been obligation of results. For example, to reduce emission by some specified amount- which allows States to choose what policies they will use to achieve the required result. One exception to this general rule is IMO work to limit emissions from Maritime shipping under the international convention for the prevention of pollution from ships (MARPOL), which has involved the adoption of mandatory Energy Efficiency is standards for vessels.[15]

Fourth issue is whether to give states flexibility in deciding where to reduce emissions? Market mechanisms such as emissions trading (ET) allow countries to implement their own mitigation commitments through emissions reductions in another country. The FCCC contained an embryonic Market mechanism which allows party to undertake activities jointly to reduce emissions. The Kyoto Protocol has employed its own market mechanism much more extensively, by allowing state to (1) credits for undertaking emissions project in another country through the clean development mechanism (CDM) and joint implementation; and (2) trade emissions allowances with other parties, selling to countries with higher mitigation cost and buying from countries with lower costs.[16]

Fifth issue is the extent to which commitments of states should be tailored to their differing capabilities and responsibilities, and how?

The issue of differentiation in mitigating has been one of the most controversial in the context of mitigation. The principle of common but

---

[13] The Montreal Protocol on Substances that Deplete the Ozone Layer (the Montreal Protocol) is an international agreement made in 1987. It was designed to stop the production and import of ozone depleting substances and reduce their concentration in the atmosphere to help protect the earth's ozone layer.

[14] Daniel, B., Jutta, B., et al., (2017). *International Climate Change Law* (Vol. 13). London: Oxford University Press.

[15] *Ibid.*

[16] *Ibid.*

differentiated responsibilities and respective capabilities (CBDRRC)[17], first articulated in UNFCCC, represents a departure from the traditional approach of international agreements, namely, to define a common set of obligations for all parties it gives expression to the profound equity concerned raised by climate change challenges, by providing that the climate change commitments of party should be differentiated, based on their different responsibilities and capability. Although there is a Universal support what the principle of CBDRRC, there is very little agreement on its rational, core content, and application in particular situations.[18] With respect to the rational for differentiation, developing countries have tended to focus on the term 'responsibilities,' which they understand to be the function of 'historical emissions,' where like US in particular have focused on term, if the different historical contributions of countries to the climate change problem provide the basis for differentiation as developing countries content, then differentiation will change relatively slowly. In contrast, if capabilities provide the basis for differentiation, then countries obligation could evolve more rapidly, as it develops and gains greater financial technological and administrative capabilities. The principal of CBDRRC preserves the position of both sides in this debate by including both responsibilities and their respective capabilities as base for differentiation.[19]

3.6. METHODS OF MITIGATION OF CLIMATE CHANGE

There are various methods of mitigating climate change like reducing Greenhouse Gas Concentrations in Atmosphere, creating carbon sink, Individual emission cut, carbon capture storage (CCS), etc.

3.6.1. Reduction in Greenhouse Gas Concentrations in Atmosphere

Climate change has become well identified fact and its impacts are also being observed. Climate change is anthropogenic and it can be mitigated only if

[17] The principle of Common but Differentiated Responsibilities and Respective Capabilities (CBDRRC) establishes the common responsibility of states for the protection of the global environment. But in addition it also lays down different standards of conduct for developed and developing nations.

[18] See Lavanya, R., (2006). *Differential Treatment under International Environmental Law*. London: Oxford University Press.

[19] Daniel, B., Jutta, B., et al., (2017). *International Climate Change Law* (p. 13). London: Oxford University Press.

there will be reduction in greenhouse gas concentration in atmosphere. And reduction is possible if we opt for green energy that is renewable energy. By creating carbon sink also greenhouse gases can be reduced to the large extent. Uses of energy efficient equipment, cutting down individual emission, using bio fuels in place of fossil fuels can help in reducing greenhouse gases.

The International Panel on Climate Change provides and distinguishes three main options for the mitigating atmospheric CO_2 concentrations by the agricultural sector: (1) reduction of agriculture-related emissions, (2) creation and strengthening of Carbon sinks in the soil, and (3) production of biofuels to switch fossil fuels. Options for sustained sequestration of Carbon in the soil through adapted management of land resources are being reviewed in order to reduce greenhouse gas concentrations in the atmosphere. Enhanced sequestration of atmospheric CO_2 in the soil, ultimately as stable humus, may well prove a more durable solution than (temporarily) sequestering CO_2 in the standing biomass through reforestation and afforestation. Such actions will also help to reverse processes of land degradation, thus contributing to sustained food productivity and security for the people in the regions concerned[20].

3.6.2. Creation of Carbon Sinks

Climate change is anthropogenic and it is due to the large-scale emission of $CO_{2,}$ for mitigation we create carbon sink to absorb the CO_2 to large scale it will help in mitigation to great extent. Carbon sinks are very important for our environment, because they act like sponges to soak up the carbon compounds that are playing such an enormous role in global climate change. Basically, carbon sinks are holding tanks for carbon or carbon compounds, like carbon dioxide (CO_2).[21] The main carbon sinks are oceans, fossil fuel deposits, the atmosphere, and the Earth itself where the element is present in rocks and sediments, in swamps, wetlands, and forests, and in the soils of forests, grasslands, and farmland which store carbon in themselves. A stock that is taking up carbon is a sink. Oceans, soils, and forests all have some potential for carbon sequestration to large scale. But forests help in the greatest immediate mitigation promise. Unlike many plants and most crops,

[20] Batjes, N. H., (1998). Mitigation of atmospheric CO_2 concentrations by increased carbon sequestration in the soil. *Biology and Fertility of Soils*, 27, 230–235. https://link.springer.com/article/10.1007/s003740050425 (accessed on 07 July 2023).

[21] *What are Carbon Sinks?* https://www.conserve-energy-future.com/carbon-sinks.php (accessed on 07 July 2023).

forests are accumulating carbon over decades and centuries.[22] Their potential for accumulation is large enough that significant amounts of carbon may be sequestered in a matter of decades. Fortuitously, forests that people manage for timber, wildlife, and recreation sequester carbon as a matter of course.[23] Forests may also be managed strictly as sinks.[24]

3.6.3. Transportation Sector

Greenhouse gas emissions from transportation is increasing faster than any other sector, accounting for roughly 29% of total greenhouse gases emission in the world in 2019.[25] Mitigation in transportation sector can be possible by using this Low carbon electric vehicle by replacing fossil diesel or petrol vehicle.[26] Electricity is not itself a clean energy source of energy but it is crucial to compare the wearing environmental impacts of energy sources used to produce electricity as a single source of polluters electric plants are easier to regulate the millions of vehicles. Carpooling, using public transport, metro, cycling is another method of mitigation.[27]

3.6.4. Carbon Budget

To keep the planet from overheating we can make a budget that is called carbon budget. In a way just like a household budget is the amount of a household, which I can spend while sticking to certain savings goals and achieving the desired lifestyle, and calorie budget determines, how much we can eat while sticking to a certain activity level and a desired weight. A global carbon budget tells humanity what amount of greenhouse gases

---

[22] See Bas, V. W., & Eric, F. L., (2001). Carbon Sinks and Conserving Biodiversity. *Science, New Series, 294*(5549), 2094–2095.

[23] See Chaturvedi, R. K., Tiwari, R., & Ravindranath, N. H., (2008). Climate change and forests in India. *The International Forestry Review (Special Issue: The Indian Forest Sector – Current Trends and Future Challenges), 10*(2), 256–268.

[24] Roger, A. S. *Harvesting the Benefits of Carbon Sinks.* http://www.rff.org/files/sharepoint/WorkImages/Download/RFF-Resources-133-carbonsinks.pdf

[25] Global transport emissions increased by less than 0.5% in 2019 (compared with 1.9% annually since 2000) owing to efficiency improvements, electrification and greater use of biofuels. Nevertheless, transportation is still responsible for 24% of direct CO_2 emissions from fuel combustion, https://www.iea.org/topics/transport

[26] See Balasudarsun, S. P. N. L., (2020). Review on emergence of electric mobility and feasibility of electric mobility adoption in India. *Aegaeum Journal, 8*(9), 1407–1414, http://aegaeum.com/gallery/agm.j-4143.144-f.pdf

[27] Elizabeth, B., (2007). Multilateral climate change mitigation. *U. S. F. L. Rev., 41,* 373.

we can emit while having a reasonable chance of a staying within a safe temperature increase. As one can imagine coming up with this number is not easy task. scientist must look at all the process that emit CO_2 and observe it from the air then they need to look at how CO_2 levels after with temperature and finally agrees on a set amount of CO_2 that can be emitted to have a good chance of keeping the temperature rise limited to 1 degree centigrade. the budget to keep the temperature rise to 1.5 degree centigrade is frighteningly small uncertain and is quickly being exhausted so much so that in 2016 we have about five years before we blew through the entire budget which means that even with China's breakneck adoption of renewable it is highly unlikely that our Earth is going to warm beyond 1.5 degree centigrade from 19th century temperatures.[28]

We can't stop this warming, the task is too great into short time to achieve. our existing economic, social and political system will not allow us to make the changes required in nearly five years can we dial up the 'safe' temperature knob a little bit to see if we can buy more time? We have about 20 years from 2017 before we exhaust the budget for the 2-degree centigrade limit. We will have to make comprehensive change to our way of life if we want to cut our emissions enough to stay within this limit, but this would still leave us with a one-in-three chance of our planet warming above 2°Centigrade and these are the best odds that we have. In fact, our 'time' to zero emission might be significantly shorter if we considered the role of non-CO_2 greenhouse gases. The budget is a pie to be share buy all countries however this is easier said than done. Anyone who has worked in the departmental context knows the aggressive jogging that takes place while allocating budget. Imagine this happening on a planetary scale, with Trump[29] lade America in the mix adding to this difficulty is the uncertainty surrounding these numbers because of the wholesale nature of change required in our lives many of whom are powerful today stands to lose a lot because of the proposed changes. As such what is straightforward-but-as-yet-not-fully understood-science becomes a highly politicized topic with the uncertainty being used as a weapon to delay action. Given the stakes (humanities future), we need to ask: why the uncertainty, what are the political compulsions that weigh in while fighting against the climate

[28] Mridula, R., (2018). *The Climate Solution: India's Climate-Change Crisis and What We Can Do About It* (p. 28). New Delhi: Hachette India.

[29] In June 2017, a few months into his presidency, Trump withdrew from the accord, saying that the deal was "draconian," while disagreeing with the standards for carbon emission reductions.

change? And why even the face of uncertainty, do we need to act NOW?[30]

3.6.5. Agriculture

In agriculture sector improved crop and grazing land management will increase soil carbon storage which will help mitigating. Restoration of cultivated peaty soils and degraded lands, improved rice cultivation techniques and livestock and manure management will reduce CH_4 emissions. Use of improved nitrogen fertilizer application techniques will help in reducing N_3O emissions. Dedicated energy crops should be used to replace fossil fuel. Use improved energy efficiency will help in mitigating climate change.

3.6.6. Building

By following methods mitigation in building is possible: if we use efficient lighting and day lighting in a building it will help in low power consumptions in building. Use of electrical appliances and heating and cooling devices which are more efficient will help in low consumption of electricity. Use of improved cook stoves, improved insulation, passive and active solar design for heating and cooling in daily basis will ultimately lead towards mitigation.

3.6.7. Mitigation Through Carbon Taxation

One more way in which mitigation can be achieved is by imposing a 'carbon tax'[31] on the carbon emissions or fossil fuel use if it exceeds a certain limit. The idea of carbon taxation is to encourage business to operate in a more environment-friendly way and is in line with the Polluter Pays principle[32] that has been applied by the Supreme Court in several decisions. From July 2010, India introduced nationwide carbon tax of 50 rupees per metric ton of coal produced and imported into India. Similar carbon taxes have been imposed in many other countries such as Australia, UK, Japan, Netherlands, Sweden, Norway, Denmark and South Africa.[33]

---

[30] Mridula, R., (2018). *The Climate Solution: India's Climate-Change Crisis and What We Can Do About It* (p. 29). New Delhi: Hachette India.

[31] See Vinod, T., (2020). *The Benefits of a Carbon Tax.* The Hindu, https://www.thehindu.com/opinion/op-ed/the-benefits-of-a-carbon-tax/article32709677.ece

[32] The 'polluter pays' principle is the commonly accepted practice that those who produce pollution should bear the costs of managing it to prevent damage to human health or the environment.

[33] Usha, T., (2016). *Climate Change Law Policy and Governance* (p. XV). Lucknow India: Eastern Book Company.

In other situations, a 'green tax' or 'energy tax' may also be levied on goods or pollutants in general that add to environmental degradation. For instance, the Regional Transport Office in Kanpur, India has started collecting 'green tax' from vehicles over 15 years old from January 2015. Spain has reduced taxes for cars that emit less carbon dioxide while increasing taxes on more fuel consuming cars. UK has in place a 'climate change levy' which is a tax on energy delivered to non-domestic users. Its aim is to provide an incentive to increase energy efficiency and to reduce carbon emissions. [34]

In a similar way, the European Union launched an ET System in 2005 under which a maximum limit is imposed on the total amount of greenhouse gases that can be emitted by all participating installations, i.e., factories, power stations, etc. Allowances for emissions are auctioned or allocated and can subsequently be traded. The participating installations must monitor and report their carbon dioxide emissions. If emission exceeds what is permitted by its allowances, the participating installation must purchase allowances from others. If the installation has performed well and reduced its emissions, it can sell its surplus credits[35].

In this context there is an interesting decision of the Income Tax Appellate Tribunal (Hyderabad) Home Power Ltd v. DCIT (151 TTJ 616), wherein income from sale of 'carbon credits 'received by the assessee on account of switching from fossil fuel (naphtha and diesel) to biomass was held to be a non-taxable capital receipt. Referring to the Kyoto Protocol, which permits transfer of carbon credits, the Tribunal observed that:

"Carbon credit is in the nature of "an entitlement" received to improve world atmosphere and environment reducing carbon, heat and gas emissions... Carbon credit is not an offshoot of business but an offshoot of environmental concerns. No asset is generated in the course of business but it is generated due to environmental concerns. Credit for reducing carbon emission or greenhouse effect can be transferred to another party in need of reduction of carbon emission. It does not increase profit in any manner and does not need any expenses."

Other kinds of environmental taxes imposed around the world include the following:

- Severance taxes on extraction of mineral, energy and forestry products.

34 *Ibid.*
35 *Ibid.*

- Taxes on hazardous waste disposal.
- Site value taxes on the unimproved value of land.

Though these measures may be perceived as harsh or regressive by businesses and consumers, they are undoubtedly necessary. Tax policy concerns are a big factor in shaping business operations. Unless tax policies are tailored to cover concerns for environmental damage that affects people at large, businesses may not necessarily follow eco-friendly ways of conducting their operations. It is true that some businesses were conscientious of ecological conservation even before the Government imposed environmental taxes. However, the need of the hour is large-scale involvement of individuals and corporate houses in the mitigation of climate change, this phenomenon can no longer be overcome by the efforts of a select few. We need to involve everyone in this process if we truly wish preserve our environment, and imposing environmental taxes is one way of ensuring that everyone plays a role[36].

3.6.8. Role of NGO in Mitigating Climate Change

NGOs have played a significant in mitigating climate change at large or those without a voice in the process. In international environmental protection, environmental NGOs have played a quasi-ombudsman role in representing, or acting as "the guardian of," the environment itself. Although various techniques have been used, environmental NGOs have particularly played this role by effectively mobilizing international public opinion. Greenpeace's activities provide several examples of this role, in its campaign to protect whales and to have Antarctica declared a national park. NGOs have already taken a strong lead in calling international attention to the problem of global warming. For example, the conference on global warming and climate change: perspective from developing countries acknowledge the "important role in the international process" played by NGOs in 'sensitizing public opinion and policy makers to the dimensions of the climate problem and needed responses.' Furthermore, NGOs play an important and large role in providing education on the global warming issue. For example, the Cairo compact stated NGO involvement in increasing "public understanding of climate change and its implications" was one of the items at 'the top of the world's agenda.' Thus, there is widespread acknowledgment of the important education function that NGOs serve. One example among many

[36] Usha, T., (2016). *Climate Change Law Policy and Governance* (p. XVI). Lucknow India: Eastern Book Company.

is Greenpeace which has produced a wealth of literature in a number of languages, including global warming: the Greenpeace report which contains both scientific information and policy responses to global climate change from a variety of scientific sources and policy perspectives.

However, public interest NGOs do not simply educate, they advocate on behalf of the public interest. NGOs are generally highlighting the danger of global climate change and putting pressure on governments to take action. This can be done through the form of public campaigns, which Greenpeace has so effectively used in the past, or via direct lobbying on international legislation. Since there is presently no effective international law in this area. NGOs international activities are presently directly towards putting pressure on states to adopt international legislation on the global warming issue.[37]

3.6.9. Forest Conservation

Forests covers about one third of the earth's surface and provided services and resources supporting human substance and wellbeing. Forest ecosystem are central to the livelihoods, economic development and cultural values of many citizens. Forest plays an important role in carbon cycle through "photosynthesis." Because of deforestation, forest dependent poor in many regions of the world faces increasing difficulties in meeting their basic needs[38].

Country like India and other which has abundance of forest they do not need to invest more in mitigation and adaptation strategy. They can mitigate by conserving forest. Forest conservation will not only help in mitigation of climate change but it will also help to conserve biodiversity, protect and safeguard the livelihood of forest dependent people, and to ensure production of round wood for industrial and commercial need. The progressive conservation of forest will contribute in reduction in CO_2 emission, stabilization and improvement of carbon stock, and conservation of biodiversity.[39]

[37] Robin, C., David, F., et al., (1991). *International Law and Global Climate Change* (p. 100). London: Graham & Trotman.

[38] Siddhant, S. K., (2016). UNFCCC kyoto protocol& developing countries: human solidarity in a divided world. In: Usha, T., (ed.), *Climate Change Law Policy and Governance* (p. 24). Lucknow India: Eastern Book Company.

[39] See Rawat, V. R. S., & Kishwan, J., (2008). Forest conservation-based, climate change-mitigation approach for India. *The International Forestry Review, 10*(2), 269–280.

3.6.10. Nuclear Power

The production of energy generation is mainly responsible for emission of greenhouse gases and if we opt for substitute then we find alternative source of energy as renewable energy on which we cannot rely totally due to many reasons. India needs to meet future energy insecurity and the most important thing is that energy should be carbon free. Countries like India need to think about the alternates, and for meeting future energy requirement nuclear energy is viable option along with renewable energy. But there is a lot of difficulties which India may face in getting nuclear energy. Such as availability of uranium, environmental clearance, opposition by local people, disposal of radioactive waste at the time of decommissioning, nuclear accidents etc.[40] the question of India's energy insecurity empowerment through nuclear energy cannot be answered without considering the large-scale impact of its safety, environmental concern and economic issues. All the international convention mentioned in chapter IV find a mention in the decision of the Hon'ble Supreme Court in *G Sunderrajan v. Union of India*[41]. Holding that the setting up of a nuclear power plant Kudankulam, Tamil Nadu was not contrary to public policy, the Hon'ble Supreme Court referred to a report of the International Atomic Energy Agency which highlighted that the world is still dependent on burning coal, oil and natural gas for 80% of its energy supply surging energy use cause surging emissions of greenhouse gases disrupting the climate with potentially catastrophic results. The report further stated that nuclear energy is a readily expandable source of low-carbon base load electricity and in the future might also help to meet other energy need. Such as hydrogen production and water desalination.

3.6.11. Individual Emission Cut

If we cut greenhouse gases collectively and individually. It will benefit all by improving air quality and mitigating climate change and would save millions of life.[42] At the end of the century. Greenhouse gas emission cause causes air pollution and climate change. Due to air pollution heart and lungs dieses is caused. Authors think that if we reduce emission then it would not only mitigate climate change but also improve public health.[43]

---

[40] See Nico, B., & Robert, J. B., (2012). Economics of nuclear power and climate change mitigation policies. *Proceedings of the National Academy of Sciences of the United States of America, 109*(42), 16805–16810.

[41] (2013) 6 SCC 620.

[42] Erin, W., (2013). *Emissions Cuts Would Save Millions of Lives, 184*(8), 16. Science News.

[43] See Pulak, D., & Himanshu, S. K., (2011). CBM: A clean energy option for India to address climate change issue. *Asia-Pacific Journal of Social Science, SI*(2), 11–16.

3.6.12. Carbon Capture and Storage

Developing country like India in which coal is the main source for energy production may opt for CCS. Carbon capture and storage (CCS) is the only technology able to deliver significant emissions reductions from the use of fossil fuels.[44]CCS can reduce emissions not only from power generation, but also from industrial sectors such as iron and steel, refining, petrochemical, and cement manufacturing.[45]CCS carbon is captured at power plant and then injected underground into reservoir deep solire, aquifers for long term storage. Although storage supply is large in demand and many regulatory and economic factor will play an important role in determining the degree to which this storage supply can be utilize.[46] Efficient nitrogen management for mitigating climate change: nitrous oxide also contributes to climate change.[47]

3.6.13. Technology Transfer

Technology development and transfer are included as priorities in both the United Nations Framework Convention on Climate Change and its Kyoto Protocol for mitigating climate change. Art. 4.1 of the UNFCCC requires all parties to promote and co-operate in the development, application, transfer of Green House Gases (GHGs) mitigation and technology. The key to protocol requires all parties to co-operate in the development, application, diffusion and transfer of environmentally sound technologies[48].

From the beginning of the United Nations Framework Convention on Climate Change, China has been one of the primary countries pushing for effective implementation of technology transfer commitments. In the first meeting of the SBSTA in September 1955 China identified the need

---

[44] See IPCC special report on carbon dioxide capture and storage prepared by working group III of the intergovernmental panel on climate change, https://www.ipcc.ch/site/assets/uploads/2018/03/srccs_wholereport-1.pdf (accessed on 07 July 2023).

[45] *Carbon Capture and Storage: The Solution for Deep Emissions Reductions*. https://www.iea.org/publications/freepublications/publication/CarbonCaptureandStorageThesolutionfordeep-emissionsreductions.pdf (accessed on 07 July 2023).

[46] See Michael, L. S., & Christopher, W. M., (2012). Lifetime of carbon capture and storage as a climate-change mitigation technology. *Proceedings of the National Academy of Sciences of the United States of America, 109*(14), 5185–5189.

[47] Raghvendra, G. R., (2007). Mitigation of global warming: Policy options for India. In: Usha, A., (ed.), *Environment and Climate Change Policy and Law* (Vol. 186, 187). Hyderabad India: The ICFAI University Press.

[48] Kyoto protocol, (1997), art. 10.

for renewable energy and technologies and the need for identification of adaptation technologies. Access to technology protected by the intellectual property rights was raised by almost immediately as concern by the Alliance of Small Island States (AOSIS)[49].

Transfer of technology can be considered as the transfer of symmetric knowledge for the manufacture of product for application process. Following the adaptation of Bali Road Map, the issue of intellectual property as a potential barrier for accessing climate change technology has come the fare as controversial. Proposal submitted by developing countries have raised concern about intellectual property and suggested the need to address it in systematic manner. The United Nations Framework Convention on Climate Change expert group on Technology Transfer has been considering the matter with greater attention. Industrialized countries have suggested indicators measuring the degree of intellectual property protection and developing countries have emphasized on technology sharing[50].

3.6.14. Climate Financing

Financial support to developing countries for mitigation, adoption technology development and transfer is an important issue in the climate change. The reason behind this is that the country which cannot meet their own requirement of basic things how they would invest in mitigation policy which itself if costly and have economic impacts.[51] Financial resources to address climate in developing countries are provided through:

i. Financial mechanism of United Nations Framework Convention on Climate Change.
ii. The mechanism of Kyoto Protocol,
iii. Paris Agreement

---

[49] *Kyoto Protocol System Key for Mitigation Ambition, Say Developing Countries* (2012). Third World Network.

[50] Christian Figures, The Contribution of Clean Development Mechanism Under the Kyoto Protocol to Technology Transfer, United Nations, Executive Secretary (November 2010).

[51] See *Climate Governance is Likely to Fail Without a Discourse on Climate Finance,* (2020). Economic and Political Weekly Engage.

Finance is one of the most urgent issue that needs to be addressed in order to achieve a compressive post-Kyoto climate agreement. Paragraph 8 of the Copenhagen Accord[52] provided, "sealed up new and additional, predictable and adequate funding as well as improved access to developing countries for mitigation, adoption, technology transfer. Paragraph 8 of the Copenhagen Accord also refers to the need of developing countries."[53]

Art. 9, 10 and 11 of The Paris Agreement reaffirms the obligations of developed countries to support the efforts of developing countries to build clean, climate-resilient futures, while for the first time encouraging voluntary contributions by other Parties. The Paris agreement also provides the Financial Mechanism of the Convention, including the Green Climate Fund (GCF)[54], shall serve the Agreement. International cooperation on climate-safe technology development and transfer and building capacity in the developing world are also strengthened: a technology framework is established under the agreement and capacity building activities will be enhanced through, inter alia, enhanced support for capacity building actions in developing country Parties and appropriate institutional arrangements.[55]

3.6.15. Emission Trading

Carbon trading system is a system in which power plants and manufacturers will be allowed to emit a certain number of tons of carbon dioxide. If they exceed that amount they must buy credits from the companies that pollute less than their allowance. The key to a successful tradable permits program is to determine initial emission accurately and allocate permit wisely. Carefully designed tradable permits system can allocate permit in an equitable manner and lower the overall cost of climate stabilization.[56]

The Kyoto target commitment represents an initial allocation of 'rights to emit' or emission quotes that can then be traded between countries for example, if Australia purchased emission quotes from Russian Federation,

---

[52] The Copenhagen Accord, (2009).

[53] Stephen, M. G., et al., (2010). Ethics and climate change: An introduction. *WIREs Climate Change, 1*, 54–66.

[54] The Green Climate Fund (GCF) – a critical element of the historic Paris Agreement – is the world's largest climate fund, mandated to support developing countries raise and realize their Nationally Determined Contributions (NDC) ambitions towards low-emissions, climate-resilient pathways.

[55] The Paris Agreement, (2015). https://unfccc.int/sites/default/files/english_paris_agreement.pdf (accessed on 07 July 2023).

[56] Elizabeth, B., (2007). Multilateral climate change mitigation. In: Usha, A., (ed.), Environment *and Climate Change Policy and Law* (p. 83). Hyderabad India: The ICFAI University Press.

it would be conceptually equivalent to paying the Russian Federation to include some of Australia's Green House Gases emission in its inventory.[57] The Kyoto Protocol allocates emission limitations (quotes) to each of the Annex B countries.[58]

Different concepts are founded under emission trading like abatement costs, protocol costs, quotes, permits. A country effort to reduce its emission of Green House Gases (GHGs) will normally decrease the net national income in the country. The reduced national income is called the abatement costs. Protocol costs have a broader meaning than the costs connected only to national efforts.[59] A basic prerequisite for emission trading to take place is that a number of emitters by some sort of a legal instruments like Kyoto Protocol are restricted to keep their emission within limit. The emission quotas are the goods being the object to trade.[60]

3.6.16. Clean Development Mechanism (CDM)

The CDM is a manifestation of the global commitment to address the impact of climate change and achieve Sustainable Development (SD)[61]. In CDM host countries in line with the objectives of the Kyoto Protocol.[62] To achieve this target, each host country is expected to comply with the CDM international rules and in addition ensure that the CDM functions within an effective legal

---

[57] Matthew, C., (2002). Prospects and pitfalls of Kyoto protocol to the United Nations framework convention on climate change. *Melbourne Journal of International Law, 3*(1), 165.

[58] Frank, J., (2005). Developing countries and future of the Kyoto protocol. *Global Change, Peace & Security, 17*(1), 77–86.

[59] Bjart, H., et al., (1998). Emission trading under Kyoto protocol. *Centre for International Climate and Environmental Research,* 11.

[60] Siddhant, S. K., (2016). UNFCCC, Kyoto protocol & developing countries: Human solidarity in a divided world. In: Usha, T., (ed.), *Climate Change Law Policy and Governance,* (p. 21). Lucknow, India: Eastern Book Company.

[61] The concept of sustainable development was described by the 1987 Bruntland Commission Report as "development that meets the needs of the present without compromising the ability of future generations to meet their own needs."

[62] According to Article 12 of the Kyoto Protocol, "the purpose of the clean development mechanism shall be to assist Parties not included in Annex I in achieving sustainable development and in contributing to the ultimate objective of the Convention, and to assist Parties included in Annex I in achieving compliance with their quantified emission limitation and reduction commitments under Article 3." This means Annex I and non-Annex I countries facilitate projects that would support low-emission development, which would reduce potential increase in emissions in the non-Annex I country and which would count towards the Annex I country's emissions reduction. In effect, contributing to the prevention of increased emissions in one country would be treated as an effort towards stabilization of GHGs to 1990 levels. A list of CDM projects underway in India is available at See http://cdmindia.nic.in.

and institutional framework provided by that CDM host country. CDM is one of the most innovative tool of the Kyoto Protocol. India is the country most similar to China with respect to development of CDM. India's basic national development including economic development, population, and the impact of climate change and projection of carbon emission are similar to China. Due to the surging economic growth energy demand in future both India and China are the focal point[63]. The Indian Government framed the National Action plan on climate change 2007, identifying 8 national missions that aim to promote sustainable development. The Plan states that;

"India is determined that its per capita greenhouse gas emissions will at no point exceed that Off developed countries even as we pursue our development objectives."

Gloria Erinma Orie[64] demonstrates[65] that in the case of Nigeria there is a huge potential for CDM project activities as a result of abundant solar energy, vast reserve of fossil fuel, oil and solid minerals, good forestry reserve, a promising market for gas Utilization, and generally a high mitigation potential. These advantages are indicative of high potential for gaining carbon credits and ultimately socio-economic development benefits. In spite of these potentials, she examines that Nigeria still grapples with several legal impediments that undermine its level of compliance with the CDM international rules. legal issues that Nigeria must address in complying with the CDM rules which include the domestication of the Kyoto Protocol, legal recognition for the Designated National Authority, absence of policy on CDM and level of awareness. [66]

3.6.17. Mitigating Climate Change by Educating People on Climate Change

Due to the complexity of the science of climate change and the underlying uncertainties, there are a lot of misconceptions with regard to climate change in general, and issues like the greenhouse effect and ozone layer in particular. The general level of "science illiteracy" in our country

[63] Antony, D., et al., (2007). *The Clean Development Mechanism and the International Diffusion of Technologies: An Empirical Study* (p. 105). *NOTA DI LAVORO.*

[64] Gloria, E. O. Practicing Advocate, Nigeria.

[65] In: *International Conference on Mitigation of Climate Change: Law, Policy and Governance,* (2014). Campus Law Centre University of Delhi.

[66] Usha, T., (2016), *Climate Change Law Policy and Governance* (p. XVI). Lucknow India: Eastern Book Company.

further complexes this fact. Efforts for creating awareness and science education may make a significant difference in literary on climate-related issues. Few years back, Gowda conducted a study among school students in America. Similar misconceptions also exist among the public and the students in India. The best estimate of the IPCC is that, to date, human activities have increased mean air temperatures by about 0.3°-0.6°C above the natural increase. According to their "business as usual" scenario, the IPCC estimates a 1.5°C increase in the next 50 years, with an uncertainty range 1°-2.5°C. In spite of these figures, the general perception is that the warming is much higher. People in the north and high mountains believe that lack of snow at Christmas time is an evidence of global warming. Yet in the climate sense, evidences for the region may show little change over time. People also exhibit a lack of clarity about global climate change and its relationship to chlorofluorocarbons (CFCs) and the ozone hole in the stratosphere. Nevertheless, people prominently identify CFCs and ozone layer depletion as being the primary causes of Climate change. Hence, the use of air conditioning, CFCs, and aerosol cans are often cited as an evidence of global warming. Significant confusions exist among the educated mass, and to some extent among science students as well, regarding these two atmosphere related environment problems. Generally, people's perceptions regarding causes give rise to their responses to the problem.[67]

The common sources of information for the public today are television, newspapers, magazines, and books, Facebook. Often, the wrong perception of the people is enhanced by the frequent misrepresentation of facts about global warming, leading people to believe that they can also sense climate change. People tend to correlate multiple unrelated environmental issues. For example, the common belief is that Climate change is caused by a broad variety of environmentally harmful activities, such as garbage disposal, water logging, water and air pollution, and so on. These are not significant causes although it is true that such environmentally harmful methods change the chemical processes in the atmosphere, and hence, indirectly affect the climate.[68]

3.6.18. Role of Environmentalists in Mitigation

Nowadays, a number of people facing difficulties in getting fresh air, pure

---

[67] Sushil, K. D., (2017). *Climate Change an Indian Perspective* (pp. 226–230). Ahmedabad: Centre for Environment Education.

[68] *Ibid.*

water, and basic facilities for a healthy life. Public grievances in matters related to the environment have given rise to environmentalism, which has become a common phenomenon in most parts of South Asia, including India. In India, public intolerance to environmental degradation started from the stage of voluntary, decentralized, and people centered resistance, such as the silent Valley in Palakkad district of Kerala, the well-known forest movement in the Himalayas called the *Chipko Andolan* in 1976[69], various subsequent Public Interest Litigations (PIL)[70] filed in the Supreme Court on issues related to the environment. However, over the decades, more centralized and organized activities have grown, with good financial strength, in some instances. They are quite effective in bringing valuable information on concerned issues to the notice of public and the government. As far as the issue of climate change is concerned, the India environmentalists have been successful in protecting the interests of the nation.

There are variety of stake holder in the global environmental meet search as national government, United Nations, UN agencies. Multinational company (MNCs), individual consultants, and NGOs. Inter-governmental negotiations are always difficult. NGOs have contributed to the national interest in global platforms by providing a great deal of information basis on research and analysis. To combat environmental degradation and conserve nature, a comprehensive action plan is needed, based on the short coming in the present system, incomplete involvement in implementing it. PILs definitely help in getting clean air in protecting the Taj Mahal and other search momentum of national pride. However, it is a time consuming and gigantic Taj to approach the Supreme Court for every issue of environment. What is needed is the right policy on crucial environmental issues being faced by the people, and the easy implementation of the policy. This can be done only with public realization, which comes for awareness. It is true that longer term global issues in environment like global warming, see level rise, ozone hole, and melting of glaciers need to be addressed. However, interesting often highlighted global issues, one should not ignore the more down to earth local environment challenges related to the survival and wellbeing of the average citizen of the country.

---

[69] The 'Chipko Movement' or 'Chipko Andolan,' was a forest conservation movement in India. It began in 1973 in Uttarakhand.

[70] Public interest litigation is the use of the law to advance human rights and equality, or raise issues of broad public concern. It helps advance the cause of minority or disadvantaged groups or individuals. Public interest cases may arise from both public and private law matters.

We should start believing that our culture tradition of environmental conservation will help find ways to address future environmental challenges. The idea of environmental conservation wall built into India's traditional culture, as per the Vedas. People should be convinced about the benefits of traditional culture.[71]

3.6.19. Effective Science Policy Interface for Mitigating Climate Change

There is need of Effective Science Policy Interface for mitigating climate change the need is because of the seriousness of the global climate change those international treaties are periodically reviewed and supported by research updates. Up to date scientific knowledge and state of the art capacity building at the national level plays a great role in international debate, negotiation and decision making. For capacity building, the key factors are: nationally assessing the different aspects of climate change, building up skills and using the skills in coordinated manner. In the past, there has been effort through various international programs, such as the International Geosphere Biosphere Programs (IGBP), to build capacity in global environmental changes in developing countries. It is well understood that the consequences of climate change will affect all sector of human activity, as well as our national environment. On the other hand, the information base at the micro level is generally unusual in our country. System which can assemble information related to all sector influenced by climate change such as hydrology and water resources, agriculture, coastal zone, ecosystems and human health is needed. Today there are very few comprehensive documents available to guide policy makers on issues of Climate change impacts, mitigation and adaptation. Since science and technology support for policy issues relating to climate change involved multiple sectors and several organization, it is very essential to have continuous interaction and exchange of ideas between these organizations. Climate change policies are passed by various department of the government, such as the Ministry of Environment and Forestry (MOEF), DST, finance, industry, energy and the like. However, expertise in various topic is very essential to deal with impact assessment and the formulation of mitigation adaptation strategies for climate change. Thus, creating the manpower of deal with climate change needs special efforts.

[71] Sushil, K. D., (2007). *Climate Change an Indian Perspective* (p. 233). Ahmedabad: Centre for Environment Education.

It is seen that different communities deal with natural adversities in different manner depending upon issues peculiar to their region. Generally, the response of a particular community depends and the local resources available and the social and cultural backgrounds, including the value system they are familiar with. Therefore, adaptation and mitigation strategies should be formulated with knowledge of the communities for whom they are designed. Hence, dialog with locals is vital for effective policies.[72]

3.6.20. By Living in Equal Minimum of the Essentials

The making of a per capita calculations implicitly assumes that equal shares per capita are at the very least a relevant benchmark, is not a moral requirement. That all person is fundamental equal has to imply that they have important equal entitlements, but it need not imply that they have equal entitlement in everything or even in most things. Something should not, and cannot, be distributed centrally. And consequently, cannot be guaranteed any practical kind of distribution. Something good lost their value if they were equally distributed. And so forth. Believe in fundamental human equality still leaves open what shape, if indeed any particular shape, the distribution of many things should have. So, any position on the distribution of something as particular as greenhouse gas emissions requires specific, faculty informed argument.[73]

And it is argument that is entirely dependent on distinctive features of the stage of economic history in which we find ourselves. It happens to be the case that a single from of industrial society has spread throughout the globe and that this dominant form of industrialization is overwhelmingly dependent upon energy obtained from burning fossil fuels, which of course emits carbon dioxide in huge amounts. The poor and powerless comfort industry licensed or industrializing societies in which engaging in enough economic activity to support oneself and one's family requires the emission of carbon. In the fossil fuel dependent economies into which poor people are born, people can survive only through activities that produces carbon. Significant carbon emissions are a necessity of life, given the energy source made available to the poor by those who control energy policy. For as long as these emissions are necessity of life for as long as those who control the economy, and energy policy in particular, choose to rely heavily on fossil

[72] Sushil, K. D., (2007). *Climate Change an Indian Perspective* (p. 235). Ahmedabad: Centre for Environment Education.

[73] Henry, S., (2014). *Climate Justice: Vulnerability and Protection* (p. 98). London: Oxford University Press.

fuel-everyone should be allowed at least an equal minimum amount of emissions sufficient for at least a decent life.[74]

3.6.21. Mitigation Through Climate Friendly Life Style

There are some specific ways through which we can decrease greenhouse gas. This is as follows:

Walking or carpooling for reaching on your way to Work should be used. Cars burns fossil fuel, so smaller, more fuel-efficient cars emit less carbon dioxide, particularly hybrid cars. Make sure car is properly tuned up which allows it to generate it fewer harmless gases. Turning lights, computers, TV and other electronic appliances off when you are not using them should be in habit. Even though a light doesn't generate greenhouse gas, the power plant that generates electricity used by the light probably does. Switch from incandescent bulbs to fluorescent bulbs, which use less energy and last longer. Recycle garbage that doesn't get recycled ends up in a landfill, generating methane. Recycled good need less energy than products made from scrapes. People should not burn garbage, because burning of garbage releases Carbon dioxide and hydrocarbons into the atmosphere. People should avoid using plastic bags or bottles. People should Use only reusable things and should try to returns the plastics bags in the Shop for reuse. The plastics bags thrown out releases methane gas which causes diseases like cancers. Planting trees is a best way to mitigate[75]. The reason behind this is plants absorb carbon dioxide and release oxygen. If people really want to cut the emissions of greenhouse gases, they need to develop no- fossil energy sources. Hydroelectric power, solar power, hydrogen engines and fuel cells could all create big cuts in greenhouse gases if they were to become more common. For our energy requirements, the world should depend on theses renewable resources of energy. These forms of energy do not pollute the emissions from vehicles, or land or water pollution due to increased amounts of waste.[76]

3.6.22. Imposition of Green Tax

Green tax is also called as the pollution tax or environmental tax. The tax is an excise duty on goods that ultimately results into the environmental

---

[74] *Ibid.*

[75] See Cornelis, V. K. G., Sabina, L. S., et al., (2002). Mitigating climate change by planting trees: The transaction costs trap. *Land Economics, 78*(4), 559–572.

[76] Tumpa, M., (2010). Climate change and sustainable development. In: Sanjay, K. S., (ed.), *Environment Law and Climate Change* (p. 144). New Delhi: SBS Publishers & Distributors Pvt. Ltd.

pollutants. It is thought that if taxes will be charged on emissions causing pollution will lower the environmental impairment in a cost-effective manner. The tax will encourage the behavioral changes in households and firms that are required to reduce the pollution. The main aim of green tax is to ensure that polluters are duly punished for their pollution creating activities. Revenue collected from the green tax are kept in a separate account for tackling the problem of pollution.

This green tax in India is relatively new trend. However, the RFID tags are being given and CCTV cameras have been deployed at border entry points in the major cities of India. With the help of CCTV cameras, the commercial vehicles that enter the city are monitored for emissions. In the cities like Delhi, Environmental Compensation Charge is imposed on pollutants depending upon the vehicle's size. Under the Green Tax norms, the transport vehicles older than eight years are charged with the green tax at the time of renewal of fitness certificate. They will be charged at the rate of 10 to 25% of road tax. While, the public transport vehicles like city buses will have to pay paying lower Green Tax. However, the vehicles used in agriculture purpose like tractor and harvester will be exempted from the tax.[77]

3.6.23. The Fly Ash Utilization Policy

This policy makes mandatory to use only fly ash/ fly ash-based products in construction of building, roads and reclamation / compaction of land within a radius of 100 km from a coal or lignite based thermal power plant, thus displacing the cement use. This policy also mandates utilization of Fly Ash for backfilling or stowing of the mines. Thus, it helps in reduction of pollution.[78]

3.6.24. Flexible Fuel Vehicles (FFVs)

Flexible Fuel Vehicles (FFVs) are the modified version of the vehicles which can run both on the gasoline and the doped petrol with varying levels of ethanol blends. These vehicles are at present being used in Brazil successfully. FFVs will give people the option to switch the fuel based

[77] Green tax to be imposed on older vehicles soon, to clean up the environment by phasing out unfit and polluting vehicles Revenue collected through the tax will be used for tackling pollution, https://pib.gov.in/PressReleasePage.aspx?PRID=1692228 (accessed on 07 July 2023).

[78] *Ash Utilization*, https://www.ntpc.co.in/en/environment/ash-utilisation (accessed on 07 July 2023).

in the price and convenience. Majority of the vehicles sold in Brazil are FFVs. It is also called as the "dual-fuel vehicle" or a flex-fuel vehicle. FFV is an alternative fuel vehicle which comprises of the internal combustion engine which can run on more than one fuel. It usually run on the gasoline blended with either ethanol or methanol fuel. Both of the fuels are stored in same common tank. The engines of the vehicle are capable of burning any proportion of the resulting blend in its combustion chamber. The fuel injection and spark timing are automatically adjusted in accordance with the actual blend detected by the fuel composition sensor. The FFV is different from the bi-fuel vehicles. In the bi-fuel vehicle, two fuels are stored in the separate tanks and engine runs on one fuel at a time.[79]

3.7. RENEWABLE ENERGY SOURCES

We have seen the effect of climate change repercussion and to mitigate it, there must be reduction in the greenhouse gas emissions which are released from the burning of fossil fuels. It needs to be stopped and it can be stopped when new source of energy, which is cost effective, green and, renewable will be used for meeting energy needs. Renewable energy sources are solar power, wind power hydropower, tidal power, geo thermal energy, and biomass; which emit only small amount of GHGs or even none at all and are essential alternative to fossil fuels. Besides climate change issue renewable energy is needed to meet the secured energy demand of future. Exporting renewable energy will ring new business opportunity.[80] "If our nation wants to reduce global warming, air pollution and energy instability, we should invest only in the best energy options. Nuclear energy isn't one of them."[81]

3.7.1. The Importance of Renewable Energy

In present there are limited number of energy source such as oil gas and coal. As we know coal has adverse impact on climate and dependency on the oil and gas is increasing concern for a country which is importing it from another country. And due to these reasons, countries like India are spending

[79] *Flexible Fuel Vehicles*, https://afdc.energy.gov/vehicles/flexible_fuel.html (accessed on 07 July 2023).

[80] Clarissefrass-Ehrfeld, (2009). *Renewable Energy Sources A Chance to Combat Climate Change* (p. 110). The Netherlands: Wolters Kluwer Law & Business, Kluwer Law International.

[81] Mark Z. J. https://www.brainyquote.com/quotes/quotes/m/markzjaco551093.html?src=t_global_warming

more for importing oil and gas and less for domestic economic activities. These are the sufficient reason to look for renewable energy which will provide secure energy supply and reduce dependence on importing oil and gas for energy as well as it will save money of government which could be used for generating own energy sources[82]. The importance of renewable energy can be summarized as follow:

- Fossil fuels are finite in nature. They will not be available in future to meet energy demand and by the time passes the demand of energy and its price will also increase.
- The renewable energy is unlimited in nature. So, there will be secured energy supply through this.
- It is environment friendly and will help to mitigate climate change; and will also increase the opportunity of new business
- Renewable energy sources compete with conventional fuels and it will replace the conventional fuels in power generation, hot water and space heating, transport fuels and off-greed energy.[83]

3.7.2. Difficulties in Generating Renewable Energy

There may be administrative problem in development of renewable energy because high number of authorities are involved and there may be lack of coordination between different authorities which may cause the delay in permission. Financial problem includes lack of trust among bankers and investors. Grid related problem may arise due to insufficient grid capacity, non-transparency of grid procedure of grid connection and cost of grid connections[84]

3.7.3. Definition and Kind of Renewable Energy Sources

Renewable energy is generated through natural process. Renewable energy sources may be defined as new and renewable energy sources developed by new technology and new material, different from conventional fossil fuel resources as they can be generated continuously, obtained endlessly, can be

---

[82] Clarissefrass-Ehrfeld, (2009). *Renewable Energy Sources a Chance to Combat Climate Change* (p. 126). The Netherlands: Wolters Kluwer Law & Business, Kluwer Law International.

[83] See Klaus, B., & Pushpa, K. L., (2018). "The Atmosphere as a Global Commons and Cleansing it with New Energy Options. In Usha, T., (ed.), *Energy Law and Policy*. New Delhi: Oxford University Press.

[84] Clarissefrass-Ehrfeld, (2009). *Renewable Energy Sources a Chance to Combat Climate Change* (p. 113). The Netherlands: Wolters Kluwer Law & Business, Kluwer Law International.

renewed and supplied after use, produce little or no pollution and have no harm to the environment.[85] Renewable energy as defined by international atomic energy and European Union directive includes wind power, solar power geo thermal power hydropower wave power tidal power biomass biogas including land field and sewage gas.[86]

3.7.4. The Advantages and Environmental Impact of Renewable Energy

The advantages of renewable energy are; these are everlasting, the installation is easy and very low cost of generation. If India opts for renewable energy it will reduce fossil fuel consumption; which ultimately leads to millions of CO_2 emission per year which will help in mitigating climate change and economic boost to country[87]. Renewable energy is environment friendly and helps in controlling exhaustion of fossil fuels. On the other side it has some disadvantages; it requires high-cost construction and huge investment, it is not cost effective; per unit cost is double of that generated through fossil fuel, it will need heavy subsidy from the government.[88]

3.7.5. Solar Power

Solar energy involves capturing and harnessing the sun's energy[89]. The raw material needed for solar energy is of renewable and unlimited in nature and that will never be exhausted. Along with this there is no emission and no noise pollution from generating energy from solar power. It is good for remote areas which is not connected to power grid.[90] It is growing like a new large array of photovoltaic put considerable demand on land. They weaken land aesthetics. The use of chemicals in the manufacture of solar panels and the use of lead acid barriers also causes many adverse impacts

[85] See Wang, X., (2005). Strategy, policy and law promoting renewable energy resources in China. In Adran, J., B., et al., (eds.), *Law of Energy for Sustainable Development* (pp. 302, 303). New York: Cambridge University Press.

[86] Clarissefrass-Ehrfeld, (2009). *Renewable Energy Sources a Chance to Combat Climate Change* (p. 114). The Netherlands: Wolters Kluwer Law & Business, Kluwer Law International.

[87] *Ibid* 125.

[88] See Renewables 2018 global status report (2018): "A comprehensive annual overview of the state of renewable energy. *Renewable Energy Policy Network.* https://www.ctc-n.org/sites/www.ctc-n.org/files/resources/global_status_report_2018.pdf (accessed on 07 July 2023).

[89] Clarissefrass-Ehrfeld, (2009). *Renewable Energy Sources A Chance to Combat Climate Change* (p. 115). The Netherlands: Wolters Kluwer Law & Business, Kluwer Law International.

[90] Advantages of Solar Energy, https://www.nachi.org/advantages-solar-energy.htm; https://www.sciencedirect.com/science/article/abs/pii/S1364032116304841 (accessed on 07 July 2023).

on environment. It is costly also.[91] There are two major types of solar power plants. These are Floating Photo Voltaic Solar Power Plant and Concentrated Solar Power Plant. The concentrated solar power plants use mirrors to redirect the solar power. On the other hand, the floating solar power plants are mounted on a structure that floats on a water body like lake, reservoir pond. The advantage of floating solar power plants is they are compact, and their installation and decommissioning are much easier. No land is required for this and the cooling system of floating solar power plants is simpler as it is natural and Environment friendly.[92]

3.7.6. Wind Power

Wind energy is unlimited, inexhaustible, best alternative for conventional energy. As we know. Wind is available everywhere in this world and use of it for energy will not reduce it. It is environment friendly, clean energy and security of supply can be ensured because wind is free along with this unlike solar energy agriculture can be continued around this.[93]Apart from advantages it has some disadvantages just like it creates noise pollution, it may obstruct the radio signals and it may have impact on the movement of bird just like Siberian crane.

3.7.7. Biomass Energy

Biomass as a renewable energy source covers organic matter that can be processed to produce electricity, heat, or can be converted into other fuels. Three way of using biomass Compose of biomass energy sector: biomass for heating purpose, biomass for electricity production, and biomass for transport fuels. The benefits of biomass is it is of very low cost and can be stored and used on demand. It is carbon free thus environment friendly and contributes to secured energy supply and agriculture growth.[94] The main problem with the biomass is the collection and transportation of biomass at the place of utilization.[95]

---

[91] see Atmaja, G. B., (2019). *How can India Become a Global Leader in Solar Power Generation?* (Vol. 54, No. 7). Economic & Political Weekly Engage. https://www.epw.in/engage/article/india-global-leader-solar-energy-g20 (accessed on 07 July 2023).

[92] Alok, S., Neha, Y., et al., (2016). Floating photovoltaic power plant: A review. *Renewable and Sustainable Energy Reviews, 66*, 815–824.

[93] Clarissefrass-Ehrfeld, (2009). *Renewable Energy Sources a Chance to Combat Climate Change* (p. 117). The Netherlands: Wolters Kluwer Law & Business, Kluwer Law International.

[94] Clarissefrass-Ehrfeld, (2009). *Renewable Energy Sources a Chance to Combat Climate Change* (p. 118) The Netherlands: Wolters Kluwer Law & Business, Kluwer Law International.

[95] *Ibid* 119.

3.7.8. Hydro Power

Hydroelectricity power is the energy derived from flowing water in the river of man-made installation where water flows from high level of reservoir down through a tunnel and away from dam. The electricity generated from hydropower is environment friendly and there is no carbon emission from it. [96]

3.7.9. Geothermal Power

This type of renewable energy involves the exploitation of different grades of thermal energy stored within the earth. In some cases, hot water and steam resources exist in ground that can be piped to the surface and used to drive turbine generator- called hydrothermal resources. The so called hot-dry rocks technology on the other hand take advantages of heat roof deep masses of rock that contain little or no steam or water. To exploit these resources, water from the surface through must be pumped through hole to extract heat from the rock. Once the water is heated up it can be pumped to the surface hole and may be used to drive turbine generator. Hydrothermal resources are already used commercially for electric generation. Hot- dry rock technology is only applied in demonstration project as boring two holes seems to be very expensive.

3.7.10. How Can the Mitigation be Achieved by Renewable Energy

Renewable energy sources compete with conventional fuels and it will replace the conventional fuels in power generation, the renewable energy is unlimited in nature. So, there will be secured energy supply through this. It is environment friendly and will help to mitigate climate change; and will also increase the opportunity of new business. It can be achieved through promoting energy efficiency policies consistent with sustainable development. By creating a viable framework that induces energy producers and consumers to use energy economically, efficiently and eco-friendly to the extent possible. By preserving the dwindling reserves of fossil fuels from further unnecessary waste due to unsustainable patterns of energy production and consumption. To act as a framework for harmonious and sustainable economic and social development.

---

[96] *Ibid* 120.

3.8. DIFFICULTY IN MITIGATING CLIMATE CHANGE

We the human have known that our action has resulted in unnatural warming of the earth. Even after knowing this fact no concrete action has been taken from our side and due to our inaction, it is increasing day by day. Our inaction towards mitigation has accelerated climate change. As Robert Gifford[97] puts it, we too face 'dragons of inaction'- Mental barriers that stymie action. These are in our mind, but are no less real for that. Indeed, they are harder to fight. Because they are part of our self. These barriers are ignorance bias and doubt.

Ignorance comes in two part the first par begins with not knowing whether problem exists. In 2012 survey 41% of Indian were ignorant of global warming saying that they had never heard of it. A much larger section of the population is ignorant about what causes the global warming and what to do about. To overcome this problem the communicator must be simpler. Biasness is another form of hindrance at personal level until or unless man is not affected personally and has personal damage, he will not act immediately. Most Indian thinks that global warming affects future generation more than present one, and that they and their immediate surroundings would not be affected as rest. This result in discounting that makes them less willing to act. Understanding climate change is an exercise in probabilistic thinking-how to respond in the face of uncertain facts; but most of us are notoriously poor at this. Most of us ignore climate change until the next flood, drought, or heat waves hits us. Another barrier is doubt in under which we think that we are contributing towards climate change mitigation but in fact we don't. We think that our policy maker and industries would be acting in good faith for mitigating climate change but in reality, they are not taking any action for this.[98]

The debate on climate change has clearly shifted in recent years, and the current focus is on mitigation and adaptation. To stop worsening and speeding-up of climate change. And to cut global GHG emissions by at least half by 2050 prevention. Mitigation, and especially adaptation is crucial and need to be done on a global basis on public and private\corporate level. Reduction measures must be implemented immediately and sweeping

---

[97] Robert Gifford is professor of Psychology and Environmental Studies at the University of Victoria (British Columbia, Canada). His main research interests are environmental psychology, social psychology, and personality psychology. He has worked on nonverbal behavior and on climate change behavior barriers.

[98] Mridula, R., (2018). *The Climate Solution: India's Climate-Change Crisis and What We Can Do About It* (pp. 131–135). New Delhi: Hachette India.

actions have to be prepared to profoundly transform the world's production and consumption patterns and systems.[99]

3.9. ADAPTATION TO CLIMATE CHANGE

Adapting to climate change involves taking the right measures either to reduce the negative effects of climate change or to exploit the positive ones by making the appropriate adjustments and changes. In general, many options and opportunities exist for countries to adapt, with adjustments and changes required at every level community, private, national, and international.

The UNFCCC refers to adaptation in several of its official Articles (especially Article 4). And the COP to the Convention has made several decisions in regard to the adaptation to climate change. The UNFCCC provides that 'all parties are to formulate, implement, publish, and regularly update national and where appropriate, regional programmers containing measures to (…) facilitate adequate adaptation to climate change' (Article 4.1(b)). Referring to Article 4.8, specific domains especially in need of adaptation are coastal zones, water resources, agriculture, and areas affected by drought and desertification, floods, small island countries, countries with forest areas liable to forest decay, countries prone to natural disasters, and countries with fragile ecosystems, including mountain ecosystems.

3.9.1. Adaptation

Adjustment in natural or human systems in response to actual or expected climatic stimuli or their effects, which moderates harm or exploits beneficial opportunities[100]

The IPCC distinguishes several types of adaptation:

- **Anticipatory Adaptation**: Adaptation that takes place before impacts of climate change are observed. Also referred to as proactive adaptation.
- **Autonomous Adaptation**: Adaptation that does not constitute a conscious response to climatic stimuli but is triggered by ecological changes in natural systems and by market or welfare

[99] Mridula, R., (2018). *The Climate Solution: India's Climate-Change Crisis and What We Can Do About It* (pp. 131–135). New Delhi: Hachette India.

[100] *Glossary of Terms and Definitions on Climate Change and Adaptation*, http://www.mrcmekong.org/assets/Publications/glossaries/Glossary-of-Terms-n-Definitions-on-CCA-Eng-04072013.pdf (accessed on 07 July 2023).

changes in human systems. Also referred to as spontaneous adaptation.

- **Planned Adaptation**: Adaptation that is the result of a deliberate policy decision, based on an awareness that conditions have changed or are about to change and that action is required to return to, maintain, or achieve a desired state.
- **Private Adaptation**: Adaptation that is initiated and implemented by individuals, households or private companies. Private adaptation is usually in the actor's rational self-interest.
- **Public Adaptation**: Adaptation that is initiated and implemented by governments at all levels. Public adaptation is usually directed at collective needs.
- **Reactive Adaptation**: Adaptation that takes place after impacts of climate change have been observed."

3.9.2. Appropriate Adaptation Strategies Involve

1. A synergy of the correct assessment of current vulnerabilities to climate change impacts.
2. The use of appropriate technologies ranging from technological options such as increased sea defenses or flood-proof houses on stilts. To behavior change at the individual level. Such as reducing water use in times of drought and using insecticide-sprayed mosquito nets.
3. Information on traditional coping practice. Diversified livelihoods and current and local interventions.
4. Early warning systems for extreme events.
5. Better water management.
6. Improved risk management.
7. Various insurance options and biodiversity conservation.

Adaptation should be implemented in the context of national sustainable development efforts, with the international community, identifying resources, tools, and approaches to support this effort. The parties shall take full account of the specific needs and special situations of the least developed countries (LDCs) in their actions with regards to technology[101]. Not every society has

[101] United Nations Framework Convention on Climate Change, 1992, art. 4, para 4

the same capacity to adapt to climate change. In general adaptive capacity is lined to social and economic development. And is influenced by a society's productive base like capital assets, social network, governance, national income, etc., therefore adaptive capacity is unevenly distributed across and within societies. Nevertheless even societies with adaptive capacity on the highest level remain vulnerable to climate change, variability, and extremes.

3.9.3. Need of Adaptation in India

The adverse impacts of climate change on the developmental prospects of the country are amplified enormously because of the existence of widespread poverty and heavy dependence of a large proportion of the population on climate sensitive sectors for livelihood like agriculture, water, etc. Hence for India adaptation becomes inevitable and an imperative for the development process. Adaptation for India is of immediate importance and requires action right now because of Vulnerabilities in India that differ among states, among regions and among different groups of people within the same region due to substantial variations in topography, climatic conditions, ecosystems as well as diversity in its social structures, economic conditions and needs of different communities.[102] A wide range of actions have been introduced to address it. Out of the eight National Missions on Climate Change five mission focus on adaptation in sectors like agriculture, water, Himalayan ecosystems[103], forestry, Capacity building and Knowledge management.

3.9.4. Water

Water is the most critical and important component of life support system. The adaptation strategies in the water sector focuses on enhancing efficient use of water, ensuring access and tackling the adverse impact of Climate Change. Conservation of water, minimizing wastage and ensuring its more equitable distribution both across and within States through integrated water resources development and management" is one of the options to enhance water use efficiency. As we know that Groundwater is the major component of the total available water resources in India. Rapid expansion of groundwater uses in India for last three decades has resulted in a steep decline in the

---

[102] See Kavi, K. K. S., Priya, S., et al., (2010). *Economics of Climate Change Adaptation in India* (Vol. 45, No. 18, pp. 25–29). Economic and Political Weekly.

[103] See Nishikant, G., Arabinda, M., & Nand, K. A., (2018). *Interstate Cooperation for Climate Change Adaptation in Indian Himalayan Region* (Vol. 53, No. 12, pp. 35–38). Economic & Political Weekly.

groundwater table in vast areas of the country. Rainwater harvesting is the best promising solution to replenish and recharge the groundwater. Other Adaptation Strategy are water storage, water reuse, and desalination, drip irrigation. Key constraints to implementation in these strategies are financial problem, human resources and physical barriers.[104]

3.9.5. Agriculture

Agriculture is the source of livelihood for nearly two-thirds of the population in India who are living in rural area. It is predominantly dependent on rain water and Droughts and floods are frequent and this sector is already facing high degree of climate variability. The performance of agriculture sector has a direct bearing on food supplies and food security in the country. India is going to become the most populated country by 2030 and will need to produce an additional 100 million tons of food grains to feed the large population. Adaptation options in agriculture sector focuses is on enhancing food security and protection of resources such as land, water, biodiversity and genetics. New technologies and practices are used in cultivation, genotypes of crops that have enhanced CO_2 fixation potential, which are less water consuming and more climate resilient. Tree plantation in complementarity and integrated manner with crops and livestock will help to protect and stabilize ecosystems, and promote resilient cropping and farming systems to minimize the risk during extreme climatic events. This will also complement achieving the target of increasing forest/ tree cover. Restrictions which may be faced in implementation of adaptation in agriculture sector are Technological and financial constraints and access to new varieties.[105]

3.9.6. Coastal Regions & Islands

India has a long coastline border of 7517 km including island territories. India is one of the countries that has been identified as one of the countries which are most vulnerable to the impact of accelerated sea level rise due to

---

[104] See Veena, S., (2019). Climate adaptation in the water sector in India. In: Navroz, K. D., (ed.), *India in a Warming World: Integrating Climate Change and Development* (pp. 498–517). New Delhi: Oxford University Press.

[105] See Kavi, K. K. S., & Brinda, V., (2019). Mainstreaming Climate Change Adaptation: Agriculture. In: Navroz, K. D., (ed.), *India in a Warming World: Integrating Climate Change and Development* (pp. 519–536). New Delhi: Oxford University Press.

global warming.[106] Adaptation option are Relocation, seawalls and storm surge barriers, land acquisition and creation of marshlands\wetlands as buffer against sea level rise and flooding, protection of existing natural barriers. Obstruction in implementation Financial and technological barriers, availability of relocation space.[107]

3.9.7. Human Health Action Plans

As we know that Climate change will increase health related stress from extreme weather-related disasters such as wider spread of vector-borne diseases as malaria and dengue and increasing frequency of heat and cold waves. Adaptation strategy to protect human health is emergency medical services, improved climate-sensitive disease surveillance and control, safe drinking water and improved sanitation in the country. Key Constraints to implementation for adaptation to protect human health is Limits to human tolerance (vulnerable groups), knowledge limitations and financial capacity.

3.10. CONCLUSION

We the human being are actually responsible for climate change. Climate change is not seen but already happening. Disregarding the fact of climate change does not make it less true and it is going to get extreme worse. Climate change will make food system more unstable. It will also exacerbate health problems and will result into displacement of people. Due to unexpected climate change infrastructure may be weaken and may lead to fuel conflict. The entire ecosystem will be forced to move. It will be touching every area of life. The repercussion of climate change could be worse than which we are predicting. We are not left with any option to avoid the repercussion of climate change except opting for mitigation and adaptation. There is incumbent need of mitigation policy and laws and rules. Mitigation and adaptation both are complementary and supplementary of each other, so both are needed. Mitigation will help in lessening the future effect of climate change and adaptation will help in ensuring survival ship and adjustment to natural environment.

---

[106] See Rohan, A., (2019). Shoring up: Climate change and Indian coasts and Iceland. In: Navroz, K. D., (ed.), *India in a Warming World: Integrating Climate Change and Development* (pp. 537–559). New Delhi: Oxford University Press.

[107] Richard, J. T. K., & Robert, J. N., (2001). Technological options for adaptation to climate change in coastal zones. *Journal of Coastal Research, 17*(3), 531–543.

In various method of mitigation opting for renewable energy is best one as we know energy is sine qua none for the growth of any country. It would not be hyperbole that Today energy is as important as clean air and clean drinking water. But to meet this energy requirement we can't deny the fact of environment degradation. For meeting our energy requirement, we are exploiting natural environment. Energy is needed for development and comfort but at the same time we cannot forget about the natural environment. And energy should be generated in such a way that it could not harm natural environment and climate. And for this the most viable option is renewable energy which will never be exhausted, endless in nature, raw material is free and pollution free. But with growing population renewable energy is only an alternative to and could meet some portion of energy demand. Along with creating energy whether from conventional source or non-conventional source for meeting future requirement India should concentrate on saving of energy through efficient energy appliance. Renewable energy has advantages and disadvantages both advantage is it is carbon free and will help in mitigating climate change. Disadvantage is it require heavy investment and it is too costly. In India son wind depend on monsoon and it won't be possible to generate energy from solar power and wind power in disturbed monsoon. Renewable energy will help in mitigating climate change but this does not mean that it does not affect the environment. Hydel projects lead to displacement of the many people but it was observed in Narmada Bachao Andolan v. Union of India[108];

"*Giving the option of generating eco-friendly electricity and substituting it by thermal power may not, therefore best option. Perhaps setting up of a thermal power plant may not displace as many family as a hidel projects may but at the same time the pollution caused by the thermal plant and the adverse effect on the neighborhood could be far greater than the inconvenience caused in shifting and rehabilitating the oustees of reservoir.*"

India should enact law for renewable energy and other energy law should be amended with concern of environment protection. All projects dealing with Renewable energy requires heavy investment and for sustainable energy government should provide financial incentives for renewable energy projects. Indian government should provide subsidy on solar panels. The population in India is increasing. Unless this population growth is controlled Indian government would not be able to supply enough energy to its citizen. Along with the generation of energy through renewable energy the focus should be on energy conservation. And it could be possible through high

[108] (2000). 10 SCC 664.

level of efficient electrical appliances. As environment as a subject is thought at school level. Student should also be thought also for energy conservation since school times because children are more sensitive towards anything. Metro projects should use renewable energy. The recent example is Nagpur metro which is using solar energy and along with this they are planting on pillars to create carbon sink to absorb emissions. Solar energy should be used for non-productive purposes such as water heating in place of using energy produce from conventional source. Scientific and technological research should be encouraged for renewable energy production. To meet energy requirement new technology should be introduced. Along with this focus should be on forest conservation, creation of carbon sink, educating people about climate change; causes and impacts. For mitigation effort should be at every level; local, national, international.

Chapter

4

CLIMATE CHANGE MITIGATION: INTERNATIONAL LEGAL ORDER

Contents

4.1. INTRODUCTION

The Term 'climate change' has become point of political discussion as well as scientific discussion. Never the less this term not a pick, the climate has been changing since the beginning of the mankind. Today everybody knows that climate is changing too fast, that the earth is warming rapidly, and that therefore climate change may be the one of the greatest threats facing the planet. In general climate change problem is related to changes in the concentration or the greenhouse gases. Which trap infrared radiation from the earth surface and thus cause greenhouse effect. This effect is neither dangerous nor-unknown it is a natural phenomenon that helps to maintain stable temperature and climate on earth, without which people would not be able to live on this planet. But the human activities like fossil fuel combustion, deforestation, and some industrial process causes emissions leading to an unnatural increase in greenhouse gases concentration.[1]

The climate change issue has been known about for decades; scientists, policy-makers economists, and others have been discussing global climate change that could result from the accumulating build-up of GHGs in the atmosphere for a long time. The concern for human-induced change of the earth's atmosphere has become a priority for environmental officials around the globe. In the 1970 vigorous scientific debates on climate change started, leading to many different actions against this threat.

Climate change has been part of a greater problem of environmental degradation on the globe. The attempts of global community with respect to climate change are being made politically, socially, and legally. The attempt to address climate changes through International regulations, convention and protocol has been shown broadly into category: general and specific. In general, the environmental concern begins to emerge as global problem in early 1970 when United Nation organized its first UN Conferences on Human Environment at Stockholm in 1972[2] and United Nation Environmental Program[3] was set up which continue to act as Global catalyst for action to protect environment. In 1983, the United Nation set up World Commission

[1] Clarissefrass-Ehrfeld, (2009). Renewable *Energy Sources A Chance to Combat Climate Change* (p. 5). The Netherlands: Wolters Kluwer Law & Business, Kluwer Law International.

[2] The 1972 United Nations Conference on the Environment in Stockholm was the first world conference to make the environment a major issue. The participants adopted a series of principles for sound management of the environment including the Stockholm Declaration and Action Plan for the Human Environment and several resolutions.

[3] The United Nations Environmental Program (UNEP) was founded in June 1972 as a result of the Stockholm Conference on the Human Environment. The UNEP is the coordinating body for the United Nations' environmental activities.

on Environment and Development (WCED)[4], which advocated the concept of "*Sustainable Development*" under the reports 'Our Common Future' in 1987. Later the issue of Environment protection was gradually integrated with the process of development in 1990, and the United Nation Conference on Environment and Development, 1992 popularly known as "Earth Summit" held in Rio de Janeiro (Brazil)[5]. The all such efforts have been environment oriented, but the impact of pollution population and production of waste has been considered as cause of degradation climate change as well.

Specifically, the environmental concern over the climate change emerged in the series of negotiations conferences and debates between 1985 and 1987 at the global level. Legally speaking United Nation General Assembly on December 6, 1988 adopted resolution 45/53 on the conservation of global climate for present and future generation of mankind. It is stated that climate change is a common concern of mankind and that it has necessary to adopt properly the necessary measures to deal with it within global framework. after 2 years on 21st December, 1990 the United Nation general assembly adopted another resolution for global climate protection with desire for an intergovernmental negotiating committee (INC) to prepare a general and effective convention on climate change in view of the United Nations Framework Convention on Climate Change was negotiated and adopted on May 9th, 1992.

The UNFCCC has been main legal instrument at global level providing Framework with 26 articles defining objective mechanism and global consultation and guiding principle on the mitigation of Climate Change it defines climate change as "modification of climate that it attributed directly or indirectly to human activity which after the composition of global atmosphere and that is in addition to National climate variability observed over comparable time period." The basic objective has been to stabilize the concentration of all greenhouse gases in the atmosphere at a level that will prevent dangerous anthropogenic interference with climate system. To

4 The General Assembly, in its resolution 38/161 of 19 December 1983, inter alia, welcomed the establishment of a special commission that should make available a report on environment and the global problematique to the year 2000 and beyond, including proposed strategies for sustainable development.

5 The United Nations Conference on Environment and Development (UNCED), also known as the Rio de Janeiro Earth Summit, the Rio Summit, the Rio Conference, and the Earth Summit (Portuguese: ECO92), was a major United Nations conference held in Rio de Janeiro from June 3 to June 14, 1992

achieve this objective there has to follow precautionary measures to anticipate prevent or minimize the cause of Climate Change and mitigate its adverse effect. The most important principle of this convention states that the parties would protect climate system for the benefit of present and future generation of mankind on the basis of equity and in accordance with their common but differentiated responsibility and irrespective capabilities.[6] Accordingly, the developed countries have been directed to take the lead in combating climate change and the adverse effects. The basic obligation to all the parties is provided as such: developing, updating, and publishing periodically and making available National incentives of anthropogenic emissions and sinks. formulating and implementing National and regional program, policies and law containing measures to mitigate in climate change promoting the application of process that control emissions of greenhouse gases, transfer of Technology, establishing integrated plant for coastal zone management. For this convention recognizes necessity for returning to earlier level of anthropogenic emission of carbon dioxide and other greenhouse gases by set out no is specific time table and targets for limiting such emissions however it is tabulation different obligations among the parties as provided in annex 1[7] placing to adopt national and regional policies and take corresponding major to mitigate climate change.

For proper implementation of principles and objectives of this convention also in December, 1997 as a legal instrument to limit greenhouse gases up to 2°C to the parties under the convention. It has been developed and developing countries fusion of the main features of the production target accepted by industry development of greenhouse gases and their inclusion in the target the possible creation of bubbles and the reduction of greenhouse gases emission trading the protocol has been structured with the relevant principles rules and guidelines of such a force of mitigating the climate change in addition it also outlines clean development mechanism (CDM) the task of which is to assist developing countries in achieving sustainable development and in contributing to the ultimate objective of the convention.

The list of congresses and conferences on Climate Change is long; the most important ones are the following:

---

[6] United Nations Framework Convention on Climate Change, 1992, art. 3.

[7] Within the UNFCCC regime, the term "Annex I countries" refers to a list of countries that includes members of the Organization for Economic Co-operation and Development (OECD) as well as economies in transition. Annex I countries are subject to additional obligations under the UNFCCC, obliging them to take the lead in combating climate change.

1. Stockholm Declaration 1972;
2. The First World Climate Conference (FWCC) in February 1979 during which policymakers and economists found out that it is highly important for all nations:
   a. To take full advantage of man's present knowledge of climate;
   b. To take steps to improve that knowledge significantly;
   c. To foresee and to prevent potential man-made changes in climate that might be adverse to the well-being of humanity.
3. The Eight World Meteorological Congress in 1979, which set up the World Climate Program (WCP); an international, interagency and inter- disciplinary effort to be able to foresee the future climate changes.
4. The Villach Congress that took place in 1985.
5. The establishment of the IPCC in 1988.
6. The Rio Earth Summit in 1992; since this congress, the world has moved towards regulating emissions of GHGs.
6. The UNFCCC, which was set up in March 1994. To start action against climate change and to be able to consider what can be done to reduce global warming.
7. The Kyoto Protocol (i.e., a protocol of the UNFCCC) entered into force on 16 February 2005.[8]
8. Paris agreement in November 2015.

4.2. HISTORICAL BACKGROUND OF COMBATTING CLIMATE CHANGE

On 23rd June 1988, James Hansen,[9] then head of NASA, told the United States Senate Energy and Natural Resources Committee that the unusual heat of recent times was not a natural occurrence but a direct result of humanity's action. The earth was a warmer in 1988 than any time in the history of

[8] Clarissefrass-Ehrfeld, (2009). Renewable *Energy Sources a Chance to Combat Climate Change* (p. 6). The Netherlands: Wolters Kluwer Law & Business, Kluwer Law International.

[9] James Edward Hansen (born March 29, 1941) is an American adjunct professor directing the Program on Climate Science, Awareness and Solutions of the Earth Institute at Columbia University. He is best known for his research in climatology, his 1988 Congressional testimony on climate change that helped raise broad awareness of global warming, and his advocacy of action to avoid dangerous climate change. In recent years he has become a climate activist to mitigate the effects of global warming, on a few occasions leading to his arrest.

instrumental measurements. Altogether, the evidence that the world is warming by an amount which is just too large to be a chance oscillation… the greenhouse effect has now been detected and it is changing our climate now. [10]

The next day the *New York Time's* headline read 'Global Warming Has Begun' an expert tell senate.'

Around the same time, 300 scientists from all around the world gathered in Toronto for a conference to persuade policymakers to cut emission. Michael McElroy, conference leader and Harvard University physicists said "if we choose to take on this challenge, it appears that we can slow the rate of change substantially. We would close our eyes, hope for the best, and pay the cost when the bill comes due." The media widely covered both Hansen's testimony and the therefor the conference. Global warming had entered the mainstream consciousness and the public attention was enough to make the government to think about how to cut greenhouse gas emission. But now came the rub: the atmosphere is a shared global resource curtailment of greenhouse gases intended global action, which needed an global scientific consensus this came through the formation of the Intergovernmental Panel on Climate Change (IPCC) – a conglomeration of the world's Government and the world's leading scientists, in 1988.[11]

Spencer Weart[12] make available a wonderful view of the entire process in his book, *the discovery of global warming,* describing the IPCC's finding as 'the result was not "mainstream" science so much as conservative, lowest-common-denominator science. The conclusion was neither the findings of scientific expert not the political statement of the governments- they were the statements that the scientist agreed were scrupulously accurate and that the governments found politically acceptable. So, when the IPCC finally announced its conclusion, they had solid credibility.

The IPCC's is first report came out in 1990. It is said: "we are certain of the following: there is a natural greenhouse effect which already keeps the Earth warmer than it would otherwise be emission resulting from

---

[10] Mridula, R., (2018). *The Climate Solution: India's Climate-Change Crisis and What We Can Do About It* (p. 35). New Delhi: Hachette India.

[11] *Ibid.*

[12] Spencer R. Weart is a noted historian specializing in the history of modern physics and geophysics. Until his retirement in 2009 he was Director of the Centre for History of Physics of the American Institute of Physics (AIP) in College Park, Maryland, USA, and he continues to be affiliated with the Centre.

human activities. Enhance the greenhouse effect, resulting on average in an additional warming of Earth's surface."[13]

The report proved to be an effective call to action which saw the birth of the United Nations Framework Convention on Climate Change (UNFCCC) in 1992, that would coordinate international negotiation on climate change.

The 1990 saw climate research becoming a high-stakes political arena, with representatives of the energy industry and NGOs like Greenpeace entering the IPCC process. The energy industry was recognized as biggest threat in the climate change announcement and had swung into full lobbying mode. The consensus building process was proving to be a double-edged sword-on the one hand, it gave tremendous power to the words in the report because every government's scientist had agreed to it. On the other hand, so little was definitive, so much was vague, due to this it made the second report a difficult read. But the threat was clear enough to get the world to work together towards the Kyoto protocol. The Kyoto protocol was adopted in Kyoto, Japan, on 11 December 1997 and entered into force on 16th February 2005. For the first time, all the countries were responsible for meeting binding emissions targets. The Kyoto protocol committed developed countries for reducing their overall greenhouse gas emissions by at least 5% below 1990 levels during the five-year commitment period 2008 – 2012. These modest targets left many feeling it was far too little- a bit like hunting an elephant with peashooter. But it was a start. The Kyoto protocol also took a stab at managing global emissions by requiring actual emission to be monitored, reported by developing market-based mechanism to meet reduction targets and including sinks activities, such as forest management. Emission cut commitments were differentiated-reflecting both contribution to a planetary warming, and the ability and willingness to cut. For instance, the EU committed to reducing its emission by 8% while Australia could increase hers. Eventually, 191 countries ratified the treaty. Notably, the US did not.[14]

4.3. NEED OF INTERNATIONAL CLIMATE CHANGE LAW

there is need of international climate change law because it is important to treat the atmosphere as a shared natural resource, similar like international

---

[13] Mridula, R., (2018). *The Climate Solution: India's Climate-Change Crisis and What We Can Do About It* (p. 36). New Delhi: Hachette India.

[14] Mridula, R., (2018). *The Climate Solution: India's Climate-Change Crisis and What We Can Do About It* (p. 37) New Delhi: Hachette India.

river, which must be utilized in an equitable manner by all those states within whose sovereignty if falls. The principle of equitable utilization is commonly recognized as a basic of international water law, and it implies both substantive and procedural limitations on the freedom to pre-empt the use or quality of such a resource. Equitable utilization is conceptually deficient in addressing environmental problems only from the perspective of states which share sovereignty over the resources.[15]

Another reason for need of international climate change law is because it is Common property, or *res communes,* refers to areas, such as the high seas, which are open for legitimate uses by all the states, and which may not be appropriated to the sovereignty of any individual state.as we have seen, airspace above the high seas is in this sense common property, because it is available for aerial navigation by all nations. But, like sovereign airspace, common property is fundamentally a spatial dimension, insufficiently inclusive to deal with the whole Atmosphere as a global unity. On the contrary, the atmosphere will necessarily occupy both sovereign airspace and common property. However, else it is defined, it is not possible to view the global atmosphere as "common property" in the normally accepted sense of an area beyond the jurisdiction of contiguous or subjacent states.

In 1985, the Vienna Convention for the protection of the ozone layer[16] defined that area as the layer of the atmospheric ozone layer in the same manner, without referring whether it is above the high seas or land territory. Thus, it is not a legally distinct zone, interposed between airspace and outer space, but falls partially within areas of national sovereignty, and partly into areas of common property. In this rather different sense, it is appropriate to view the common ozone layer as a common area, regardless of who enjoys sovereignty over the airspace which it occupies.

The most important reason for need of international climate change law is Common heritage. Common heritage implies that a resource must be exploited and conserved for that benefit of mankind as a whole, that no state enjoys sovereignty over the resources and that it is open for use by all the states without discrimination. It differs from common property, however, in

[15] Robin, C., David, F., et al., (1991). *International Law and Global Climate Change* (p. 8). London: Graham & Trotman/M. Nijhoff.

[16] The Vienna Convention for the Protection of the Ozone Layer is a multilateral environmental agreement signed in 1985 that provided frameworks for international reductions in the production of chlorofluorocarbons due to their contribution to the destruction of the ozone layer, resulting in an increased threat of skin cancer.

that the benefits must be shared among all states, including those unable to participate in exploitation, such resources will usually require some form of international regulation and administration in order to control the allocation of exploitation rights and benefits. Concept of common heritage has been applied by treaty to the deep seabed and the celestial bodies, and it has been suggested that it may also govern activities in Antarctica, although this is questionable.[17]

4.4. COMMON CONCERN OR COMMON INTEREST

In 1988, the UN general assembly declared in resolution 43/ 53 bad global climate change was the common concern of mankind. The same terminology was subsequently used by the Noordwijk Declaration of the Conference on Atmospheric Pollution and Climate Change,[18] and it has been adopted by UNEP's governing council in decisions on the subject. "Common concern" is not a concept previously employed in international law. The choice of language was itself the outcome of a political compromise agreed on after Malta's original proposal to treat the atmosphere as a part of the common heritage of mankind encountered predictable opposition, but it indicates a status for the global atmosphere distinctively different from any of the other concepts considered so far.

Resolution 43/53 does not make the global atmosphere common property beyond the sovereignty if individual states, but like the ozone layer, it does treat it as a global unity in so far as injury in the form of global warming or climate change may affect the community airspace under the sovereignty of a subjacent state or not.

4.5. THREE PERSPECTIVES ON THE CLIMATE CHANGE PROBLEM

Climate change problem can be understood in many ways such as a scientific, technological, or even it can be considered as religious problem.

[17] Robin, C., & David, F., et al., (1991). *International Law and Global Climate Change* (pp. 9–11). London: Graham & Trotman/M. Nijhoff.

[18] The Noordwijk Ministerial Conference was a critical milestone on the road to international emission targets for carbon dioxide. Held in November 1989 in Noordwijk, The Netherlands, the Conference was attended by representatives of 67 countries, 11 international organizations, and the Commission of the European Community (EC). The conclusions of the resulting Noordwijk Declaration on Atmospheric Pollution and Climate Change1 have often been referred to in subsequent climate debates.

But mainly three perspectives have dominated the international policy response to climate change. Firstly, European countries have tended to see it as an environmental problem, which is reflected in their representation for many years at the UN climate negotiations by their environment ministries. Small island states of the world have not surprisingly, been even more environmentally minded given the existential threat posed by climate change to their existence. In contrast, many non-European developed countries (in particular the US) have tended all, from the start to see climate change through an economic lens in which economist plays a major role in formulating policies meanwhile many developing countries understand climate politics as part of larger pattern of historical and economic injustice- a continuation of the 1970s debate about the 'new international economic order,' which grew out of decolonization movement. In their view, developed countries not only bear the primary historical responsibility for combating climate change, but should also support developing countries in their efforts to do so. In some developing countries even claim 'compensation' and argue that developed countries must discharge their 'ecological debt.'[19]

4.6. PRINCIPLES OF THE CLIMATE CHANGE REGIME

Article 3 of the UNFCCC sets out a number of international environmental principles applicable to the regime. The preamble to the UNFCCC also contains references to international environment principles. The principles are to be used to guide the implementation of the instrument and assist in meeting the ultimate objective in Article 2.[20] The principles contained within the UNFCCC are applicable to the Kyoto protocol and all other instruments of the regime seeking to implement the ultimate objective of the convention.

The international environmental principles referred to within the UNFCCC are sourced from earlier international instruments, binding acts of international institutions and customary international law. Most of the international environmental principles referred to within the UNFCCC have not reached the status of customary international law. As such, it is necessary to analyze the text of the UNFCCC in order to understand the manner. Scope and application of the principles within the climate regime.

[19] Daniel, B., Jutta, B., et al., (2017). *International Climate Change Law* (p. 5). London: Oxford University Press.

[20] United Nation Framework Convention on Climate Change, (1992).

There is no exhaustive list of international environmental principles, seven principles have been identified:

1. The obligation reflected in Principle 21 of the Stockholm Declaration and Principle 2 of the Rio Declaration, namely that states have sovereignty over their natural resources and the responsibility not to cause trans boundary environmental damage;
2. The principle of preventive action;
3. The principle of cooperation;
4. The concept of sustainable development (encompassing the concepts of sustainable use, inter-generational equity, intra-generational equity and integration);
5. The precautionary principle;
6. The polluter-pays principle; and
7. The principle of common but differentiated responsibility.[21]

The principles listed above are different from substantive legal rules that are in and of themselves enforceable. Principles are used within regulatory framework to guide the interpretation and implementation of the obligations within the source of law under consideration. The difference between a legal rules and principle was examined in the Gentini case, where it was stated that:

"A 'rule' ... is essentially practical and, moreover, binding... There are rules of art, as there are rules of government, while a principle 'expresses a general truth, which guide our action, serves as a theoretical basis for the various acts of our life, and the application of which to reality produces a given consequence."

The inclusion of a set guiding principles within the text Convention was controversial during the drafting the agreement. The United States of America, along with other developed countries, did not want to include open-ended principle within the agreement due to concerns that their inclusion would lead to the creation of additional commitments beyond those clearly spelled out within the Convention. Developing countries felt that it was essential to include a statement on principles within the articles of the text to guide the implementation of the text.[22] The final text of the agreement adopts the

---

[21] Philippe, S., (2003). *Principles of International Environmental Law* (2nd edn., p. 231). London: Cambridge University Press.

[22] Daniel, B., (1993). The United Nations framework convention on climate change: A commentary. *Yale Journal of International Law, 18*, 501.

developing country perspective, thus creating within Article 3 a normative framework to support the implementation of the UNFCCC. The principles contained within Article 3 are therefore not directly enforceable, but can be used to inform policy development and implementation modalities within the broader climate change regime. The seven international environmental general rules and principles identified by Sands will now be analyzed in the context of the entire text of the UNFCCC.[23]

4.6.1. Principle of Preventative Action

The *principle of preventative action* requires states to prevent damage to the environment and to reduce, limit, or control activities which might cause or risk such damage. While the principle of preventative action is not included as principle within Article 3 of the UNFCCC, the principle of preventative action is encapsulated the objective clause within Article 2, which requires parties to prevent dangerous anthropocentric interference with the climate system. While similar to the principle of sovereignty and responsibility, the principle of preventative action can be distinguished from the principle of sovereignty and responsibility in two ways. Firstly, the principle of preventative action requires a certain objective to be fulfilled: that of reducing environmental damage. Secondly, the preventative principle can operate to prevent a state from damaging the environment within its jurisdiction. The Kyoto Protocol in Article 3(1) applies the principle of preventative action by requiring parties to reduce their overall emission of greenhouse gasses by at least 5% below 1990 levels during the first commitment period. Such a provision seeks to reduce environment further within their jurisdictions.[24]

4.6.2. Principle of Cooperation

The principle of cooperation sometimes referred to as 'good neighborliness' is defined in principle 27 of the Rio Declaration as requiring that "states and people shall co-operate in good faith and in a spirit of partnership in the fulfillment of the principle embodied is this Declaration and in the further development of international law in the field of sustainable development." This general principle of cooperation has evolved to include more concrete duties such as information sharing and participation in decision-

[23] Jimmy, A. F., & Pilar, R. M., (2016), *Climate Change Mitigation: Greenhouse Gas Reduction and Biochemicals* (p. 92). Florida: CRC Press Taylor &Francis Group.

[24] Jimmy, A. F., & Pilar, R. M., (2016), *Climate Change Mitigation: Greenhouse Gas Reduction and Biochemicals* (p. 94). Florida: CRC Press Taylor & Francis group.

making processes. The UNEP Draft Principles recognize this evolution of the principle by stating in principle 7 that "exchange of information, notification, consultation and other forms of cooperation regarding shared natural resources are carried out on the basis of the principle of good faith and in the spirit of good neighborliness."[25]

The principle of cooperation is found in many instances within the UNFCCC. Such repeated inclusion of this principle demonstrates an understanding of the necessity of global cooperative action in addressing climate change. The preamble in paragraph 6 calls for "the widest possible cooperation by all countries and their participation in an effective and appropriate international response." This call for cooperation could be interpreted in two ways. Firstly, it could be interpreted as requiring all parties to adopt commitments and act in implementing mitigation and adaptation policies and activities. Or secondly it could be interpreted as merely requiring all parties to the UNFCCC to take part in the negotiation process. The first interpretation would clearly place a much heavier onus on parties, and would likely be well received many who are frustrated by the lack of good neighborliness conduct at recent COP negotiations. The second interpretation, only requiring participation at negotiating session, is however reflective of state practice at COP negotiation. The ambiguity of this statement and the potentially onerous obligations that it could impart likely explain the inclusion of this statement within the preamble.[26]

Article 4 of the UNFCCC creates binding and more specific cooperation duties.[27] The parties are requested to cooperate on a number of tasks including to:

- Promote and cooperate in the development, application and diffusion, including transfer, of technologies, practices and processes that control, reduce or prevent anthropocentric emissions[28];
- Promote and cooperate in the full, open and prompt exchange of

[25] Draft Principles of Conduct for the Guidelines of States in the Conservation and Harmonious Exploitation of Natural Resources Shared by Two or More States, United Nations Environment Programme Governing Council, XII Plenary Meeting, UN Doc. UNEP/GC/101and Corr.1, 9 To 25 May 1978

[26] Jimmy, A. F., & Pilar, R. M., (2016). *Climate Change Mitigation: Greenhouse Gas Reduction and Biochemicals* (pp. 95, 96). Florida: CRC press Taylor &Francis group.

[27] Also note that Kyoto Protocol creates additional cooperation obligation for parties to this agreement in Article 1 (b) and article10 (c),(d),(e).

[28] United Nations Framework Convention on Climate Change, 1992, art. 4 (c).

relevant scientific, technological, technical, socio-economic and legal information related to the climate system[29];

- Promote and cooperate in education, training and public awareness related to climate change and encourage the widest participation in this process, including that of non-governmental organizations[30];
- Communicate to the Conference of the Parties information related to implementation in accordance with Article 12[31].

4.6.3. The Concept of Sustainable Development

Right to life implies the right to live in pollution free and ecologically balanced environment. Whether the environment can really be protected these days when almost all the countries in South-East Asia are still developing? It is quite obvious that the development is achieved through industrialization, but at the same time we should not forget that the industrialization is the main cause of environmental degradation. To resolve this issue, the experts worldwide have come up with a doctrine called 'Sustainable Development,' i.e., there must be balance between development and ecology. The concept of 'Sustainable Development' can be traced back as early as in 1972 in the Stockholm Conference. So, it is not a new concept though it was given a definite shape in a report by WCED, which was known as 'our common future.' The commission, which was chaired by the then Norway Prime Minister, Ms. G. H. Brundtland defined "Sustainable Development" as:

"Development that meets the needs of the present without compromising the ability of the future generations to meet their own needs."[32]

The concept of sustainable development has grown since its inception at the international level and it has acquired different dimensions in term of economic growth, development and environment protection. However, some of the salient principle of sustainable development as called out from Brundtland Report and other international document such as Rio Declaration and Agenda 21, are as under.

1. Inter-Generational Equity
2. Use And Conservation of Natural Resources,

[29] *Ibid* 4 (h).

[30] *Ibid* 4 (i).

[31] *Ibid* 4 (j).

[32] www.un-documents.net/ocf-02.htm (accessed on 07 July 2023).

3. Environmental Protection,
4. The Precautionary Principle,
5. The Polluter Pays Principle
6. Obligation to Assist and Co-Operate,
7. Eradication of Poverty, And
8. Financial Assistance to the Developing Countries.[33]

1. **Inter-Generational Equity:** The present generation should ensure that the health, diversity and productivity of the environment is maintained or enhanced for the benefit of future generations.) Inter-Generational Equity principle talks about the right of every generation to get benefit from the natural resources. **Principle 3** of the Rio Declaration states that:

"*The right to development must be fulfilled so as to equitably meet developmental and environmental needs of present and future generations*." The main object behind the principle is to ensure that the present generation should not abuse the non-renewable resources so as to deprive the future generation of its benefit.[34]

2. **Use and Conservation of Natural Resources:** In order to meet the needs on sustainable basis, it is absolutely necessary to use the earth's natural resources carefully and the natural resources base must be conserved and enhanced. It is part of our moral obligation to other living beings and future generations. The resources must be conserved and enhanced to meet the need of growing population.[35]
3. **Environmental Protection:** The protection of environment is essential part of sustainable development. Without adequate environment protection, development, resources will be inadequate for needed investments and environmental protection will fail. The strong environmental policies complement and reinforce sustainable development.[36]
4. **The Precautionary Principle**: This principle has widely been recognized as the most important principle of *'Sustainable*

[33] *Vellore Citizens' Welfare Forum v. Union of India,* (1996) 5 SSC647.

[34] http://www.sd-commission.org.uk/pages/what-is-sustainable-development.html (accessed on 07 July 2023).

[35] Rio Declaration on Environment and Development, 1992, prin.19.

[36] Jaswal, P. S., & Nishtha, J., (1999). *Environmental Law Environment Protection, Sustainable Development and the Law* (p. 122). (Allahabad, India: Allahabad Law Agency.

Development.' **Principle 15** the Rio declaration states that:

"*In order to protect the environment, the precautionary approach shall be widely applied by States according to their capabilities. Where there are threats of serious or irreversible damage, lack of full scientific certainty shall not be used as a reason for postponing cost-effective measures to prevent environmental degradation.*"

In other words, it means:

1. Environmental measures by the state government and the local authority must anticipate, prevent and attack the causes of environmental degradation.
2. Where there are threats of serious and irreversible damage, lack of scientific certainty should not use as a reason for postponing measures to prevent environmental degradation.
3. The 'onus of proof' is on the actor or the developer to proof that his action is environmentally benign.[37]

5. Polluter Pays Principle: Principle 16 of the Rio declaration states that:

"*National authorities should endeavor to promote the internalization of environmental costs and the use of economic instruments, taking into account the approach that the polluter should, in principle, bear the cost of pollution, with due regard to the public interest and without distorting international trade and investment.*"

It is quite obvious that the object of the above principle was to make the polluter liable not only for the compensation to the victims but also for the cost of restoring of environmental degradation. Once the actor is proved to be guilty, he is liable to compensate for his act irrelevant of the fact that whether he's involved in development process or not.[38]

6. Obligation to Assist and Co-Operate: The environmental problem is not the problem of individual or that of one country. It is a global problem and it can be tackled only with assistance and co-operation of all. Principle 9 **of** Rio Declaration call for an international transfer of the scientific knowledge needed for sustainable development. Principle 10 recognizes the fact that environmental issues are best handled with the participation of

[37] *Vellore Citizens' Welfare Forum v. Union of India* (1995) 5 SSC 647 at 658

[38] Jaswal, P. S., & Nishtha, J., (1999). *Environmental Law Environment Protection, Sustainable Development and the Law* (pp 127, 128). Allahabad, India: Allahabad Law Agency.

all concerned citizens at relevant level. Principle 27 says that state and people shall co-operate in good faith and in a spirit of partnership in the fulfillment of principles embodied in the Declaration and in the further development of international law in the field of sustainable development. Thus, obligation to assist and co-operate is also one of the important principles of sustainable development.

7. **Eradication of Poverty:** The sustainable development has to address the problem of the large number of people who live in absolute poverty and who cannot satisfy even their basic needs. At the Stockholm conference in 1972, our former Prime Minister Mrs Indira Gandhi said: "of all pollutants we face, the worst is poverty." The Brundtland Report has rightly pointed out that poverty reduces people's capacity to use resources in a sustainable manner and hence it intensifies pressure on environment. Most of the developing countries are under stress of poverty. Therefore, it is necessary that the growth must revived in developing countries because that is where the link between economic growth, the alleviation of poverty, and environmental condition most operate directly.[39]

8. **Financial Assistance to the Developing Countries**: It has already been explained above that the developing countries face poverty as the worst pollutant. The people in the developing country strain their natural resources and over exploit them to meet their basic needs. The developing countries also do not have the finances and modern technology to follow the path of development which is sustainable. Therefore, the financial assistance and transfer of technology from the developed nations to the developing nation is a must if we want to achieve the goal of sustainable development.[40]

General principles rights and responsibilities for environment protection and sustainable development adopted by Brundtland commission are as under: -

1. **Fundamental Human Right:** all human beings have the fundamental right to an environment adequate for their health and well-being.

[39] Jaswal, P. S., & Nishtha, J., (1999). *Environmental Law Environment Protection, Sustainable Development and the Law* (p. 129). Allahabad, India: Allahabad Law Agency.

[40] *Ibid.*

2. **Inter-Generational Equity:** state shall conserve and use the environmental and natural resources for the benefit of present and future generation.
3. **Conservation and sustainable use**: state shall maintain ecosystem and ecological processes essential for the functioning of the biosphere, shall preserve biological diversity, and observe the principle of optimum sustainable yield in the use of living natural resources and ecosystems.
4. **Environmental Standards and Monitoring**: State shall establish adequate environmental protection standards and monitor changes in and publish relevant data on environmental quality and resource use.
5. **Prior Environmental Assessments**: state shall make or require prior environmental assessment of proposed activities which may significantly affect the environment and use of natural resource.
6. **Prior Notification Asses and Due Process**: state shall inform in a timely manner all person likely to be significantly affected by a planned activity and to grant them equal access and due process in administrative and judicial proceedings.
7. **Sustainable Development and Assistance**: state shall ensure that conservation is treated as an integral part if the planning and implementation of development activities and provide assistance to other states, especially to developing countries, in support of environmental protection and sustainable development.
8. **General obligation to co-operate**: State shall co-operate in good faith with other state in implementing the preceding rights and obligations.[41]

From the above principles, it is evident that for achieving sustainable development, economy and ecology will have to merge. Environmental concerns must become an integral part of the decision making at all level and sustainable development must become the overriding goal of all governments.

4.6.4. The Precautionary Principle

The precautionary principle seeks to provide assistance in the development and implementation of international environmental law when there

[41] Jaswal, P. S., & Nishtha, J., (1999). *Environmental Law Environment Protection, Sustainable Development and the Law* (pp. 348, 349). Allahabad, India: Allahabad Law Agency.

is scientific uncertainty. The principle includes considerations of risk prevention, cost effectiveness, ethical responsibilities towards the earth and the shortcomings of human understanding.[42] Principle 15 of the Rio Declaration provides a definition of the principle:

"*Where there are threats of serious or irreversible damage, lack of full scientific certainly shall not be used as a reason for postponing cost-effective measures to prevent environmental degradation.*"

Differing views exist as to the meaning of the principle with some believing that it provides a justification for early intervention, while others view the principle as hampering human activity and creating a system of over-regulation.[43] From a legal perspective the crucial component of the principle is a requirement to take positive action to protect the environment, prior to the existence of scientific evidence detailing specific harm. The principle is also proactive in nature and operates to prevent unsustainable or degrading environmental regulation.

The precautionary principle is included within Article 3(3) of the UNFCCC and requires parties to take precautionary measures to anticipate, prevent or minimize the causes of climate change and mitigate its adverse effects. Where there are threats of serious or irreversible damage, lack of full scientific certainly should not be used as a reason for postponing such measures, taking into account that policies and measures to deal with climate change should be cost-effective so as to ensure global benefits at the lowest possible cost.

The complicating factor in implementing the precautionary principles arises in the determination of what is considered to be a cost-effective solution. The preamble to the UNFCCC provides some guidance on cost effectiveness in paragraph 17 by stating that "*various actions to address climate change can be justified economically in their own right and can also help in solving other environmental problems.*"[44]

---

[42] Minna, P., Anne, B., & Hanna, P., (2010). The precautionary principle. In: Malgosia, P., David, O., & Panos, M., (eds.), *Research Book on International Environmental Law* (p. 203). London, United Kingdom: Edward Elgar Publishers.

[43] Philippe, S., (2003). *Principles of International Environmental Law* (2nd edn., pp. 267, 268). London: Cambridge University Press.

[44] Jimmy, A. F., & Pilar, R. M., (2016), *Climate Change Mitigation: Greenhouse Gas Reduction and Biochemicals* (p. 99). Florida: CRC press Taylor & Francis Group.

4.6.5. The Polluter Pays Principle

The polluter pays principle requires that individuals, states or corporations engaging in polluting or hazardous activities that cause damage to the environment should be held responsible for the consequences of their action.[45] The polluter pays principle is defined in principle 16 of the Rio Declaration as:

"National authorities should endeavor to promote the internalization of environmental costs and the use of economic instruments, taking into account the approach that the polluter should, in principle, bear the costs of pollution, with due regard to the public interest and, without distorting international trade and investment."

The UNFCCC does not incorporate the polluter pays principle. A number of climate change suits are being bought which in effect seek to implement the polluter pays principle. National courts have taken a cautious approach with such cases and it seems that there is a preference from both the judiciary and the literature surveying these decisions for legislative response to be created dealing with the imposition of liability for harm and the specification of the appropriate remedy for such harm. The evolution of cases in this area is a response to the lack of legislative responses to climate change at both the international and national levels giving effect to a polluter pays type obligation.[46]

4.6.6. The Principle of Common but Differentiated Responsibility

The concept of '*common but differentiated responsibility*' adopts a substantive approach to justice by recognizing that different groups before the law require different rights and responsibilities. As such, the principle recognizes:

- The common responsibility of countries to protect the environment;
- The differing contributions of countries to climate change; and
- The differing inabilities of countries to prevent, reduce and control the threat of climate change.

---

[45] Minna, P., Anne, B., & Hanna, P., (2010). The precautionary principle. In: Malgosia, P., David, O., & Panos, M., (eds.), *Research Book on International Environmental Law* (p. 215). London, United Kingdom: Edward Elgar Publishers.

[46] Jimmy, A. F., & Pilar, R. M., (2016), *Climate Change Mitigation: Greenhouse Gas Reduction and Biochemicals* (p. 101). Florida: CRC Press Taylor & Francis Group.

The principle therefore, recognizes the historical differences in the contribution of developed countries to climate change and the difference in their respective economic and technical capacity to respond to these problems. This concept was defined and brought to life through the 1992 Rio Declaration. The principle was also defined and explained in Article 3(1) of the UNFCCC[47] which states that:

"*The parties should protect the climate system for the benefit of present and future generations of humankind, on the basis of equity and in accordance with their common but differentiated responsibilities and respective capabilities. Accordingly, the developed country Parties should take the lead in combating climate change and the adverse effects thereof.*"

The Kyoto Protocol is explicitly based on the Common but Differentiated Responsibility Principle.[48] As stated by Honkonen[49] "the burden-sharing agreement under the Kyoto Protocol was a remarkable achievement, paving the way to country specific commitments in international environmental cooperation." Specifically, the Protocol demonstrated the applicability of the common but Differentiated Responsibility principle through its operational provisions, for example, by excluding non-Annex 1 countries from binding emission reduction obligations. The principle of common but differentiated responsibility also features prominently within adaptation policy of the regime. Paragraph 14 of the Cancun Adaptation Framework, the key provision of the framework defining the parameters of domestic adaptation policies and measures recognizes the application of the principle of common but differentiated responsibility in implementation of such measure. The principle of common but differentiated responsibility is a central pillar in the climate change regime.[50]

4.7. THE STOCKHOLM CONFERENCE, 1972

The earliest evidence of the need for collective action on world level on climate change emerged from the Stockholm Conference on the Human

---

[47] See United Nations Framework Convention on Climate Change, 1992, art. 4(i), which outline the national and regional development priorities, and circumstances.irt. 4(ction and Biochemicals 101, 8gdompanos merkouris (eds), research book on international environmental law 203 (edward

[48] See United Nations Framework Convention on Climate Change, 1992, art. 10

[49] Tulla, N., (2009). The principle of common but differentiated responsibility in post 2012 climate negotiations. *Rev. Eur. Comp. & Int'l Envtl. L.*, 8, 257. https://heinonline.org/HOL/LandingPage?handle=hein.journals/reel18&div=32&id=&page= (accessed on 07 July 2023).

[50] Jimmy, A. F., & Pilar, R. M., (2016), *Climate Change Mitigation: Greenhouse Gas Reduction and Biochemicals* (pp. 102, 103). Florida: CRC press Taylor &Francis group.

Environment in 1972. It acted as a catalyst to developments in the United Nations for the next decade and beyond that. In particular we can say, it established the principle that the use of the Earth's resources has to be regulated in line with the aim of maintaining developmental opportunities. The way, how this was to be achieved was not made clear at that stage, but the principle has been established. The Stockholm conference mainly adopted a non-binding Declaration of Principles for the preservation and enhancement of the human environment which was designed to "inspire and guide the peoples of the world in the preservation and enhancement of the human environment."[51]

Institutionally, it led to the formation of the United Nation Environment Program (UNEP) which was to administer an Action Plan for the development of environmental policy. This Action Plan was consisting of 109 recommendations for environmental action at the international level.

Principles 21 and 22 are the most important principles of the Stockholm Declaration are. Principle 21 drew on existing treaty and customary law to proclaim:

"States have, in accordance with the Charter of the United Nations and the principles of international law, the sovereign right to exploit their own resources pursuant to their own environmental policies, and the responsibility to ensure that activities within their jurisdiction or control do not cause damage to the environment of other States or of areas beyond the limits of national jurisdiction."

Basically, the normative force of the provision is that trans-boundary environmental harm has to be controlled. Principle 22 of the declaration requires States to cooperate in the development of international law with respect to liability and compensation for the victims of pollution and other environmental damage, insofar as this is caused by activities within the jurisdiction or control of these States into areas beyond their jurisdiction. Most of the other Principles are in the nature of policy-oriented, rather than normative in character. Responsibility towards future generations was set out in Principle 1 and Principle 2 which states that the "natural resources of the earth, including the air, water, land, flora and fauna must be safeguarded for the benefit of present and future generations through careful planning or management, as appropriate." Principles 8 to 11 of the declaration recognize

[51] United Nations, *Report of the UN Conference on the Human Environment,* Held in Stockholm from 5 to 16 June 1972, UN Doc. A/CONF. 48/14 and Corr.1 (1972); 11ILM 1416(1972). http://www.un-documents.net/aconf48-14r1.pdf (accessed on 07 July 2023).

that economic and social development is essential and that environmental policies should "enhance and not adversely affect the present or future development potential of developing countries."

The Stockholm Declaration introduced some key ideas on how environmental and developmental issues in future would be handled in international relations. It has treated environmental protection and economic development as linked subjects. The Declaration envisaged the need for cooperation among States and for coordinating activities addressing the issues identified, within and outside the UN system; and it also promoted the idea of inter-generational equity. However, two decades have passed but very little progress has been made in applying these ideas. Indeed, little progress is any how made in developing a common response for combating climate change issues for an even longer period.

4.8. THE FORMATION OF INTER-GOVERNMENTAL PANEL ON CLIMATE CHANGE (IPCC)

Although there were a number of intergovernmental conferences for dealing with climate change issues from 1979 onwards,[52] but Intergovernmental Panel on Climate Change (IPCC) was established by the UNEP and the World Meteorological Organization (WMO) in 1988. The IPCC was handed over with the task of assessing current scientific, technical and socio-economic information relevant to understanding the scientific basis of risk of human-induced climate change, as well as the potential impacts and options for adaptation and mitigation. IPCC was neither given task of conducting research, nor to monitor data or other relevant parameters. Nevertheless, it has to depend itself on the available literature. IPCC has three working groups and a task force on national greenhouse gas inventories. Its secretariat is hosted by WMO in Geneva and supported by UNEP and WMO. A main activity of the IPCC is to provide an assessment of the existing knowledge on climate change at regular intervals. The fourth of assessment reports of the IPCC was published at the end of 2007. The first report was published in 1990 and this report was used by the INC as the scientific basis for the United Nations Framework Convention on Climate Change (UNFCCC). Subsequently the second report was published in 1995 included socio-

[52] The Principal initiatives were the WMO's World Climate Conference in 1979 and the UNEP/WMO/ICSU Conference in Villach, Austria: "Assessment of the Role of Carbon Dioxide and of Other Greenhouse Gases in Climate Variations and Associated Impacts."

economic aspects of climate change and made an influential impact in the negotiations on the Kyoto Protocol. The fourth report was built on previous work and declared with greater confidence that "the warming of the climate system is unequivocal."[53] From time to time, special reports and technical papers have been published by the IPCC on specific issues such as emissions scenarios.

The Intergovernmental Panel on Climate Change published its Fifth Assessment report in 2014 by summarizing the work of thousands of scientists across the world. The message was, in the panel's own words, "unequivocal." Concentrations of carbon dioxide and other greenhouse gases are now higher than they have been for nearly a million years, long before human society began. Report made clear that the burning of fossil fuels is the main reason behind this increase. Without strong action, temperatures are very likely to exceed the two-degree centigrade target that all the governments of the world have committed to, resulting into serious consequences including sea level rises, heat waves, loss of snow and ice cover, disruptions to agriculture and food production, and greater extremes of drought and rainfall.

The Intergovernmental Panel on Climate Change (IPCC) has published its special report on the implications of a 1.5°C rise in average global temperature.

The report states Climate change from human-induced causes is already taking place and is both accelerating and intensifying. Average global temperatures have risen about 1°C since the start of the Industrial Revolution. The harmful effects are visibly evident like the extreme weather, rising sea levels and diminishing Arctic Sea ice, among other changes.

The key findings are Even if the temperature rise could be limited to the current level by countries efforts, the changes would not only continue but will intensify because of the strong feedback loops. Temperature rise, even of 1.5°C, may result in disastrous and irreversible changes. 70–90% of coral reefs across the world would die at 1.5°C temperature rise. At 2°C rise, none would be left. According to report Global warming so far may have led to the loss of over 40–50% of the coral reefs. The loss of fish populations due to loss of coral reefs will have an impact on food security which in turn affects the human health. There are some parts of the planet which are already past the 2°C threshold. Places like Arctic temperature has

[53] IPCC, Fourth Assessment Report, (2007). *Climate Change 2007: Synthesis Report; Summary for Policy-Makers.*

already risen by 3°C. According to the report, once the ice begins to melt, they will release vast quantities of methane that lie locked in the ice. This would further fasten the speed of climate change. The report highlights that the techno-optimism which leads us to believe that before disaster strikes, the wonders of technology such as artificial intelligence, machine learning, and quantum computing and geo-engineering will be able to find answers to the challenges of climate change. But this is delusion and there are no technical fixes to deal with the scale and rapidity of ecological change.[54]

For the first time Intergovernmental Panel on Climate Change (IPCC) had released a report which solely focuses on the land sector and its impact on climate change. This special report of IPCC talks about the contribution of land-related activities to global warming, i.e., how the different uses of land, like agriculture, industry, forestry, cattle-rearing, and urbanization, was affecting emissions of greenhouse gases makes the following observations:

- "If pre-production activities like cattle rearing and post-production activities like transport, energy and food processing are considered, then food production could contribute as much as 37% of all greenhouse gas emissions every year.
- About 25% of all food produced is either lost or wasted and even the decomposition of the waste results in releasing emissions.
- The report talks about the manner in which even existential activities like food production contributes to global warming and is also affected by it."

As we know that land acts as both the source as well as a sink of carbon, Land use, and changes in land use forms an integral part of the conversation on climate change. On one hand activities like agriculture and cattle rearing are a major source of emission of methane and nitrous oxide. Both the gases are hundreds of times more dangerous than carbon dioxide as a greenhouse gas. On the other hand, soil, trees, plantations, and forests absorb carbon dioxide for the natural process of photosynthesis, thus reducing the overall carbon dioxide content in the atmosphere. As a result, large-scale land-use changes, like deforestation or urbanization, or even a change in cropping pattern have a direct impact on the overall emissions of greenhouse gases. Land and ocean together absorb approx. 50% of greenhouse gases emitted every year through natural processes in the carbon cycle. Hence the importance of land, or ocean, as a carbon sink, thus cannot be undermined

[54] IPCC Special Report, *Global Warming of 1.5°C*. https://www.ipcc.ch/site/assets/uploads/sites/2/2019/06/SR15_Full_Report_High_Res.pdf (accessed on 07 July 2023).

in the global fight against mitigating climate change. Hence afforestation and reduction in deforestation are vital approaches in a global strategy to combat climate change.[55]

Intergovernmental Panel on Climate Change (IPCC) Working Group 1 has recently published its sixth assessment report (AR6) titled "Climate Change 2021: The Physical Science Basis" on 9 August 2021. IPCC in its assessment report confirmed that, Indian Ocean was warming at a rate higher than other oceans on earth. According to this AR6 report, warming of Indian Ocean will result into rise in sea levels causing more frequent and severe coastal flooding across world having low-level areas. This will also result into intense and frequent heat waves and humid heat stress in the 21st century in South Asia. The AR6 Report highlights, even if the temperature is limited 1.5°C from pre-industrial levels, extreme weather events will be witnessed and Warned developed countries to undertake immediate, deep emission cuts and decarbonization.[56]

The IPCC is currently busy in its sixth assessment cycle, in which the IPCC is producing the Sixth Assessment Report (AR6) with contributions by its three Working Groups (in which one is published on 9 August 2021) and a Synthesis Report, three Special Reports, and a refinement to its latest Methodology Report. The Synthesis Report will be the last of the AR6 products, currently due for release in 2022.[57]

4.9. THE RIO CONFERENCE, 1992

After 20 years of the Stockholm Conference, the UN Convened a new conference intended to build upon the principles it adopted in 1972. The UN Conference on Environment and Development, 1992 (UNCED) included two novel features. Firstly, nongovernmental organizations (NGOs) were permitted to play a significant role in preparatory committees. Secondly, it was sponsored not only by States, but also by major companies and foundations. Eventually, at the Earth Summit in Rio de-Janeiro in Brazil on June 14, 1992, a Declaration on Environment and Development and the

---

[55] IPCC Special Report, (2019). *Climate Change and Land.* https://www.ipcc.ch/srccl/ (accessed on 07 July 2023).

[56] IPCC sixth assessment report, *Climate Change 2021: The Physical Science, Basis.* https://www.ipcc.ch/report/ar6/wg1/downloads/report/IPCC_AR6_WGI_Full_Report_smaller.pdf (accessed on 07 July 2023).

[57] The Sixth Assessment Report is underway, https://www.ipcc.ch/assessment-report/ar6/ (accessed on 07 July 2023).

UNFCCC were agreed by delegates from the 176 Participating States.[58] The Rio Declaration Contained 27 Principles which set out the basis on which States and people had to cooperate and further develop "international law in the field of sustainable development." Although it is non-binding it has more significance than its Stockholm predecessor. Its core lies in Principles 3 and 4, which declare that the right to development needs to consider the role of environmental protection in the development process. Principle 10 sets out a number of important procedural requirements such as public participation in environmental decision-making, adequate public access to information, and access to any relevant judicial proceedings, etc. In these respects, the provisions of the Rio Declaration are much more specific than those in the Stockholm Declaration. Both the key Principles of the Stockholm Declaration, i.e., Principles 21 and 22 have been replicated into the Rio Declaration in its 2nd and 13th Principles. Principle 12 also supports an open trading system by requiring that trade restrictions for environmental purposes are not to constitute "a means of arbitrary or unjustifiable discrimination or a disguised restriction on international trade."

The different contributions of States to regional and global environmental degradation are recognized in Principle 7 with the notion of "common but differentiated responsibility." The United Nation Framework Convention on Climate Change (UNFCCC) is another important achievement of the Rio Conference. It has been the center piece of the international community's efforts to combat climate change since its adoption.[59]

The UN General Assembly approved the commencement of negotiation in December 1990 through resolution 45/212, and a Convention was finalized on May 9, 1992- all in the short period of 15 months. The aim was to secure universal participation and achieve a consensus among all countries. The UNFCCC was a framework for a process of negotiating further agreements and taking specific measures to address climate change.[60] It was signed by 154 countries in Rio the following month and has the objective of "stabilization of greenhouse gas concentrations in the atmosphere at a level

[58] There was also agreement on a Convention on Biological Diversity and an ambitious Agenda 21, but these are less relevant to our purposes in this chapter. See also Philippe, S., (2003). *Principles of International Environmental Law* (2nd edn.). London: Cambridge University Press.

[59] The Convention was adopted on May 9, 1992 and opened for signature in June 1992. It entered into force on March 21, 1994, after deposit of the fiftieth instrument of ratification.

[60] See Chandrashekhar, D., (2019). Present at the creation: The making of framework convention on climate change. In: Navroz, K. D., (ed.), *India in a Warming World: Integrating Climate Change and Development* (pp. 142–156). New Delhi: Oxford University Press.

that would prevent dangerous anthropogenic interference with the climate system."[61] The countries listed in Annex I of the FCCC (the industrialized countries) agreed to work to return greenhouse gas emissions to 1990 levels and to demonstrate a reversal in the trend toward growing emissions before the year 2000. The FCCC established several institutions, including the Conference of Parties (COP)[62] which meets annually as the governing body of the FCCC and deals with the issues related to it. It is the main supervisory institution of both the FCCC and the Kyoto Protocol. Table 1 mentioned in the sub chapter 'Kyoto Protocol' sets out the various meetings up to end 2007 and the main developments in each case.

The FCCC also stated that the industrialized countries "may implement policies and measures jointly with other Parties." This short reference in an otherwise long document sowed the seeds for the so-called flexible mechanisms of Joint Implementation (JI), the CDM and emissions trading (ET). However, the FCCC is a framework agreement that depends upon supplementary protocols and agreements for implementation. The Kyoto Protocol is an example of this process at work.

It was in the second half of the 1990s that the international response to climate change started to gather real momentum. In Berlin in 1995 (at the first Conference of the Parties, known as CoP-1), it was agreed that the emissions reductions targets provided for in Articles 4(2) (a) and (b) of the FCCC were insufficient, that new commitments would be needed for the post-2000 period and that these commitments should be laid down in "a protocol or another legal instrument" at the CoP-3 in Kyoto, Japan, in December 1997. However, it was also agreed that no new commitments would be applied to developing States.

In December 1995, the IPCC's Second Assessment Report stated that "the balance of evidence suggests that there is a discernible human influence on global climate." In July 1996 at CoP-2, negotiators agreed that the IPCC report justifies "action by Annex I countries to limit and reduce emissions of greenhouse gases." During the 16 months that followed, there was intense negotiation to meet the schedule for completion of a new instrument at CoP-3 in 1997.

[61] United Nations Framework Convention on Climate Change, 1992, art. 2.

[62] COP serves as the formal meeting of the UNFCCC Parties (Conference of the Parties, COP) to assess progress in dealing with climate change, and beginning in the mid-1990s, to negotiate the Kyoto Protocol to establish legally binding obligations for developed countries to reduce their greenhouse gas emissions.

4.10. THE UNITED NATIONS FRAMEWORK CONVENTION ON CLIMATE CHANGE, 1992 (UNFCCC)

It has now been over 29 years that the world has been debating climate change. In 1988, the WMO and the United Nations Environment Program (UNEP) set up the Intergovernmental Panel on Climate Change (IPCC) for looking into climatologists' claim of the threat of climate change. The First Assessment Report published by the IPCC in 1990 established this threat and urged the international community for taking action to mitigate it.

The UNFCCC was an outcome of negotiations which was started in the United Nations General Assembly in December 1990. An INC was set up for writing the framework, which took 15 months to be prepared. The text of Convention was presented for signing at the Earth Summit in Rio de-Janeiro, Brazil in June 1992 and it was signed by 154 countries. The UNFCCC came into force after receiving ratification by the governments of 50 nations on March 21, 1994. At present, the Convention has universal membership, having been ratified by 192 UN member nations.[63]

The main objective of the UNFCCC was to achieve stabilization of greenhouse gas concentrations in the atmosphere at a level that would prevent dangerous anthropogenic interference with the climate system in a time frame that would allow ecosystems to naturally adapt to climate change.[64]

Crucially, the Convention asserted that developed countries should lead in taking action for mitigation, by recognizing climate change as a global problem, and that all nations have a responsibility towards its mitigation. Hence, it recognized the principle a "common but differentiated responsibility"[65] for member nations. This implied, and was later reiterated in the Kyoto Protocol, that developing countries need not to take any commitments on emission reductions, as their economic growth may be adversely affected. Annex I countries listed under the Convention are those who have been made responsible for the problem.

The Convention set up two bodies, a Subsidiary Body for Scientific and Technological Advice (SBSTA) and a Subsidiary Body for Implementation (SBI), for assessing the progress of efforts and to provide advice on

---

[63] Centre For Science and Environment, "Climate Change: Politics and Facts" (2009).

[64] United Nations Framework Convention on Climate Change 1992, art.2, Objective, at 4, FCCC/INFORMAL/84 GE. 05-62220 (E) 200705.

[65] United Nations Framework Convention on Climate Change 1992, art. 3.1.

methodologies for projects taken up under the Convention. It also established financial mechanisms for projects under it by suggesting them to be financed by the Global Environment Fund (GEF), the International Bank for Reconstruction and Development (IBRD or the World Bank) and the UNDP.

The member countries meet annually to assessing progress and to negotiate further rules and protocols to meet the objectives set by the Convention. These meetings are called the 'Conference of the Parties' or COP.[66]

4.11. CONCEPT OF CONFERENCE OF PARTIES (COP) AND MEETING OF PARTIES (MOP)

The UNFCCC was one of the significant achievements of the Earth Summit, 1992. It has been the center piece of the international community's efforts for combating climate change since its adoption. This convention was adopted on May 9, 1992, opened for signature in June 1992 and it entered into force on March 21, 1994 after submission of the fiftieth instrument of ratification.

The Framework Convention on climate change (FCCC) established a number of institutions, including the COP which meets annually as the governing body of the FCCC and deals with the issues related to it. COP is the main supervisory institution of both the FCCC and the Kyoto Protocol. Since 1995 the COP to the UNFCCC have been meeting annually for assessing progress in dealing with climate change. Successive decisions taken by the COP make up a detailed set of rules for practical and effective implementation of the Convention. The COP also serves as the meeting of the Party (MOP) to the Kyoto Protocol, which also adopts decisions and resolutions on the implementation of its provisions. The first meeting of parties was held in 2005 with 10th COP when Kyoto Protocol came into force. This annual meeting is referred to as the Conference of the Parties serving as the meeting of the Parties to the Kyoto Protocol (MOP).

However, Parties to the Convention that are not Parties to the Protocol are able to participate in the MOP as observers, but without the right to take decisions.

---

[66] United Nations, *Report of the UN Conference on the Human Environment,* Held in Stockholm from 5 to 16 June 1972, UN Doc. A/CONF. 48/14 and Corr.1 (1972); 11ILM 1416(1972), http://www.un-documents.net/aconf48-14r1.pdf (accessed on 07 July 2023).

4.12. COP /MOP PRESIDENCY

The office of the Conference President rotates annually between the five United Nations regional groups. Following the procedural rules of the Conference, it is customary for the COP and MOP to elect a minister as President from the host country.

4.13. IMPORTANT CONFERENCES OF PARTIES (COPS)

After enforcement of the UNFCCC, the concept of COP has been institutionalized and every year all the member States started discussing various issues in such Conference. Some important conferences of parties have been discussed under the following sub heads:

4.13.1. COP 1-Berlin, 1995

In the first 'Conference of the Parties,' the question of the ability of the Annex I (industrialized) countries to meet the commitments of the Convention was raised and voiced in a UN declaration called the 'Berlin Mandate.' This asked for negotiations on actions to be taken by the countries to fight climate change, and set the path for negotiations on a protocol with a schedule for emission reductions by Annex I (developed) countries by 1997.[67]

4.13.2. COP 3-Kyoto, 1997

On December 11, 1997, the Conference of the Parties adopted the 'Kyoto Protocol to the UN Framework Convention on Climate Change.' This put binding restrictions on greenhouse gas emissions on the developed countries, requiring a decrease of 6–8% from the 1990 levels of emissions in the period 2008–12.[68]

The Protocol to the United Nations Framework Convention on Climate Change, agreed to in December 1997 at Kyoto in Japan, envisages legally binding steps for achieving reductions in emissions of greenhouse gases. Many of these heat-trapping gases are produced by carbon-intensive industries, so implementation of the Protocol was to impact upon the energy

[67] Conference of The Parties, United Nations Framework Convention on Climate Change, *Report of the Conference of the Parties on its First Session.* Held in Berlin From 28 March To 7 April 1995, https://Unfccc.Int/Resource/Docs/Cop1/07a01.pdf (accessed on 07 July 2023).

[68] See Igor, S., Romain, M., & Valentin, B., (2016). Compliance of the parties to the Kyoto protocol in the first commitment period. *Climate Policy, 16*(6), 768–782.

industry. Some of the impacts was to be positive, creating opportunities for new products and services to assist companies that need to reduce their greenhouse gas emissions (through trading schemes and investment funds). Other impacts were likely to be negative, such as those affecting electricity generating companies that rely upon coal and oil for much of their power generation, or heavy industry and manufacturing industries that use coal or oil in their heating systems, as well as those industries that are energy intensive.

The heart of the Kyoto Protocol is its legally binding character, its core idea being that without this element of multilateral legal constraint the sovereign states of the globe are unlikely to make any significant progress in addressing the problem of climate change. Amid the considerable body of scientific data, the shifting sands of negotiating positions and the complex economic arguments over methods of implementation, the simplicity of this feature cannot be overstated. Legal constraint on the exercise of sovereign power is an essential element in the solution of the climate change problem, and Kyoto is a first – and far from perfect – step in the development of a legal response on a multilateral basis.

The Third Conference of the Parties (COP 3) at Kyoto marked a significant step forward in the UNFCCC process. With the conclusion of a Protocol to the FCCC in December 1997, the UNFCCC process moved in a distinctly practical direction. The Kyoto Protocol imposes legally binding commitments, and offers radical new mechanisms, to combat climate change.

4.13.2.1. Main Features of the Kyoto Protocol

The main features of the Kyoto Protocol may be summarized as follows:

i. Signatories to the Protocol have undertaken to make a legally binding commitment to achieve its objectives.
ii. The developed countries listed in Annex I to the Convention (Annex I Parties) have each signed up to binding individual emission limitations target to be achieved by the years 2008 to 2012 (the EU, which ratified the agreement as a bloc in May 1998, is a special case). Collectively, the Protocol provides for differentiated Annex I targets which amount to an average reduction of 5.2%. This affects six GHGs.[69]. The base year is

[69] *Carbon Dioxide, Methane, Nitrous Oxide, Hydrofluoro Carbons, Perfluoro Carbons and Sulphur Hexafluoride are Called Greenhouse Gases.*

1990 for cuts in the three most important gases: carbon dioxide, methane and nitrous oxide.

iii. Most important of all is probably the inclusion of so-called flexible mechanisms of Joint Implementation (JI), CDM and ET. These are market-based mechanisms which can be used by industrialized countries to lower their compliance costs. The CDM was a real surprise and emerged from the negotiations very much at the last minute. Included for similar reasons was a limited category of carbon sequestration activities or "sinks."

iv. No quantitative commitments are included for developing countries, reflecting the guiding principle of common but differentiated responsibility, under which the industrialized countries are required to take the lead.[70]

4.13.2.2. Main Issues in the Negotiations of Kyoto Protocol

There were three main issues in the negotiations leading to the conclusion of the Protocol. The first of these was the level of GHG reductions for the developed countries. The second concerned the role of the developing countries and the countries in transition (i.e., the post-Communist countries of Central and Eastern Europe) in limiting GHGs. The third issue was the mechanisms to implement GHG reductions, such as the flexible mechanisms under the third bullet above. The other two issues concerned the inclusion of man-made GHGs and the use of forests as sinks to offset emissions from fossil fuels. These were issues of a much lesser order than the first three. of all of them, the one that turned out to be the most contentious was the role of developing countries in limiting GHGs. The issue highlighted a paradox because the major polluters were the developed countries who were signing up to legal obligations. But by the year 2010 the developing countries have become the major producers of GHGs in the world and yet under the Kyoto regime their legal obligations are either distinctly weaker or non-existent. For some Annex I Parties this represents a serious shortcoming in the final text signed at Kyoto.

Kyoto left open this crucial question of participation by the developing countries. Developing countries (or at least some of them, such as China, India and Brazil) are expected to be the major contributors of GHGs by 2010, so that efforts by the industrialized countries alone will be inadequate to deal

[70] *Kyoto Protocol to the United Nations Framework Convention on Climate Change,* (1997). https://unfccc.int/resource/docs/convkp/kpeng.pdf (accessed on 07 July 2023).

with the problem. In such cases, there may also be problems of implementation even where the legal obligations have been accepted by governments, due to weak legal and institutional frameworks, as well as in some cases a high degree of centralization and little input from industry into decision-making. This is a highly unsatisfactory context for the implementation of long-term projects such as ET schemes or indeed a CDM project. In addition, the application of controls on Annex I countries may result in industries relocating to developing countries to avoid the restrictions. However, for some developing countries or countries in transition, the prospect of joining in an ET scheme is an attractive one. By exporting quotas, they could generate hard currency income.[71] For other developing countries, there is some resentment that a problem which has been created over the past 200 years largely through industrialization by the developed world is one that they are now being expected to pay for.

Many of the participating countries at Kyoto accepted that the rules for implementing the Protocol needed to be worked on. For example, the Article on ET (Article 17) could be interpreted with a wide discretion by states even though the basic ideas were ones that most participants were agreed upon. However, the parties agreed on the basic principles and postponed the details to a later COP. The Kyoto Protocol remains a landmark environmental agreement which, once ratified and complied with, offered the potential for a transformation in the way energy is produced and used, and in the way transport systems are utilized. The socio-economic impact could be considerable – and it would be global. In addition, the legal implications would be very wide ranging.

4.13.2.3. Flexible Mechanisms under Kyoto Protocol

Kyoto Protocol being an international agreement-under the UNFCCC that sets binding reduction targets for greenhouse gas emissions for developed nations (Annex 1 under UNFCCC). It was adopted by COP-3 held in Kyoto, Japan in 1997 and came into force on February 16, 2005.

The Kyoto Protocol defines 'flexible' methods which may be adopted by developed nations to meet their emission reduction targets. These methods include ET, CDM and Joint Implementation (JI) which are discussed as below:

---

[71] Christian, V., (1999). *The Buenos Aires Climate Conference: Outcome and Implications*. RIIA Briefing Paper New series.

4.13.2.4. Emissions Trading

The Kyoto Protocol sets the guidelines for calculation[72] of and certification[73] of Emission Reduction Units (ERU), commonly referred to as 'carbon credits.' These are created by countries that have emissions to 'spare': this means the country has released emissions less than the amount which was allowed to them by the Kyoto Protocol. This difference created a new market commodity. The Protocol states these credits can be traded[74] amongst Parties to meet their reduction caps, as under Annex B to the Protocol. These credits may also be generated by CDM or JI projects and bought/sold as per market value.

4.13.2.5. Clean Development Mechanism (CDM)

Clean Development Mechanism or CDM is a flexible tool under Kyoto protocol which was created to help developed countries meet their emission reduction targets. It was established by the Kyoto Protocol with the objective of reducing greenhouse gas emissions at a lower cost by financing sustainable development projects in developing counties rather than in developed countries. These projects would result in emission reductions, which may be certified (CERs) and then bought by industrialized countries to meet their emission reduction requirements as established under Annex B to the Protocol.

CDM projects are under the authority of the COP to the UN Framework Convention on Climate Change (or MoP to the Protocol) and are supervised by the Executive Board (EB) of CDM created by the Protocol. Prior to applying for registration by the RB, CDM projects must be approved by Designated National Authorities (DNA) as sustainable development projects that would be beneficial for the local community. For ensuring transparency and accountability, projects must also be audited and verified by an independent agency, which is known as Designated Operating Entities (DOE).

A proposed CDM project must show emission reductions lower than those in the absence of the project, or lower than the baseline adopted. This has come under fire from various agencies in the past as being vague and

[72] UNFCCC (1998). Kyoto Protocol to United Nations Framework Convention on Climate Change, art. 3.7.

[73] UNFCCC (1998). Kyoto Protocol to United Nations Framework Convention on Climate Change, art. 3.3.

[74] UNFCCC (1998). Kyoto Protocol to United Nations Framework Convention on Climate Change, art. 17.

ambiguous. The risks of false credits, and passing off a non-sustainable development project as a CDM one, are quite high, as many CDM watchdogs have shown in the last few years.

4.13.2.6. Joint Implementation (JI)

Joint Implementation is a strategy proposed in the Kyoto Protocol by which Parties under Annex 1 of the Convention may cooperate and meet their emission reduction requirements together. Countries proposing JI projects must show a greater decrease in emissions due to the project than would occur without it. The key difference between CDM and JI is that JI occurs between the Annex 1 countries having emission caps in Annex B to the Kyoto Protocol.

JI projects are overseen by the Joint Implementation Supervisory Committee (JISC). Equivalent to DNA in JI projects is the Designated Focal Points (DFP), which are the national level agencies responsible for overseeing JI projects. These projects are also independently audited and verified by Accredited Independent Agencies (AIA).

4.13.3. COP 6 – BONN, 2001

At COP 6, held at The Hague in 2000, negotiators were unable to come to an agreement due to major political disagreements over carbon sinks, penalty for non-compliance and financial mechanisms for developing countries to cope with adverse effects of climate change. The talks were resumed in Bonn in 2001. To the surprise of many observers, agreements were reached on most of the issues. The US delegation acted as observers to the meeting since the Kyoto Protocol had been rejected by the US government.[75]

4.13.4. COP 11- Montreal, 2005

In December 2005, the COP held in Montreal was significant for two reasons. It was the first COP since the Kyoto Protocol came into force in February 2005; hence, it was also the first 'Meeting of the Parties to the Kyoto Protocol' (MOP). Also, the 'Montreal Action Plan' was agreed upon, which called for negotiations on emission reduction requirements beyond 2012. After the expiration of the Kyoto Protocol.

---

[75] Conference of the Parties, United Nations Framework Convention on Climate Change, *Report of the Conference of the Parties on the Second Part of its Sixth Session, Held At Bonn From 16 To 27 July 2001* Fccc/Cp/2001/5 25 (September 2001), https://unfccc.int/resource/docs/cop6s-ecpart/05.pdf (accessed on 07 July 2023).

4.13.5. COP 13- BALI, 2007

In December 2007, the Parties agreed on setting a timeline for negotiations on a second commitment period agreement to be adopted at COP 15 in Copenhagen in December 2009. The 'Bali Action Plan' was based on enhanced action on four pillars: adaptation, mitigation, technology transfer and finance. It is on these principles the agreement in Copenhagen was based.

4.13.6. Bali Action Plan

The Bali Action Plan was agreed upon by negotiators at the 13th Conference of the Parties held in Bali, Indonesia in December 2007. This Plan sets guidelines and a timeline for up to and beyond 2012[76] agreement on climate change mitigation and greenhouse gas emission reductions. This Plan sets a two-year time-table for talks and calls for an agreement to be adopted at the 15th Conference of the Parties in Copenhagen, Denmark.

The Bali Action Plan is set on enhanced action on four pillars: mitigation, adaptation, technology and financing. All these measures, it states, must be shared by all Parties by the principle of "common but differentiated responsibilities" Negotiations in the UNFCCC are currently happening under two tracks: the Ad Hoc Working Group on Long-term Cooperative Action (AWG-LCA) and Ad Hoc Working Group under Kyoto Protocol on Further Commitments of Annex I Nations (AWG-KP).

The AWG-LCA was established by the 13th Conference of the Parties (COP 13) held in Bali, Indonesia in December 2007, as a follow-up process to the "Dialog on long-term cooperative action to address climate change by enhancing Implementation of the Convention." This flew subsidiary body was mandated to ensure a comprehensive process to enable the full, effective and sustained implementation of the Convention through long- term cooperative action up to and beyond 2012. The AWG-LCA is to complete its work by the time COP 15 is held in Copenhagen in 2009. The AWG-KP was established by COP 11/MOP 1 in Montreal, Canada in December 2005. Under the decision accepted by the Parties, the Group was to work on further commitments of the Annex I party and ensure that there is no time gap between the first and second commitment periods under the Kyoto Protocol.[77]

---

[76] UNFCCC, (2007), Bali Action Plan, art. 1., UN Doc FCCC/CP/2007/6/Add.1.

[77] United Nations, *Report of the UN Conference on the Human Environment,* Held in Stockholm from 5 to 16 June 1972, UN Doc. A/CONF. 48/14 and Corr.1 (1972); 11ILM 1416 (1972), http://www.un-documents.net/aconf48-14r1.pdf (accessed on 07 July 2023).

4.13.7. COP 15- Copenhagen Summit, 2009

The United Nations Climate Change Conference, 2009, commonly Known as the Copenhagen Summit was held at the Bella Centre in Copenhagen, Denmark between 7 December and 18 December, 2009.[78] The Conference included the 15th Conference of the Parties (COP 15) to the United Nations Framework Convention on Climate Change (UNFCCC) and the 5th Meeting of the Parties (COP/MOP 5) to the Kyoto Protocol.[79] Actually, in the Bali Road Map held in December 2008 itself, a framework for climate change mitigation beyond 2012 was to be agreed there.[80]

Connei Hedegaard was president of the conference until December 16, 2009, handing over the chair to Danish Prime Minister Lars Lokke Rasmussen in the final stretch of the conference, during negotiations between heads of state and government. On December 18, 2009, the final day of the conference, international media reported that the climate talks were "in disarray." Media also reported that in lieu of a summit collapse, solely a "weak political statement" was anticipated at the conclusion of the conference.[81]

4.13.8. COP 16 – Cancun Climate Summit, 2010

In 16th COP was held in Cancun (Mexico) in December 2010 in which delegates from 193 countries agreed to cut carbon emissions and to help developing countries to tackle climate change as part of an historic deal to stop global warming. They also agreed on a GCF to share new technology, help conserve forests and ease the stresses on the poorest countries – the ones that emit the lowest per-capita share of greenhouse gas emissions. But it is not yet decided how the funds will be raised, although preferred options were a new tax on aviation or shipping, or increased carbon taxes more generally. During the negotiation the UK played a key role in helping nations to resolve the contentious issue of the Kyoto Protocol. Developing

---

[78] Conference of the Parties, United Nations Framework Convention on Climate Change, *Report of the Conference of the Parties on its Fifteenth Session, Held in Copenhagen from 7 to 19 December 2009 – Addendum Part Two: Action taken by the Conference of the Parties at its Fifteenth Session* UN Doc FCCC/CP/2009/11/Add.1 (30 March, 2010).

[79] See Generally Shyam, S., (2019). One Long Day in Copenhagen. In Navroz K Dubash (ed.), *India in a Warming World: Integrating Climate Change and Development* (pp. 157–168). New Delhi: Oxford University Press.

[80] Dhawan, T. K., (2010). *Global Warming & Climate Change* (p. 172). New Delhi: Mohini Publishers & Distributors.

[81] *Ibid.*

nations were refusing to sign up to a deal unless the existing treaty, signed by most developed nations except the US, was honored.

The outcome of this Summit was a weak and ineffective agreement but at least it gave us a small and fragile lifeline. We can say that Cancun had saved the process little bit but it did not yet save the climate.

4.13.9. COP 17- Durban Climate Conference, 2011

The 17_{th} COP was held in Durban during 28 November to 9 December 2011 by the United Nations under the scheme of UNFCCC, 1992 itself.[82] It was very much crucial for the survival of human beings because the commitment period of Kyoto Protocol was going to expire in this year, i.e., 2012 and world community has yet not agreed upon any successor of this protocol. It was really the matter of great concern for our survival. We cannot imagine our existence without the environment which is was at stake due to various reasons. The purpose of the Durban Climate Conference, 2011 was inter alia to assess solutions through the negotiation among the heads of state and other bodies. But the Conference had miserably failed to achieve its main purpose. However, there were a few modest successes. One of them was an agreement to set up a "green fund" to help poor nations deal with climate change and reduce deforestation, a major source of greenhouse gas emissions.[83] More importantly, with the United States was applying much of the pressure, China and India consented to participate in any future agreement limiting emissions and play by the same rules as everyone else. These two nations are huge producers of greenhouse gases and, until now, have disclaimed responsibility for reducing them. The latest figures showed that China and India each had emission increases of about 10% since 2009. Since 1990, China's emissions had tripled; India's nearly so.

4.13.10. COP 19- Warsaw Climate Conference, 2013

The Warsaw conference – well-known as the Nineteenth Session of the Conference of the Parties to the U.N. Framework Convention on Climate Change (UNFCCC), or COP 19 – marked the midpoint in a round of talks launched two years ago in Durban. This Conference tried to keep all governments on a track towards a universal Climate agreement which

82

83 See Lavanya, R., (2012). The Durban platform for enhanced action and the future of the climate regime. *International and Comparative Law Quarterly, 61*(2), 501–518.

will take place in 2015. Marcin Korolec was the President of the COP 19 Conference. He had incorporated new decisions for cutting emissions from deforestation and having international mechanism for loss and damage. The Conference had set a pathway for governments to work on a draft text of a new universal climate agreement.[84] This was an essential step to reach a final agreement in Paris in 2015. All countries had decided to initiate or intensify domestic preparation for their intended national contributions towards this universal climate agreement which will come into force from 2020. Parties have to submit clear and transparent plans by the first quarter of 2015 well in advance of COP 21.[85] This Conference also decided to establish an international mechanism to provide most vulnerable populations with better protection against loss and damage caused by extreme weather events and slow onset events such as rising sea levels.[86]

4.13.11. COP 20- Lima Climate Conference, 2014

The 20th session of the Conference of the Parties and the 10th session of the Conference of the Parties serving as the Meeting of the Parties to the Kyoto Protocol took place in Lima, Peru from 1st to 14th December 2014.[87] An unprecedented number of world leaders attended the Summit, including thereby 100 Heads of State and Government and they were joined by more than 800 leaders from business, finance and civil society. This Summary details their most significant announcements. The main purpose of the 2014 Climate Summit was to raise political momentum for a meaningful universal climate agreement in Paris in 2015 and to galvanize transformative action in all countries to reduce emissions and build resilience to the adverse impacts of climate change.

In this Conference, world leaders agreed that climate change is a major issue of our time and that bold action is needed today to reduce emissions and build resilience and that they would lead this effort. Leaders further also acknowledged that climate action should be undertaken within the context of efforts to eradicate extreme poverty and promote sustainable development.

---

[84]https://unfccc.int/files/press/news_room/press_releases_and_advisories/application/pdf/131123_pr_closing_cop1 pdf (accessed on 07 July 2023).

[85] *Ibid.*

[86] See Lavanya, R., (2014). The Warsaw climate negotiations: Emerging understandings and battle lines on the road to the 2015 climate agreement. *International and Comparative Law Quarterly, 63*(3), 721–740. https://doi.org/10.1017/S0020589314000311.

[87] http://unfccc.int/meetings/lima_dec_2014/meeting/8141.php (accessed on 07 July 2023).

4.14. PARIS CLIMATE CONFERENCE, 2015

UN secretary general Ban Ki-Moon categorized the 2015 Paris Agreements, adopted after years of deeply argumentative multilateral negotiations, as a 'monumental triumph.' Others also regarded Agreement as 'historic,' a 'landmark,' the world's greatest diplomatic successes, and a 'big, big deal.' To the extent these claims are true, it is not because the Paris Agreement either decisively resolves the climate crisis or is novel in its approach, but because the agreement represents a considerable achievement in multilateral diplomacy. However, the fact is that world reach long-term, balanced and virtually universally accepted agreement, despite the many crisscrossing red lines of parties, was not a foregone conclusion.[88]

The Paris Agreement sets an ambitious direction for the climate change, and complements this direction with a set of common core obligations for all countries, including legally binding obligations of conduct in relation nationally determined mitigation contributions, and an expectation of progression over time.[89] Is also establishes a common transparency and accountability framework and an iterative process, in which parties take stock, every five years, of their collective progress and put forward emission reduction contributions for the next five-year period. The Paris Agreement, moreover, commands universal or near universal acceptance, and is applicable to all. As of 20 January 2017, over 190 countries representing roughly 99% of global emission had put forward intended nationally determined contributions (INDCs).[90]

4.15. CHARACTERISTIC OF PARIS AGREEMENT

4.15.1. Legal Form of the 2015, Agreement

The legal form of the Paris agreement 2015 was at issue from the beginning. The European Union (EU), the United States (US), and other developed countries collectively consistently favored a global and comprehensive legally binding agreement under the FCCC. The developing nations like Brazil, China, and India, concerned about the constraints of a new legal

[88] Daniel, B., Jutta, B., et al., (2017). *International Climate Change Law* (p. 209). London: Oxford University Press.

[89] See Ashok, L., (2019). Reaching Agreement in Paris: A negotiator's perspectives. In: Navroz, K. D.., (ed.), *India in a Warming World: Integrating Climate Change and Development* (pp. 169–186). New Delhi: Oxford University Press.

[90] Daniel, B., Jutta, B., et al., (2017). *International Climate Change Law* (p. 210). London: Oxford University Press.

agreement on their development projects, were initially reluctant to endorse the call for a legally binding instrument, but, in the final hours of the 2011 Durban conference, which launched the negotiations for the 2015 agreement, only India remained firm in its oppositions to such an instrument. India feared that a legally binding instrument would contain binding mitigation commitments that would pose challenges for its development ambitions. Ultimately, the EU prevailed upon India to accept a compromise that called for the development of a 'protocol, another instrument or an agreed outcome with legal force under Convention applicable to all parties.' In India's view, this wording was sufficiently open-ended that it preserved the option of an instrument other than a treaty.[91]

Parties to the Paris agreement chose not to decide the legal form of the 2015 agreement and the legal character of its constituent provisions until the end of the four-year negotiating process. However, by the time parties reached in Paris, there was emerging consensus that the 2015 agreement would take the form of a legally binding instrument. The US was willing to accept a legally binding instrument, despite significant domestic political constrains, so long as development and developing countries were equally bound by the agreement. India, despite its historical reluctance to accept a legally binding instrument, had become softer at its stance. Other developing countries were more worried with particular of the agreement than its legal form, the softening of positions in relation to legal form can be drawn to at least three developments.

First, an influential political momentum had built up over time, due to the efforts of the EU and many vulnerable countries, toward adoption of a legally binding instrument. [92]

Second, the reluctance of many countries across the developed-developing country divide to take on internationally-negotiated commitments had led to the emergence and gathering traction of the notion of 'NDCs' an approach that, by privileging sovereign autonomy, respecting national circumstances, and permitting self-differentiation, significantly reduced the sovereignty costs of a legally binding instrument.[93] Third, due to the efforts

[91] *Ibid* 211.

[92] Lavanya, R., (2019). Understanding the 2015 Paris agreement. In: Navroz, K. D., (ed.), *India in a Warming World: Integrating Climate Change and Development* (pp. 208–210). New Delhi: Oxford University Press.

[93] See Lavanya, R., & Jutta, B., (2017). The legality of downgrading nationally determined contributions under the Paris agreement: Lessons from the US disengagement. *Journal of Environmental Law, 29*(3), 537–551.

of the US and other, there was increasing recognition and acceptance by states of the distinction between the legal form of the instrument (i.e., could be binding) and the legal character of national determined contributions.

Two key points are important of noticing about the Paris Agreement. First, it is a treaty, as defined in the Vienna Convention on the Law of Treaties (VCLT). It was titled as the 'Paris Agreement' rather than the Paris Protocol, in deference to US political sensitivities, and was not explicitly adopted under FCCC Article 17, which governs the adoption of 'Protocol.' However, the nomenclature of an instrument is legally irrelevant.[94]

Second, the Paris agreement is an agreement 'under the United Nations Framework Convention on Climate Change.' As such, the provisions of the FCCC that apply to 'related legal instrument' apply to the Paris Agreement, including the FCCC's ultimate objective. In addition, Article 2 of the Paris Agreement links the 'purpose' of the agreement with 'enhancing the implementation of the Convention,' which some parties argue ensures the centrality of the conventions in the evolution of the climate regime. The Paris Agreement also makes use of many of the FCCC's instruments, including the COP and financial mechanism.[95]

4.15.2. Legal Character of the Provisions of Paris Agreement 2015

Parties decided to a legally binding 2015 agreement on the understanding that it would contain a variety of provisions, some with greater legal force and authority than others, the legal character of a provision rest on a variety of factors including-location, subjects, normative content, obligations, language, precision, and what institutional mechanisms exist for transparency, accountability, and compliance. Taking these factors into account, the Paris Agreement contained provisions that span the spectrum of legal character. It formulated provisions based on their nature and subjects. At one end of the spectrum are provision that create rights and obligations for parties and lend obligations, framed in mandatory themselves to assessments of compliance and non-compliance. This is, for instance, the case with individual obligations, framed in mandatory terms, with clear and precise normative content, and no qualifying or discretionary

[94] Daniel, B., Jutta, B., et al., (2017). *International Climate Change Law* (p. 212). London: Oxford University Press.

[95] See also Lavanya, R., (2019). Understanding the 2015 Paris agreement. In: Navroz, K. D. (ed.), *India in a Warming World: Integrating Climate Change and Development* (pp. 205–221). New Delhi: Oxford University Press.

elements. Such provisions can be characterized as 'hard law.' In the middle of the spectrum are provisions that indemnity actors ('each Party' or 'all parties') and set standards, but include qualifying or discretionary elements or are formulated in hortatory or advisory terms ('should' or 'encourage'). These provisions can be characterized, in varying ways, as 'sift law.' At the other end of the spectrum are provisions lacking normative content that capture understandings between parties, provide context, or offer a narrative regarding the need for the provision or its location in the broader picture. Even when these provisions are found in the operational part of a legally binding instrument, they are contextual or descriptive and thus might be characterized as 'non-law'-a purely descriptive term that should not be interpreted as denigrating their critical importance in the Paris Agreement. These three categories of provisions-hard law, soft law, and non-law-are fluid, and there are no bright lines between them. Each provision of the Paris Agreement contains a unique blend of elements, and thus occupies its own place in the spectrum.[96]

4.15.3. Architecture of Paris Agreement

A second issue with the agreement that vexed negotiators was the issue of architecture-whether the Paris agreement should reflect a top down or bottom-up approach. In contrast to the mandate for the Kyoto Protocol Negotiations, which called the negotiation of quantitative emission restriction and reduction objective, the Durban platform did not offer any concert guidance on the issue of architecture.

The Paris agreement crystallized this emerging hybrid architecture in which bottom of substance was to promote participation (contained in parties' NDCs) was combined with a top-down process to promote ambitions and accountability (content in the agreements internationally determined provision on progression and highest possible ambition accounting transparency stocktake, and compliance). The bottom-up component of the Paris agreement hybrid architecture was nearly complete by the time the Paris conference started. Over the courses of 2015 virtually every state submitted an INDC. The Paris conference focus on the other half of the Hybrid equation: the development of strong international rules to promote ambition and accountability. The Paris agreement did not include a number of important proposals, such as that NDCs be quantified or quantifiable and

[96] Daniel, B., Jutta, B., et al., (2017). *International Climate Change Law* (p. 213). Oxford University Press; See also. Daniel, B., (2016). The legal character of the Paris agreement. *Review of European, Comparative and International Law, 25*(2), 142–150.

include an unconditional element, and that proposed and NDCs be subject to a formal process of *ex ante* review to consider their ambition comparability, and fairness. On the other hand, the so-called 'friends of rules' group of countries in Paris ultimately proved successful in including actively strong rules on transparency accounting and updating. The rules that were successfully incorporated into the agreement are at the outer edge of what was politically achievable given the experience of the previous four year of negotiation.[97]

4.15.4. Scope of Paris Agreement 2015

A third predominant issue was that negotiators grappled with through the courses of the four-year negotiation process was the scope of coverage of the 2015 agreement. The Durban platform decision required the consideration of 'mitigations, adaptation, finance, technology development and transfer, transparency of action, support and capacity building.' While discussing the content of the 2015 agreement, developed countries sought to focus on mitigation, transparency, and market instruments, while some developing countries sought to bring the focus across the other pillars of the Bali action plan, namely adaptation finance and technology development.[98]

The submission and intervention of parties in the Ad Hoc working group the Durban platform (ADP) negotiations was being continued these debates. The US argued that 'mitigation is the main issue that needs updating,' Australia believed that 'mitigation must be central to the 2015 agreement,' and the independence association of Latin America and the Caribbean (AILAC) that 'mitigation will necessarily be at the core of the 2015 agreement.' China and India among others were disagreed. India argued that, hence action under the Durban platform is related not just to the mitigation but to other pillars of climate action decided upon in the Bali action plan and subsequent COP decision. China was arguing that all the elements should be address 'on an equal footing and in a holistic, balanced and co-ordinate manager' and, with India, that unresolved issue from the Bali process---such as equitable access to sustainable development, trade and unilateral measures, and technology related intellectual property rights---should also be addressed. The African

[97] See Daniel, B., & Lawanya, R., (2018). The evolution and governance architecture of the United Nations. In: Detlef, F. S., et al. (eds.), *Global Climate Policy: Actors, Concepts, and Enduring Challenges* (pp. 26–31). Massachusetts: MIT Press.

[98] Daniel, B., & Lawanya, R., (2018). Evolution and governance architecture of the climate change regime. In: Detlef, S., & Urs, L., (eds.), *International Relations and Global Climate Change: New Perspectives* (pp. 13–65). Massachusetts: MIT Press.

group, the least developed countries (LDCs), and AILAC were insisting that there should be political parties between the treatment of adaptation and mitigation. In addition, some other parties suggested that the 2015 agreement addresses loss and damage, and compliance. Given the lack of specificity in the Durban platform and the divergence in positions of States, it was thus vague at the start of the four-year negotiation process what attention, if any, different elements in the Durban platform would ultimately receive in the 2015 agreement. It was also indistinct if elements not explicitly identified in the Durban platform would feature in the 2015 agreement.[99]

As noted above, the Warsaw conference in 2013 invited all the parties to prepare and submit INDCs in 2015. In the course of the following year, however, it became clear that the carefully negotiated language of the Warsaw decision rise further issue, including:

- Weather INDCs submitted by parties will cover only mitigation or also adaptation, finance, technology and capacity- building
- Whether mitigation would be a compulsory or optimal component of a party's INDC.
- Whether parties could submit conditional INDCs conditioned on the provision of support for an action by other parties or only unconditional ones.

There were a range of views on their issue cutting across developed-developing country lines, which some States insisting that contribution should cover only mitigation and others arguing that mitigation and adaptation should be accorded legal and material parties. In addition, many developing countries, including the Like-Minded Developing Countries (LMDCs), argued that if they were required to submit mitigation contributions, there had to be a corresponding increase in the provision of technical and financial support. This, in their view, could best be insured by requiring developed countries to submit contributions on finance. Needless to say, this proved difficult to achieve. The Lima outcome therefore merely repeated Warsaw language investing parties to communicate their INDCs. It encouraged parties to consider including and adaptation component in their contributions, but was silent on a financial component. Parties were therefore free to offer a full and diverse range of contributions, and they did so: some focused-on mitigation alone, while other covered all area; some work unconditional, while other contained both conditional and unconditional elements.

[99] Lavanya, R., (2019). Understanding the 2015 Paris agreement. In: Navroz, K. D., (ed.), *India in a Warming World: Integrating Climate Change and Development* (pp. 205–221). New Delhi: Oxford University Press.

The Paris agreement 2015 addresses all of the elements listed in the Durban decision in a comprehensive manner: 'mitigation, adaptation, finance, technology development and transfer, transparency of action, find support and capacity- building. In addition, it also addresses loss and damage, any issue of importance to small all island states and LDCs, as well as compliance, of inter interest to a border coalition developed and developing countries. Also, the agreement is broad in its coverage, its treatment of these issues and the legal character of provisions in different areas vary.[100]

Article 3 of the Paris agreement requires parties to submit and INDCs only in relation to mitigation. This issue proved countries, with the LMDCs insisting until the final hours of the Paris negotiation that party should be able to offer and INDCs in areas other than mitigation, such as adaptation and means of implementation (finance, technology, and capacity- building). Their concern was addressed in Article 7.11 which recognized that and adaptation communication can be submitted as part of a state's NDC. It is also addressed through Article 3 insofar as the ambiguous reference to 'contributions' in that Article can be interpreted as cutting across issue areas. the specific Article that address mitigation, adaptation, finance, capacity-building, technology and transparency were introduced to ensure that the obligation in each issue area would be determined by these articles, and the Article 3 would not create any new (potentially conflicting or confusing) publications in relation to these areas. The one sab-stative new element which was added in Article 3 is that the progression requirement applies beyond mitigation to areas such as adaptation and support. However, this provision applies to 'all parties' not 'each party,' indicating that it could be intercepted as a collective rather than an individual requirement. Further, it used the auxiliary verb 'will' rather than the imperative 'shall' and thus sets strong expectations (rather than obligations) of more ambitious action over time. The provisions of mitigation and transparency under Paris agreement create several new legal obligations for individual parties ('Each Party Shall'). In contrast, the provision of finance generally continues existing obligations rather than create new substantive obligations. Finally, in the areas of adaptation, technology, capacity-building, and loss of damage, the provisions primarily recommended, encourage or set aspirations, rather than band parties.[101]

---

[100] See Jayaraman, T., & Tejal, K., (2016). *The Paris Agreement* (Vol. 51, No. 3, pp. 10–13). Economic & Political Weekly. http://www.epw.in/system/files/pdf/2016_51/3/CM_LI_3_The%20Paris%20Agreement.pdf (accessed on 07 July 2023).

[101] Daniel, B., Jutta, B., et al., (2017). *International Climate Change Law* (pp. 215–219). London: Oxford University Press.

4.15.5. Differentiation

The most argumentative predominant issue in the Paris agreement negotiation was the issue of differentiation. Differentiation in international climate regime is founded on the principle of 'Common but Differentiated Responsibilities and Respective Capabilities' (CBDRRC). The principle of common but differentiated responsibilities has been operationalized in several ways, most sharply especially in the Kyoto protocol, which established quantitative emission reduction targets for developed countries but not for developing countries. Even though the Kyoto model of differentiation was, from the outset, contentious, it began to seriously erode only with negotiation of the Bali action plan which was held in 2007.

Developed countries contended that any reference to 'common but differentiated responsibilities' be qualified with a statement that this principle must be construed in the light of current economic realities. The 2015 agreement, the EU argued that it must contain a border spectrum of differentiation in the obligation among parties that is the case under the convention. However, these proposals were unacceptable to developing countries. India argued that qualifying CBDRRC in this manner would be tantamount to amending the FCCC. The compromise reached was to omit any reference to CBDRRC in the Durban platform and instead simply provide that the 2015 agreement would be 'under the convention' thereby implicitly engaging the FCCC's principals, including the principle of CBDRRC. Many developing countries believed that this reference to the FCCC would hold at bay efforts to reinterpret and qualify CBDRRC. But the debate over CBDRRC in Durban, and its eventual resolution, reflected a recasting of differentiation in the 2015 climate regime.[102]

Another indicator of the shifting approach for differentiation can be found in the Durban platform's provision that the 2015 agreement would be 'applicable to all parties.' Developed countries specially the US, Japan and Australia were insisting on including this language. However, the mere fact that an instrument is applicable to all those does not mean that it is applicable in a symmetrical manner universally of application does not automatically single uniformity of application. The phases 'application to all' their four head political rather than legal significance.it was a signal that the 2015 agreement would move towards symmetrical obligations, at least in so far as the nature and form of the obligations (even if not their stringency) were

[102] Mukul, S., (2016). *Paris Agreement on Climate Change Challenges of Urban Transition Remain* (Vol. 2, No. 13, pp. 21, 22). Economic & Political Weekly.

concerned. In a similar way, preambular recitals in the Durban platform decision called for 'all parties' to urgently address climate change and for the widest possible cooperation 'by all countries.'[103]

The 2012 Doha ADP decision explicitly reintroduced into the negotiation through a preambular recital, a reference to the principle of the convention. The Warsaw decision of 2013 similarly contained a general reference to 'principles' of the convention, but no specific reference to CBDRRC. The Lima call from climate action of 2014, in contract, contained an explicit reference to the CBDRRC principle, but qualified by the clause 'in the light of different nation's circumstances.' Is qualification which represents are arrived at between the US and China- arguably shapes the interpretation of CBDRRC? Although it could be argued that the principles even in its original formulation, is dynamic, because historical responsibilities and respective capabilities both evolve, the qualification of the principle by a reference to 'different National circumstances' represented a political signal of flexibility and dynamism. Moreover, given the differences in National circumstances among states, a simple categorization of States as developed or developing might not be appropriate. It is this version of the principle of common but differentiated responsibilities with the qualifier 'in the light of different National circumstances,' that feature in the Paris agreement.[104]

The issue of CBDRRC also underlays debate about the relationship of the 2015 Paris Agreement to the FCCC. Developed countries were adamant that the Paris Agreement abandon the FCCC and Kyoto Protocol's annex-based approach to differentiation, which they regarded as outdated and rigid, and they successfully excluding any reference to the annexes in the Paris Agreement. But many developing countries were equally firm that the agreement contains a hook for arguments to reintroduce what they regard as the continuing balance of responsibilities between developed and developing countries reflected in the annex-based approach differentiation. They sought to do this indirectly, through the inclusion of general language tying the Paris Agreement to the FCCC, which they believed would implicitly engage the entirety of the FCCC, including its annexes. The shadow boxing over this issue rippled through the negotiations of the entire text. Is can be seen in the debate about whether to call the agreement simply the 'Paris Agreement'

[103] Lavanya, R., (2016). Ambition and differentiation in the 2015 Paris agreement: Interpretative possibilities and underlying politics. *International and Comparative Law Quarterly, 65*(2), 493–514.

[104] Daniel, B., Jutta, B., et al., (2017). *International Climate Change Law* (p. 222). London: Oxford University Press.

or the 'Paris Implementing Agreement.' And it was in particular evidence in the negotiation over whether to include language in Article 2 stating that the Paris Agreement should enhance the implementation of the convention, as most developed countries favored. For many developing countries, implementation of the convention implies the continued relevance of all the convention's principles and provisions, in particular the balance of responsibilities reflected in FCCC Article 4, and the corresponding annexes. Developed countries, in contrast, believe the Paris Agreement represents a paradigm. Shift in which the annex-based structure of the FCCC no longer has any continuing relevance, and sought to reinforce this position by referring only to the Paris Agreement's role in implementing the convention' objective. Ultimately the parties reached a carefully balanced compromise in Article 2, which reads: 'this Agreement, in enhancing the implementing of the convention, including its objective, aims to strengthen the global response to the threat of climate change in the context of sustainable development and efforts to eradicate poverty, ….' The intense negotiations over this somewhat convoluted provision illustrate the importance placed on language in the UN climate process and the difficulties of resolving disagreements definitively.

4.15.6. Operationalizing of the CBDRRC Principle in the Paris Agreement

The Paris Agreement operationalizes the CBDRRC principle not through the FCCC and Kyoto Protocol's annex structure, but in a tailored way, which takes into account the specificities of each the Durban pillars-mitigation, adaptation, finance, technology, capacity-building, and transparency. This more nuanced approach has resulted in different approaches to differentiation in different areas.

4.15.7. Differentiation in Mitigation under Paris Agreement

The mitigation provisions of the Paris Agreement embrace a 'bounded self-differentiation.' The Warsaw decision parties to submit INDCs in the context of the 2015 agreement.[105] In submitting this contribution, as well as the information that accompanied them. In so far as parties chose their own constraints, the differentiations have come to be characterized as self-differentiation. The mitigation section of the Paris Agreement contained self-differentiation, albeit modulated, by several normative expectation placed on parties, in contrast to the Kyoto Protocol, all of the legal obligations

[105] See Warsaw decision, para 2(b)

of parties relating to mitigation with one exception,[106] are undifferentiated legal obligations, however these obligations are 'bounded' or 'modulated' be several normative expectations placed on parties. Perhaps the most important of these articles under Paris agreement, reflecting a categorical approach to differentiations, is in Article 4, which reads as: 'developed country Parties should continue taking the lead by undertaking economy-wide absolute emission reduction targets. Developing country Parties should continue enhancing their mitigation efforts, and are encouraged to move over time towards economy-wide emission reduction or limitation targets in the light of different national circumstances'.... The use of the terms in this Article 'developed country Parties' and 'developing country Parties' and the nation of leadership are reminiscent of the FCCC. The paragraph sets strong normative expectation, but does not create any new obligations for parties, given the use of the term 'should.' Indeed, it is precisely because this provision created no new obligations that the US was able to accept the Paris Agreement. This provision was at the center of the 'shall/should' controversy that nearly unraveled the Paris deal in the final hours.

The 'take it or leave it' text presented by the French contained mandatory language ('shall') in relation to develop country targets, and recommendatory language ('should') in relation to developing country mitigation efforts. In addition to the lack of parallelism in the legal character of requirements placed on developed and developing countries, the use of mandatory language for develop countries' targets posed a problem for the US. In the light of long-standing and intractable resistance to climate treaties in the Senate, the US had worked throughout the negotiations to ensure that the Paris Agreement could be accepted by it as a Presidential-executive agreement. This approach would have been put in jeopardy, arguably, if the agreement required the US to have a quantitative emission targets, since such a target is not currently part of US law. However, the LMDCs objective to changes in what had been presented as a 'take it or leave it' text. Eventually, after furious huddling in the plenary room. And high-level negatives outside it. The 'shall' was declared a typographical error and changed to a 'should' by the FCCC secretariat.

In addition to Article 4.4's differentiation between develop and developing countries, the Paris Agreement also recognizes that peaking of emission will take longer in developing countries, and that support shall be provided to developing countries for the implementation of this article.

---

[106] Paris Agreement, 2015, art. 4.5 states that 'support shall be provided' to developing countries.

Thus, the, mitigation section of the Paris Agreement operationalizes the CBDRRC principle through self-differentiation, but sets normative expectation to the types of actions developed and developing country parties should take, and recognizes the need for flexibility and support for developing countries. It also sets normative expectations in relation to progression and 'highest possible ambition' through successive cycles of contributions. Self-differentiation was broadly acceptable to states because it provides flexibility and privileges sovereign autonomy. Self-differentiation was broadly acceptable to states because it provides flexibility and privileges sovereign autonomy. Given the lack of collective agreement about states differentiated responsibility and respective capabilities, self-differentiation was a pragmatic choice intended to encourage broad participation. But the Paris Agreement includes 'modulators or boundaries to this self-differentiation that encourage ambition and recognize differences.

4.16. STRUCTURE OF PARIS AGREEMENT

Preamble of the Paris agreement identifies a series of contextual factures that could be proved helpful in the interpretation of the agreement. It include: the CBDRRC principle,[107] best available scientific knowledge,[108] special circumstances of particularly vulnerable nation,[109]special needs and situation of the LDCs,[110] equitable access to sustainable development and eradication of property,[111] food security,[112] just transition of the work force,[113] human rights,[114] conservation and enhancement of sinks,[115] ecosystem integrity,[116] climate justice, environmental education, awareness, training and participation,[117] multi-level governance, and sustainable patterns of consumption and production.

107 Paris Agreement, preambular recital 3.
108 *Ibid*, preambular recital 4.
109 *Ibid*, preambular recital 5.
110 *Ibid*, preambular recital 6.
111 *Ibid*, preambular recital 8.
112 *Ibid*, preambular recital 9.
113 *Ibid*, preambular recital 10.
114 *Ibid*, preambular recital 11.
115 *Ibid*, preambular recital 12.
116 *Ibid*, preambular recital 13.
117 *Ibid*, preambular recital 14.

4.16.1. Purpose (Article 2 And 4.1)

The main 'purpose' of the 2015 agreement is to strengthen global response to the threat of climate change.[118]

In relation to mitigation, Paris had different options for expressing the long-term goals which are worked out in

- reducing the increase of temperature which are above pre-industrial levels (2°or 1.5°C)
- As a GHG emission reduction goal (50% by 2050).

Of the above long-term goal, the temperature goal had acquired great significance in UN climate regime.

The 2°C goal was first recognized by the G-8 in L'Aquila in 2009 which was incorporated into 'Copenhagen accord' in later year and into 'Cancun Agreement' in 2010. It also features the in the Durban platform decision that launched the negotiations towards the 2015 agreement in preambular recital, that notes significant gap between parties mitigation pledges and the required emissions pathways to reach 2°C temperature goals. However, Durban platform also refers the 1.5°C as alternative goal to 2°C leaving the option to strengthen of global temperature in future. AOSIS, LDCs and the African group and among others have long argued for 1.5°C goal. For many of these countries the 2°C temperature increase has created threat in existence.

The Paris Agreement finds a solution to hold the increasing global average temperature to 'well below 2°C' above pre-industrial levels. And find a way for 1.5°C temperature limit, where the world is far from it. This pathway would dramatically pull back the remaining carbon space with troubling suggestions to country like India which has to lift large majority of population from poverty. In spite of that, the aim for 1.5°C sets a direction to travel and brings unity from the small islands states on the frontlines of climate impacts.[119]

The long-term goals are to be achieved as set out in Article 4.1.

Through global high increase of GHGs (greenhouse gases) emissions has to be controlled as soon as possible and rapid reductions. Thereafter, so as

---

[118] Paris Agreement, 2015, art. 2.1. Although the Paris Agreement articles do not have titles, Article 3 identifies Article 2 as the 'purpose' of the agreement.

[119] Daniel, B., Jutta, B., et al., (2017). *International Climate Change Law* (p. 229). London: Oxford University Press.

to bring balance between reducing human activity emissions by sources and dropping down of GHGs in the second half of the century. However, parties had proposed quantitative global emission goals with specific deadlines or percentages of reduction form 2010, in the end it was proved to achieve but not in specific given time.

4.16.2. Mitigation (Article 4); Obligations In Relation To Nationally Determined Contributions (NDC's)

The most important legal rules of Paris Agreement are found in its mitigation article. To meet the long-term temperature goals, the parties are followed with the obligations of conduct as in relation to NDC's. Which are in Article 4.2, says that:

"*Each party shall prepare, communicate and maintain the successive nationally determined contributions that intends to achieve. Parties shall pursue (complete) domestic mitigation measures, with the aim of achieving the objectives of such contributions.*"

Drafting treasures to be minded carefully in negotiated text: First, unlike the most provisions of Paris Agreement which apply to 'Paris,' the first sentence "each party" creating individual obligation. Second sentence with "shall" as a command of preparing, communicating and maintaining the records of national contributions as well as completing domestic mitigation measures.

Third, says while all these are binding obligations, they are obligations of conduct rather than result.

- As the first sentence of 'intend to achieve' gives a good faith and expectation of NDC's but stops, short of required doing so.
- The second sentence is similar to first sentence function as it requires parties to complete measures 'with aim of achieving the objectives of their contributions.'

Thus, parties are with the above obligations and also the good faith in doing contributions to NDC's.

In the lead up to Paris, many parties including EU, South Africa and small Island states argued that parties must achieve their NDC's thus resulting in obligation of result. This was opposed by China, India and US as they don't want to be legally under obligation of result.

With this occurring, The Paris agreement has postponed this action. However, to provide the good act of faith the parties should provide the information to track the progress in implementing and achieving NDC, subject parties to a 'facilitative (free from difficulty), multilateral consideration of progress' respect to implementation and achievement.

The NDC's parties had submitted in variety ways, where some are-

- qualitative-climate friendly;
- quantitative-absolute emission reduction targets;
- conditional-instance on provision of international support;
- Unconditional.

In circumstances, if one had been created as an obligation of result, are not lent itself to enforcement.

In addition to the obligations, parties are subjected to the procedural obligations. Which are,

- Each party has to communicate a contribution in every 5 years with providing the information for clarity, transparency and understanding. Where these provisions are made with mandatory terms "shall" as a binding obligation for the parties.
- Where some also oblige (desire to help) parties to act according to the 'relevant decisions' to be taken by COP meeting of parties of Paris agreement [CMA], thus giving the CMA authority to adopt legally the binding decisions. Though it's not worth but gives the parties a discretion (a quality of making decision). For instance, they are provided with several decision but not all are required to do so.
- The Paris agreement also require parties to account (to report) with NDCs in accordance with 'guidance' adopted by CMA. Although the provision is made with "shall" the word 'guidance' could be explained as CMA decisions containing accounting guidance would not bind parties.

Alternatively, it means CMA decision on accounting should be provided general rather impose detailed rules. Given these as clues, whether parties regarded as binding. For example -drafted in mandatory or discretionary (not compulsory) terms.

The strength of these provisions as well as the transparency framework, can be attributed to jointly arrange the efforts of informal group of key negotiators from developed and developing countries, including EU, Australia, New Zealand, south Africa, Switzerland, the US, and others as well as Singaporean diplomats who facilitate the formal negotiators.

The informal group called as 'friend of rules,' formed after LIMA when its members realized that the rules of game, of profound importance to principles and morality of agreement, where getting short shrift (confession) in process focused mainly on the headline political issues.[120]

4.16.3. Adaptation (Article 7)

As in relation to the 'scope' of Paris agreement, most developing countries have argued for equality between the mitigation and adaptation in climate regime for a long time. And thus, sought out to include the provisions in Paris agreement. Though they have limited success but in part adaptation had mainly local benefits. So that countries are encouraged to regardless of what other countries are doing, and making the situation for collective active and less forceful.

The Paris agreement contains one strong law and several soft laws relating to adaptation.

- Parties are required ("each party shall") to engage in planning and implement actions of adaptation.
- Parties are made ('parties should) to update and submit the adaptation.
- Communications (possibly as part of their NDCs).
- Identifying the needs and priorities for listening on public list and to strengthen the teamwork on adaptation.

Many of these provisions are qualified by definitions like 'as appropriate,' which allow the discretion. The Paris agreement also tasks the CMA with developing possible situations to recognize the adaptation efforts of developing countries.

In addition, the adaptation Article in Paris agreement includes several provisions relating to the circumstances that form the idea.

- Article 7.2 recognizes that adaptation is a global challenge.
- Article 7.5 admits that adaptation should follow country driven approach.
- Article 7.6 recognizes /identifies the importance of support for the adaptation efforts.

[120] Daniel, B., Jutta, B., et al., (2017). *International Climate Change Law* (pp. 231–233). London: Oxford University Press.

The provisions do not specify weather in mandatory, recommendatory or even in flatter terms in particular course of action. Instead, they provide conditions, construct the description, capture the understanding things and generate the mutual reassurance (the action of removing the doubts or fears) about the nature of adaptation, problems and addressing. In doing so, these are critical functions and very essential for Paris package.

A good example relating provisions is Article 7.4 in which parties identify the current need for adaptation that is important for reducing the greater levels of mitigation which are needed for additional adaptation efforts. Where the greater the adaptation needs greater the adaptation costs. Although this provision does not need the particular control from parties. It is in spite extremely important to the unsafe/vulnerable countries. Many developing countries had stressed the link between mitigation ambition (or lack thereof) and need of increase of adaptation. In fact, the African groups had proposed a qualitative (measurable) adaptation goal that would allow adaptation impacts and costs flowing from agreed temperature goal.

The proposal above was an assumption that worst effects of climate change would be mainly carried by unsafe/vulnerable countries that had contributed the little thus creating problem. And such adaptation costs should be raised and carried/borne by developed countries. But this proposal by African group proved difficult to approve/accept from developed and some developing countries.

The Paris agreement contain the quantitative adaptation goal in Article 7.1 namely to enhance adaptive capacity. In regard, however, to the concerns of many unsafe/ vulnerable countries Article 7.4 recognizes inter linkages between adaptation, mitigation and support. Although only descriptive provisions convey the important information of which is accepted by large Paris package.

Finally, it is worth nothing. The Paris agreement also declares that aim of increasing the financial resources should achieve balance between adaptation and mitigation and especially for public and grant base resources for adaptations.[121]

4.16.4. Loss and Damage (Article 8)

The issue of loss and damage has been offered to discuss since the beginning of UN climate regime (interaction of human activity with climate system),

[121] Daniel, B., Jutta, B., Lavanya, R., (2017). *International Climate Change Law* (pp. 237–238). London: Oxford University Press.

with small island states and other vulnerable countries investing considerable arrangement of capital on it.

In 2013 Warsaw conference, parties established the Warsaw international mechanism (WIM) for loss and damage associated with climate change impacts to address this in particularly vulnerable developing countries. But parties disagreed for whether loss and damage should be called as a part of adaptation or as a different issue. Many vulnerable countries believe that the topic of loss and damage is different from adaptation and surrounded with issue of liability and compensation, while developed countries says the issue can be done within the framework of adaptation.

At Warsaw conference, parties agreed to establish a new method under 'Cancun adaptation framework,' but with a review in 2016. The following year, the parties, at LIMA conference agreed to a two-year work plan for WIM as it was said for review in 2016. Many developed countries thought that issue would not be addressed in Paris but the topic is given importance to vulnerable countries they somehow pushed to include the topic (loss and damage) in Paris agreement.

Although Article 8 may be argued for greater representative/symbolic than fundamental/substantive significance which is important for two reasons.

- First, it clearly brings the issue of loss and damage with the scope of Paris agreement.
- Secondly, it is free standing article, thus, arguably, separating loss and damage from adaptation, as developing countries are looking after for long.

As the price for including Article 8, however US instead adding sentence for developing the idea to Paris COP decision, stating that 'article 8 does not provide any basis for any liability or compensation,' as a result limiting the loss and damage of some of its distinctive elements.

Some of the areas of cooperation and facilitation identified in Article 8 are, in fact,

- forms of adaptation;
- aimed at preventing damage;
- including early warning systems;
- actions prepared for emergencies; and
- dealing with the risk assessment and management.

In spite of that, Article 8 gives loss and damage a space in the regime, which developing countries are likely to use to advance the issue going forward.[122]

4.16.5. Support (Article 9, 10, 11)

1. **Finance:** In the lead-up to Paris, finance was expected to be the most difficult issue to settle with the impossible bridge of
- developing countries that sought new commitments with Paris agreement;
- developed countries could not accept for the new commitments but sought to increase donor pool.

At the end developing countries are willing to settle with the methods of

- Advancements in Paris agreement;
- making a goal to solve issue of finance.

To provide assistance to the developing countries for mitigation and adaptation the FCCC requires Annex II parties. In Copenhagen, developed countries have committed to $100 billion per year by 2020 in climate finance in order to assist developing countries in climate change. And Copenhagen had promised to surround money from public and private sources 'in the factors of meaningful mitigation actions' and transparency of implementation by developing countries. In the recent reports of organization of economic corporation and development (OEDCD) the climate finance in 2013 & 2014 is rose from 52$ to 62$ respectively. There had been an argument on these figures but because of systematic questions it was done with what counts as climate finance. Against this drawback the Paris agreement provisions on finance are somewhat modest.

- **Finance Commitments:** 'In continuing of the existing task under agreement,' Article 9 requires developed countries to assist financially to developing countries in mitigation and adaptation. It is the conclusion note that permitted the US for forming only important financial obligations. Article 9 also brings out new reports for improvement of performance of the climate finance should be improved than previous efforts.
- **Donor Pool:** By moving away from FCCC with sharp difference, Paris agreement 'encourages' the other parties to 'provide or

---

122 Daniel, B., Jutta, B., Lavanya, et al., (2017). *International Climate Change Law* (pp. 238, 239). London: Oxford University Press.

continue to support voluntarily.' This provision is much weaker than developed countries have looked after, and it encourages rather than recommending and silent as to who is willing to do. But still this provision could prove outstanding by removing the gap between the donor and recipient countries. Besides it calls on developed countries to take charge of climate change but as a 'part of global effort.'

- **iii. Mobilization Goal:** In Paris agreement, the US and other developed countries have succeeded in everything except on Copenhagen $100 billion mobilization goal. Instead, the quantitative financial goal appears to provide COP advice to increase developed countries that set goal of $100 billion by 2025 and parties should also set a new bigger goal with comparing the level of 2025 goal.

Finance, as discussed will be a part of 2023 goal. Like Copenhagen accord, the Paris agreement wants to increase the support to balance between mitigation and adaptation.

2. **Technology:** In relation to technology Paris agreement wants to create a framework for wide range of guidance to work of technology mechanism in promoting and enhancing the technological development and transfer. It also makes support to encourage and enable the innovative approaches to help in support global stocktake.
3. **Capacity-Building:** The Paris agreement makes it important to the parties for enhancement of developing countries to implement the agreement, and requires them from time to time.[123]

4.16.6. Transparency (Art. 13)

The Paris Agreement relies on a robust transparency and accounting system by providing clarity on action and support by Parties, with flexibility for their differing capabilities. In addition to reporting information on mitigation, adaptation and support, this agreement requires that the information submitted by each Party undergoes international review. The Paris Agreement also includes a mechanism that will facilitate implementation and promote compliance in a non-adversarial and non-punitive manner, and will report annually to the COP.

---

[123] Daniel, B., Jutta, B., et al., (2017). *International Climate Change Law* (pp. 237, 238). London: Oxford University Press.

4.16.7. Global Stocktake (Art. 14)

A "global stocktake," to take place in 2023 and every 5 years thereafter, will be assessing collective progress toward meeting the purpose of the Agreement in a comprehensive and facilitative manner. Its outcomes will inform Parties in updating and enhancing their actions and support and enhancing international cooperation in combating climate change.

4.16.8. Final Clauses (Articles 20–28)

The Paris Agreement consist of a standard set of final clauses. Parties to the agreement must express their consent to be bound by means of ratification, accession, acceptance, or approval. Entering into force involved a double trigger, requiring acceptance by at least 55 States that account for at least 55% of global GHG emissions. The Paris agreement was opened for signature at a high-level signature ceremony convened by secretary general in New York on 22nd April 2016 on the occasion of Earth Day. A record 175 UNFCC parties signed the agreement on this day, and the Paris agreement entered into force on 4th November 2016 less than a year after it was adopted. The Paris agreement incorporates by reference the UNFCCC's provisions on amendment,[124] adoption and amendment of annexes[125] and dispute settlement.[126] Reservations are expressly disallowed,[127] but parties may withdraw by giving 1 years notice in beginning 3 years after the agreements entry into force.[128]

4.17. ROLE OF US IN PARIS CLIMATE DEAL

In December 2015, the then-president Barack Obama envisioned we would leave today's children "A world that is safer and more secure, more prosperous, and more free." when he announced that the United States, along with nearly 200 other countries, had committed to the Paris Agreement, an ambitious global action plan to combat climate change. But even less than two years later, the then-president Donald Trump put that future in jeopardy by announcing his plan from withdrawing the United States from

[124] Paris Agreement, 2015, art. 22

[125] *Ibid*, art.23

[126] *Ibid*, art.24

[127] *Ibid*, art.27

[128] *Ibid*, art.28

the Paris Agreement on 1 June 2017[129]. Trump also indicated his willingness to renegotiate the agreement or negotiate a new one. But other countries, reaffirmed their strong support for the Paris Agreement and said that they were not open to a new negotiation. The U.S. formally initiated withdrawal from the Paris Agreement on November 4, 2019. A step in this regard that became official on November 4, 2020 as part of a larger effort to dismantle decades of U.S. environmental policy. The delay in the withdrawal was due the complex rules that were built into the Paris agreement to cope with the possibility that a future US president might decide to withdraw the country from the deal. As The Clinton administration was unable to secure Senate backing for the Kyoto Protocol, agreed in 1997 due to internal politics. Due to this reason President Obama's negotiators wanted to ensure that if there was a change in leadership then it would take time for the US to get out. Even though the agreement was signed in December 2015, and it came into force on 4 November 2016, 30 days after at least 55 countries representing 55% of global emissions had ratified it. No country could give notice to leave the agreement until three years had passed from the date of ratification. Even after lapse of 3-year time duration a member state still had to serve a 12-month notice period on the UN.

While the US is one of the biggest global greenhouse gas emitters and it is world's biggest and most powerful economy. So, when it became the only country to withdraw from a global solution to a global problem it raised questions of trust.

Trump's perspective on the Paris agreement was that it was unfair to the US, leaving countries like India and China free to use fossil fuels while the US had to curb their carbon. The withdrawal from Paris Agreement became official on 4 November 2020 – by chance, the day after Mr Trump lost the presidential election. After becoming Joe Biden as president of US; The United States has officially re-joined the Paris climate accord on February 19, 2020. The US has re-joined the deal after 107 days it left. With the re-joining agreement by of US, the world leaders are expecting that now the country will prove its seriousness as it was mostly absent for four years. The other countries are also waiting for the announcement from Washington on the US's goal to cut emissions of heat-trapping gases by 2030. The president of US, Joe Biden, had signed the executive order to reverse the withdrawal ordered by former President, Donald Trump.

---

[129] Kevin, L., & Jim, A., (2017). *Trump on Paris Accord: "We're Getting Out.*" CNN. https://edition.cnn.com/2017/06/01/politics/trump-paris-climate-decision/index.html (accessed on 07 July 2023).

4.18. COP 22

COP22 was held in Marrakech, it was treated as an opportunity to showcase progress and start the important process of turning the UN's Paris Agreement into a detailed blueprint for action by all the member countries. The Paris Agreement was a complete document that set goal and framework for international climate action. But setting out the details was a longer process, which the countries participating in COP22 had decided should be completed by 2018, with a review of progress in 2017. Stephen Cornelius, chief advisor on climate change at WWF-UK, told just before the talks concluded. "*This COP is the first stop after Paris and it needs to cover and take up the issues to make Paris work and be implemented, so that is what it's doing*," The official outcomes of the COP 22 was a collection of various documents. Notable outcomes of the COP 22 is given below:

4.18.1. Finance

All the member Countries were urged to continue scaling up their financial contributions towards the pre-agreed "$100bn a year by 2020" goal, and to achieve a greater balance between adaptation and mitigation. Some countries had hoped for stronger wording on this, since adaptation has long trailed mitigation, to the damage of the most vulnerable countries. There was also some last-minute internal strife over the Adaptation Fund.

4.18.2. Facilitative Dialog

The organization of the "*2018 facilitative dialog*" was also proved to be controversial. In Marrakech, it was decided that the presidents of COP 22 and the forthcoming COP 23 would make consultation with countries on the organization of this dialog and report back on their findings in a year's time. Article 4 of the Paris Agreement deals with both the long-term "net-zero emissions in the second half of the century" goal, as well as the need for NDCs to provide "clarity and transparency." A main theme of COP22 was debating how best to create a fair "rulebook" that all countries could share and have confidence in when assessing each other's climate pledges. The technicalities of the rulebook – baselines, methodologies, etc., – will likely be a continuing discussion into 2018.

4.18.3. Orphan Issues

Negotiators also clashed, without resolution, with the "orphan's issues" of the Paris Agreement. These were tasks for which no one was given responsibility. These includes important issues, such as common timeframes for future climate pledges, and a new goal for climate finance.

4.18.4. Loss and Damage

Countries also agreed on "loss and damage," which was to start in 2017 and will see countries start to formally address topics such as slow-onset impacts of climate change, non-economic losses (for example, culture and identity) and migration. In other words, dealing with climate impacts that are beyond adaptation.[130]

4.19. COP 23

COP 23 is the second "conference of the parties" since the Paris Agreement took place in 2015, promised to be a somewhat technical affair as member countries continued to negotiate the finer details of how the agreement will work from 2020 onwards.

However COP 23 was also the first set of negotiations since the US, under the presidency of Donald Trump, announced its intention in June 2017 to withdraw from the Paris deal. And it also was the first COP to be hosted by a small-island developing state with Fiji taking up the presidency, even though it was being held in Bonn. The key topics of the COP 23 were related to financial support, mitigation action, differentiation, and loss and damage.

4.19.1. Key Outcomes of COP 23

The Conference made progress on framing rules for implementation of the Paris Agreement. Developing countries negotiated hard to get the rich countries report, by May 2018, and their progress under the Doha Amendment. Countries were also asked for submitting their views on what should be included in the work by 31 March 2018, with possibilities including how to improve soil carbon and fertility, how to assess adaptation and resilience and the creation of better livestock management systems. There were few project-specific funding announcements were made, such as for green buildings, 'power past coal' and eco-mobility. But the key question of when coal will be phased out globally was not addressed in the

[130] United Nations Framework Convention on Climate Change, 1992, COP 22, accessed on December 31, 2021, https://unfccc.int/process-and-meetings/conferences/past-conferences/marrakech-climate-change-conference-november-2016/cop-22.

conference. However, 15 countries led by Canada and the UK formed an alliance for cutting down their use of coal by 2030.

4.19.2. Coal Phase-Out[131]

A major event which took place at the COP was the launch of the "Powering Past Coal Alliance," led by the UK and Canada. More than 20 countries and other sub-national actors joined the alliance, including Denmark, Finland, Italy, New Zealand, Ethiopia, Mexico and the Marshall Islands; as well as the US states of Washington and Oregon. It aimed to top 50 members by this time in 2018. While the alliance notes in its declaration that "analysis shows that coal phase-out is needed no later than by 2030 in the OECD and EU28, and no later than by 2050 in the rest of the world" for meeting the Paris Agreement, it does not commit signatories to any particular phase-out date. It also did not commit the signatories to ending the financing of unabated coal power stations, rather just "restricting" it.

Claire Perry, the UK's climate minister, traveled to Bonn for launching the initiative alongside Canada's environment minister Catherine McKenna. The UK had previously pledged to phase out unabated coal by 2025, while Canada has a 2030 deadline. The US did not sign onto the pledge and several other big coal countries were notable by their absence, including Germany, Poland, Australia, China and India.

4.19.3. Talanoa Dialog

An important outcome of the COP 23 is the *'Talanoa Dialog.'* 'Talanoa Dialog' is a year-long process to assess countries' progress on climate actions. Under this, it was agreed that the next two climate conferences, in 2018 and 2019, will have special 'stock-taking' sessions. The aim of this stock-take was to focus on the 'pre-2020 actions' being taken by different countries to reduce greenhouse gas emissions. It also included the progress made by developed nations in their obligations to provide finance and technology support to the developing countries. This was to bring a strong message to developed countries that post-2020 climate action as part of the Paris Agreement cannot be divorced from pre-2020 commitments.

The main concern and challenges of the COP 23 was that Parties were still far from keeping the world safe from its harmful consequences of the

[131] Coal phase-out means stopping burning coal, and is part of fossil fuel phase-out. Coal is the most carbon intensive fossil fuel, therefore phasing it out is critical to limiting climate change and keeping global warming to 1.5°C as laid out in the Paris Climate Agreement.

climate change. The second phase of the Kyoto Protocol was not ratified by a many number of countries to enter into force. The anthropogenic greenhouse gas emissions would severely affect the world's poorest countries that are the most vulnerable to the impacts of climate change. The promise relating to Mobilization of $100 billion each year by 2020 into the GCF was not seen much inflow. Expectation that all countries of the world keep up their NDCs promises was in fact far from reality. US's withdrawal from the Paris Agreement in 2017 despite being the world's biggest greenhouse gas emitter Worsen the matters. Another problem was that China being the second biggest emitter is shutting down coal plants on its soil, but building more elsewhere in the world.[132]

4.20. COP 24

The United Nations Climate Change Conference of 2018 took place between 2 and 15 December 2018 in Katowice, Poland, which was the 24th Conference of the Parties to the United Nations Framework Convention on Climate Change (COP 24). It was also referred to as the Katowice Climate Change Conference or Katowice Climate Talks. The most important outcome of COP24 was that the countries have agreed on rules for the implementation of the 2015 Paris Agreement.

4.20.1. Key Outcomes of COP 24 in Katowice

The participating countries agreed on the rules to implement the Paris Agreement that was come into effect in 2020. The rules were regarding how the member nations will measure the carbon-emissions and report on their emissions-cutting efforts. This 'rulebook' can be called as the detailed "operating manual" of the 2015 Paris Agreement. The rulebook covered a multitude of questions, such as how countries should report their greenhouse gas emissions or contributions to climate finance, as well as what rules should apply to voluntary market mechanisms, such as carbon trading. The Intergovernmental Panel on Climate Change (IPCC) in October, 2018 published special on the impacts of 1.5°C global warming which became a major source of tension at the talks. The members of the Katowice conference did not agree to "welcome" the Intergovernmental Panel on

[132] COP 23 UNFCCC, https://unfccc.int/process-and-meetings/conferences/past-conferences/un-climate-change-conference-november-2017/sessions-of-negotiating-bodies/cop-23 (accessed on 07 July 2023).

Climate Change (IPCC) report on 1.5°C. Countries like US, Saudi Arabia, Russia and Kuwait refused to "welcome" the IPCC report. The parties to the Katowice conference agreed to record the pledges in a public registry, as per the existing interim portal. It was also agreed in the COP 24 among the members that future pledges should cover a "common timeframe" from 2031. The number of years for the timeframe will be decided later. Many difficult matters could not reach an agreement and had been postponed to next year for resolution. This included questions such as ways to scale up existing commitments on emission reduction, different ways of providing financial aid to the poor nations, wording that prevents double counting and whether member nations are doing enough to cut their respective emissions.

4.21. COP 25

The 25th COP was held in 2019 in Madrid under the Presidency of Chile. The COP 25 decision, titled Chile Madrid Time for Action, emphasized the continued challenges that developing countries face in accessing financial, technology and capacity-building support, and recognized the urgent need to enhance the provision of support to developing country Parties for strengthening their national adaptation and mitigation efforts. The decision also recalled the commitment made by developed country Parties to a goal of mobilizing jointly USD 100 billion per year by 2020 to address the needs of developing country Parties. On the issue of global ambition for combating climate change, the decision adopted provides for a balanced and integrated view of ambition that includes not only efforts for climate change mitigation, but also for adaptation and means of implementation support from developed country parties to developing country parties.

The longest climate talks in the history of United Nations COP 25 held in Madrid, Spain, ended on December 15, 2019. This climate summit ended without crucial agreement on carbon market rules of the Paris Agreement. The key concerns that were aimed to get resolved at the COP 25 was the regulation of carbon market. Carbon market puts prices on carbon dioxide emission. However, no conclusion was reached and the issue was to be taken at the next summit to be held at Glasgow in 2020. This summit failed to reach an agreement on the implementation of Article 6 of Paris Agreement.

The main Issue was in COP24 that US, Saudi Arabia, Kuwait and Russia refused to welcome the IPCC 1.5°C report. This year majority of the members were not willing to reduce their emissions at the pace suggested by the conference. According to World Resources Institute NDC tracker, only 80 countries had submitted proposals to enhance their NDCs. These countries are primarily small and developing. They represent only 10.5% of the world population.

The United Nations Environment Program released its 'Emission Gap Report'[133] at the summit to measures the gap between measures being done and measures to be done to tackle climate change. The report said that the global temperature will increase by 3.2° by 2100 as compared to pre-industrial levels. The top four emitters countries of the world namely China, USA, European Union and India contributed to 55% of total emissions in the last decade. This report suggested that a minimum of 1.58 billion USD of annual investment in renewable sector is required for efficient energy use. It also says that solutions like coal phase out, decarbonization of transport and industries and increased access to electricity will help in tackling climate change. This report warned that in order to achieve the goal of 1.5°C increase in temperatures over pre-industrial levels, it is essential to reduce the emissions by 7.6% between 2020 and 2030.

The Presidency of the summit, Chile (though summit was held at Madrid, Spain), launched the Climate Ambition Alliance (CAA). Around 73 countries had joined the CAA at the COP 25. The alliance was led by Chile and was launched at the Climate Action Summit, New York in 2019. The main purpose of the alliance was to upscale 2020 targets of members of UN and to achieve Net Zero Carbon Emissions by 2050. The alliance was important to enhance the NDCs of countries all over the world.[134]

4.22. COP 26

The 26th session of the Conference of the Parties (COP 26) to the UNFCCC was scheduled to take place from 9–19 November 2020, in Glasgow, UK. But On 28 May 2020, the COP Bureau decided that it would take place from 1–12 November 2021, in Glasgow, UK. Due to the COVID-19 pandemic. At that time, the COP Bureau announced that: "In light of the ongoing, worldwide effects of COVID-19, holding an ambitious, inclusive COP26 in November 2020 is no longer possible. Rescheduling will ensure all parties can focus on the issues to be discussed at this vital conference and allow more time for the necessary preparations to take place. We will continue to work with all involved to increase climate ambition, build resilience and lower emissions." There are two main reasons why the COP 26 is so important.

---

133 https://www.unep.org/resources/emissions-gap-report-2021 (accessed on 07 July 2023).

134 *UN Climate Change Conference,* (2019). https://unfccc.int/cop25 (accessed on 07 July 2023).

Firstly, COVID-19 has refocused priorities and caused individuals and governments to pay closer attention to the environment. As many countries are looking to rebuild their economies in the wake of the pandemic, there has been a major emphasis on 'building back better' through a green recovery. Secondly, COP26 is being viewed as the successor to COP21 where the Paris Accord was signed, arguably the greatest success from the UNFCCC in recent years. COP26 is seen as the summit to address both what has and hasn't been achieved since 2015, while also setting concrete plans to reach the Paris Agreement targets. It is also the first COP to be held since the US left and re-joined the Paris Agreement, so it's likely that there will be extra eyes on US contributions to the summit.

4.23. 2021.IS MAKE OR BREAK YEAR TO CONFRONT GLOBAL CLIMATE EMERGENCY: UN CHIEF

Initial NDCs Synthesis Report has been recently released by The United Nations Framework Convention on Climate Change (UNFCCC). While commenting on the report; the Secretary-General of the United Nations, Antonio Guterres, stated that the year 2021 is a "make or break" year so as to confront the global climate emergency. The Secretary-General further stated that in order to limit the global temperature rise to 1.5°C, the global emission should be cut by 45% of the 2010 level by 2030.

Initial NDCs Synthesis Report measures the progress of national climate action plans in the backdrop of upcoming 26th session of Conference of its Parties (COP26) in November 2021 in Glasgow. This report has also insisted the other countries to initiate more ambitious plans in order to reach the Paris Agreement goals. This report covered the submissions from countries up to December 31, 2020. As per the submission, 75 countries which are Parties to the Framework Convention communicated a new or updated NDC which represents 30% of global greenhouse gas emissions.[135]

4.24. CLIMATE ADAPTATION SUMMIT 2021

The climate adaptation summit was held online and was hosted by the Netherlands government on January 25, 2021. It is being called "CAS Online." This summit also showcased the pioneering solutions of the climate emergency and sustaining momentum through UNFCCC's COP26.

[135] Hemani, S., (2021). *2021 is Make or Break Year to Confront Global Climate Emergency: UN Chief.* The Hindu. https://www.thehindubusinessline.com/news/2021-is-make-or-break-year-to-confront-global-climate-emergency-un-chief/article33954251.ece (accessed on 07 July 2023).

The summit was organized with the ambition of realizing the transitions which is required for a climate-resilient world. The summit focused on securing the new investments in order to ensure that millions of smallholder farmers could adapt to the impact of climate on food production. India had committed to provide the clean cooking fuel to 80 million rural households during the summit. It also committed to connect the 64 million households with the piped water supply.[136]

The Ministry of road transport has decided to impose the additional taxes on old vehicles which are no more fit on road. This surplus tax is being called as the "Green Tax." This decision was taken to curb pollution. It will also motivate people to switch to environment-friendly alternatives to the vehicle[137]

4.25. CONCLUSION

The problems associated with tackling climate change are essentially global in both their causes and effects, respecting no national boundaries. As such, they require a significant element of global cooperation if they are to be tackled appropriately. Few would disagree with this unremarkable statement, but the difficulties in achieving this international cooperation in the context of nation-states driven by self-interest has been conspicuous in all of the negotiations on Kyoto implementation since 1997, however, to some extent this is now a matter of history. By the end of 2007, attention had definitively shifted to the development of a successor instrument to the Kyoto Protocol for the period after 2012. The problems in design of climate change regime do not lie only in the instruments of international law that have been made to address them, but those shortcomings are more evident at national and regional levels. Development at the international level may nevertheless in due course was to achieve agreement on how the post-Kyoto principles can be designed so that the more intensified national and regional efforts are harmonized and become much more effective in achieving climate change mitigation.

The past decade shows beyond a shadow of doubt that a consensus now exists behind the central role of market mechanisms and legal frameworks (international and regional) for tackling climate change. The Paris

[136] Climate Adaptation Summit 2021, 25-26 January 2021 – Online Worldwide, https://www.cas2021.com/ (accessed on 07 July 2023).

[137]

agreement is a landmark achievement in the history of United Nation climate negotiations. Despite longstanding and seemingly intractable differences, parties harnessed the political will necessary to arrive at an agreement that is long term, rule based and applicable to all. The Paris agreement contains ambitious goals, extensive in applications and comparatively rigorous oversight. Admittedly the goals of Paris agreement are aspirational the obligations are largely procedural. Differentiation too has come to acquire a new more nuanced form. Despite many negotiations compromise the path of Paris agreement represents a hard-fought deal among 196 Nations across the developed and developing country divide made significant concessions from long held position in the closing hours of the conference does making the final agreement possible.

Many difficult issues remain unsolved and left for post Paris negotiations like the Paris did not resolve the issue of burden sharing among parties. But it must be noted that many developing countries including Brazil China and India among that brought the Paris agreement into force. However, developing country like India is going to face a lot of problem like developmental challenges. India being developing country have limited resources and it has to devote to the significant energy transformation required to bend their emissions curve in line with 2 degree centigrade or 1.5 degree centigrade scenarios. With limited resources it remains to be seen to what extent many countries even with the best intentions will be able to make the transformational changes required domestically to meet the global temperature limit for the Paris agreement. As it was assumed that all the country of the world will make effort to mitigate climate change and will help developing countries by providing financial assistance and technology support. When Paris was signed it was thought that for the first time in the world all the nations of the world were together. But in June 2017 USA who is the largest emitter of greenhouse gas withdrew. It has hit the other country.

Perhaps the more serious problem with the Paris Agreement is that its goal of striving to keep the rise in global temperatures under 1.5°C is likely to be unattainable, since the carbon budget left to the world until the end of the century is likely to be exceeded fairly soon, and certainly by 2035–50, at current rates of emissions reduction. The release of the IPCC special report on 1.5-degree warming signals this fact clearly, even though there appears to be some ambiguity in the manner in which the feasibility of restricting temperature rises to 1.5°C is discussed in the report. Given the nature of the NDCs, even the 2°C goal appears very difficult to achieve. It is not unlikely, unless there is a serious and dramatic course action is taken at some point in time in the future, the globe will have to declare a climate emergency. In such type of scenario, the challenge of equity in all aspects of climate action is likely to be unprecedentedly severe.

Sweden is the first country in the world which has passed the law to switch on green energy till 2045 in phase-out manner. France has passed the law for banning of exploration of gas and oil in country till 2040. France and the United Kingdom have legally enshrined their 2050 net-zero GHG emissions goals. The European Union aims to achieve net-zero GHG emissions by 2050. China, has announced plans to achieve carbon neutrality before 2060. Japan has also announced a goal of net-zero GHG emissions by 2050. The Republic of Korea has committed the country to becoming carbon neutral by 2050 in a speech to parliament. Canada has also indicated its intention to legislate a goal of NetZero emissions (though it is unclear if this refers to just CO_2 or all GHGs) by 2050. South Africa aims to achieve net-zero carbon emissions by 2050; Argentina and Mexico are both part of the UNFCCC CAA working towards net-zero emissions by 2050.

India clearly is in a unique position where it not only has a significant percentage of population that is extremely vulnerable to climate change but it is also projected to be one of the major GHG emitters in the world (in term of total output). As countries move nationally from a reactionary to a proactive protection strategy towards climate change and development, it is in India's best interest to focus strongly on the internal issues and create a friendly image for other developing countries which are looking up to India for its development models and diplomacy.

Chapter

5 CLIMATE CHANGE MITIGATION: INDIAN LEGAL ORDER

Contents

5.1. INTRODUCTION

India is developing country as well as a major emerging economy possess it with unique challenges in relation to climate change. India has a population consisting of over a billion people, second only to China's, and India is expected to become the world's most populated country by 2025. With most of its population living in poverty[1] and million people with no access to electricity, India being developing country, prime challenge remains poverty eradication and raising the standards of its people. Compounding these challenges is the potential impact of climate change. Nearly half a million people are dependent for their livelihood on the glacier fed Himalayan region which is one of the main freshwater's sources for the Gangetic basic. Studies have shown that in the longer term, melting Glaciers and the resulting water stress could have a crippling effect on the Indian region. Agriculture dependent economy of the India could be affected by lower crop yields as an impact of climate change. Millions of People who live in Coastal India and climate change impacts like sea level rise could have a devastating effect on them.[2]

In spite of the developmental challenges India faces, its economy has been experiencing unprecedented growth. It is currently the one of largest economy in the world and to meet its economic and development goals, the Indian government has targeted economic growth rates of about 8 to 10% a year for the next two decades. The Government of India estimates that to meet these goals at 3 – 4-fold increase in primary energy supply and a 5-to-6-fold increase in electricity generation capacity will need to occur. Both fossil fuels, Coal and oil, account for about 75% of India's energy consumption and almost 70% of electricity is generated from coal fired power plants.[3] It is expected that due to the high demand for energy, coal will remain the back bone of the Indian economy for the foreseeable future.[4]

[1] Word bank, *New Global Poverty Estimate-What It Means for India.* http://go.worldbank.org/51QB3OCFU0 (accessed on 07 July 2023).

[2] UNFCCC, (2010). *Indian Minister of Environment Jairam Ramesh: Press Conference at COP 16.* http://unfccc.int/resource/webcast/player/app/play.php?id_episode=2988 (accessed on 07 July 2023).

[3] Energy Information Administration, (2011). *Country Analysis Briefs: India.* http://www.eia.doe.gov/cabs/India/Full.html (accessed on 07 July 2023).

[4] See Michael, J., Jan, C. S., et al., (2020). The future of coal in a carbon-constrained climate. *Nature Climate Change, 10*, 704–707.

Even though India contributes only about 7% of global greenhouse (GHG) emissions, it still is ranked fourth in terms of absolute emissions.[5] Furthermore, its emissions are projected to experience one of the highest growth rates in the next few decades. India however ranks extremely low-only about the 10th of the United States' per capita emissions and third of the world average.

Streaking the right balance between these two conflicting factors--on the one hand tackling the country's development challenges and on the other being recognized as a major emerging economy in the world is the basis on which India has developed its recent approach to climate change. Historically, India's stance on climate change has been driven primary by its categorization within the United Nation Framework Convention on Climate Change (UNFCCC) as a non-annex 1 party[6] (developing country party) where it had no obligation to reduce its GHG emissions. But recently India's emerging on the world stage as a major economy cast a spotlight on the country and it has felt an increasing pressure to undertake domestic action to address climate change. This chapter explores the evaluation of climate policy making in India. The following fraction elaborates on India's historical stance of the climate issue being a first world problem and highlights relevant domestic legislation that was developed in this context.[7] India has enacted various laws at almost regular interval to deal with deal with the problems of environmental degradation. At the same time judiciary in India has played a pivotal role in interpreting the law in such a manner which not only helped in protecting environment but also promoting sustainable development. In fact, judiciary in India has created a new "environmental jurisprudence."

The tide of judicial decisions and environmental litigation in India symbolizes the anxiety of court in finding out appropriate remedies for environmental problems. The adherence to the principle of sustainable development is now constitutional requirement. How much damages to the environment and ecology have been done has got to be decided on the fact

[5] *UNEP, Emissions Gap Report,* (2020). https://wedocs.unep.org/bitstream/handle/20.500.11822/34438/EGR20ESE.pdf (accessed on 07 July 2023).

[6] Non-Annex I Parties are mostly developing countries. Certain groups of developing countries are recognized by the Convention as being especially vulnerable to the adverse impacts of climate change, including countries with low-lying coastal areas and those prone to desertification and drought; Non-Annex 1 countries have also ratified the Kyoto Protocol but do not have any emissions reduction targets. This group has 148 members and is mainly comprised of developing countries. Large GHG emitters such as China, India, Brazil and Mexico belong to this group.

[7] Patodia, R., (2013). "India's evolving climate change strategy. In: Erkki, J. H., et al., (eds.), *Climate Change and the Law* (pp. 605–607). Dordrecht: Springer.

of each case. Therefore, the court are required to balance development needs with the protection of environment and ecology. It is duty of state under constitution to device and implement a coherent and coordinated program to meet its obligation of sustainable based on inter-generational equity.[8]

It is true that in developing country there shall have to be developments but that development shall have to be in closest possible harmony with environment, as otherwise there would be development but no environment, which would result in total devastation, though, however may not be felt in present but at some future point of time, but then it would be too late in the day to control and improve environment so that both can co-exist without affecting the other.

It may be relevant to mention here that principle 10 of the Rio declaration of 1992 specifically provides for "*effective access to judicial and administrative proceedings, including redress and remedy.*"

The judiciary in India has applied the principle of sustainable development[9] while deciding the cases. There are number of cases on this point and, therefore, it will necessary to study and analyzed a few important cases in this area. It is also worthwhile to mention here that most of the environmental cases before the courts through "public interest litigation" (PIL).

Governments all over the world have taken various policy decisions to reduce the levels carbon emissions and mitigate climate change. Road transport and emission level of vehicles are being regulated to achieve the highest global standards[10]. Subsidies being provided for alternative fuel consumption projects. The Indian Government framed the National Action plan on climate change 2007, identifying 8 national missions that aim to promote sustainable development. The Plan states that "*India is determined that its per capita greenhouse gas emissions will at no point exceed that Off developed countries even as we pursue our development objectives.*" In May 2014, the Ministry of Environment and Forest was renamed as the Ministry of Environment, Forest and Climate Change (MoEFCC). This is indicative of the seriousness with which the Government proposes to deal

[8] Paramjit, S. J., (1996). *Directive Principles Jurisprudence and Socio Economic Justice in India, 543.*

[9] Sustainable development is the idea that human societies must live and meet their needs without compromising the ability of future generations to meet their own needs.

[10] The Central Government has mandated that vehicle makers must manufacture, sell and register only BS-VI (BS6) vehicles from April 1, 2020.

with the change that climate change possesses.[11]

5.2. HISTORICAL BACKGROUND OF ENVIRONMENT PROTECTION IN INDIA

"Man's paradise is on earth;
This living world is the beloved place of all;
It has the blessings of nature's bounties;
Live in a lovely spirit" ...... Atharvaveda (5.30.6)

The Indus valley civilization (3250 B.C. to 2750 B.C) explores the Urban life of India and reveals the fact that the people were best town planners at their times. The archaeological traces of construction of bathroom and toilet and underground drains in the city of Indus valley reveal the fact that the dwellers were conscious and aware of personnel cleanliness and community sanitary management. We find many hymns in the world's oldest scripture (2000 B.C. to 600 B.C.). The Rigveda portrays the beauty of the morning and a Vedic man starts his life early in the morning worshipping the glory of sun with Surya namaskars. There are many shlokas in Rigveda, where in prayerfully Calling the goddess to grant blessings of nature in bountiful to all mankind of enrich the life happily and comfortable with all its abundance. Thus, the worship itself is a form of the admiration of nature.

During the Vedic period great emphasis was put on to purify the environment as a process of religious rites and the concept of cleanliness was an integral aspect of sacred rituals, for example, performance of Yajnya was also a method to purify the surrounding air. Traditionally, Indian people hold extreme reverence towards nature. Initially Indian people worshiped the object of nature in Hindu religion some plants, trees, bird's, animals were elevated to the position of gods or goddess water is called *Jivan* (life) and air called *Vayu* and is considered as prana life itself. River also enjoyed a sacred position in religions and social life, worshipped as goddess and considered as having a capacity to purify a man from his sins and also having self-purifying ability. Fouling of the water of a river was considered a sin and it attracted punishment of different grades, which included, penance out casting, fine etc. The earth or soil also equally had the same importance and the ancient literature provided the means to purify the polluted soil.[12]

[11] Usha, T., (2016). *Climate Change Law Policy and Governance* (p. XV). Lucknow, India: Eastern Book Company.

[12] Bhaskar, K. C., (2006). Environmentalism: Indian constitution and judiciary. *Journal of Indian Law Institute, 48*, 99.

Yajnavalkya Smriti which is a historic Indian manuscript on statecraft and jurisprudence, suggested to have been written before 5th century AD, prohibited the cutting of trees and also prescribed punishment for such acts. Kautilya's Arthashastra, written in Mauryan period, gave emphasis on the need for forest administration. With reference to ancient Indian laws, during the Mauryan period, due attention was drawn to preserve and protect the trees, Forests, animals and birds. Further, punishments were also prescribed to those who harm them. Ashoka went further, and his Pillar Edicts expressed his views about the welfare of environment and biodiversity.[13] Hindu religion viewed the forests as 'Vanadevata' in a manner of worshipping the nature. The sacred groves are kept unmolested and undisturbed times immemorial. Doing any harm to those groves was believed to offending the forest spirit and deities. It is said Rishis had warned against deforestation and cutting of trees; they thought that this would result in poor rainfall. Thus, ancient India had a philosophy of environmental management enshrined in old.

5.3. LAW OF TORTS

The law of torts is mainly uncodified and also based on accepted principles. Thus, the common law principles of torts including nuisance, negligence, trespass and strict liability are the oldest form of remedies to abate the hazards of pollution. In India the law of torts has been commonly followed to provide equity, justice and good conscience. The Supreme Court has evolved the principle of absolute liability in addition to the above traditional categories as it was felt that the rule of strict liability laid down in *Ryland v Fletcher*[14] was no longer suitable in the prevailing socio-economic conditions of the country.[15]

5.4. EARLY LEGISLATIONS IN INDIA FOR PROTECTION OF ENVIRONMENT

The Shore Nuisance (Bombay-Kolaba) Act, 1893 was enacted to regulate wastes and marine water pollution. The Oriental Gas Company Act, 1857 and the Bengal Smoke Nuisance Act, 1905 were enacted to prevent or reduce atmospheric pollution in and around Calcutta. The Bombay Smoke

[13] Shanthakumar's, S., (2009). *Introduction to Environmental Law* (p. 75). Nagpur, India: Wadhwa and Company, Reprint.

[14] (1868) LR 3 HL 330.

[15] *M. C. Mehta v Union of India,* AIR 1987 SC 1086.

Nuisance Act, 1912 was passed to regulate smoke nuisance in Bombay area. For preservation and protection of forests, the Cattle Trespass Act, 1871 and Indian Forest Act 1927 were passed. The Indian Easement Act of 1882 assured property rights of riparian owners against "unreasonable" pollution by upstream users.

5.5. THE INDIAN PENAL CODE, 1860

The Indian Penal Code, 1860[16] was passed during the British rule, which contains one chapter (Chapter XIV) on offenses affecting public health, safety, convenience, decency and morals. It deals with offenses affecting public health, safety and conveyance, which covered aspects like water, air and noise pollution. Section 268 define public nuisance. It defines public nuisance as "*any act or an illegal omission which cause any common injury, danger, or annoyance to the public or to the people in general who dwell or occupy property in the surrounding area, or which must necessarily cause injury, obstruction, danger or annoyance to person who may have occasion to use any public right.*" Sections 269 and 272 of IPC deal with adulteration of food or drink for sale and adulteration of drugs respectively. According to Section 277, "*whoever, voluntarily corrupts or fouls the water of any public spring or reservoir, so as to render it less fit for the purpose for which it is ordinarily used shall be punished with imprisonment for a term which may extend to 3 months, or with a fine which may extend to Rs. 500, or with both.*" Section 278 of IPC lays down that "*whoever voluntarily vitiates the atmosphere so as to make it noxious to the health of persons in dwelling or carrying on business in the neighborhood or passing along a public way shall be punished with fine which may extent to Rs. 500.*" Sections 284, 285 and 286 deal with negligent conduct with respect to poisonous substances, combustible matter and explosive substances. Sections 428 and 429 of IPC cover mischief to animals. Other concerned provisions in the IPC are a "negligent act likely to spread infection or diseases dangerous to life"[17] and making atmosphere noxious to health.[18] Thus, even in the pre-constitutional statues like Indian Penal Code, 1860 and the criminal sanctions contained in specific environmental legislation and municipal Acts contain provisions for criminal aspects relating to environment.

16 The Indian Penal Code, 1860 (Act 45 of 1860).

17 The Indian Penal Code, 1860 (Act 45 of 1860), s.270.

18 The Indian Penal Code, 1860 (Act 45 of 1860), s.278.

5.6. CONSTITUTION OF INDIA

The post-Independence period witnessed a lot of changes in the policies of the Indian government with respect to environment protection especially after the framing of the Constitution of India. The constitutional law plays very important role in protection and preservation of environment. In fact the space for environment-related statutes in the constitution is an outcome of the human urge for development with a social face.[19] The constitution of India when adopted did not contain any specific provisions on environment and even the word environment did not find a place in the Constitution but there were certain provisions in the constitution which to a great extent had direct bearing on the environment such as improvement of public health,[20] organization of agricultural and animal husbandry on modern and scientific lines[21] and protection of natural monuments from spoliation, disfigurement etc.[22] The provision contained in Article 47 of the Indian constitution is more important as it Provides that the state shall regard the raising of the level of nutrition and the standard of living of its people and improvement of public health as among its primary duties. Protection and improvement of environment is inherently included in the improvement of public health because without it public health cannot be assured. Through these provisions it is clearly reflected that the framers of our constitution were very much conscious about the environmental concern. Constitutional Amendment, 1976 of added special provisions relating to environment. 42nd amendment of the constitution was the outcome of The UN Conference on Human Environment held at Stockholm in 1972. A National Committee on Environmental Planning and Coordination (NCEPC) was established in the Department of Science and Technology in 1972 to make necessary preparations for the Conference. The Government of India took a number of steps to implement the decisions taken at the Conference by means of amendments to the Constitution, new legislations relating to environmental protection and creation of institutions for implementing the legislations. The Constitution of India was amended by the 42nd constitutional amendment and the subject of "ecology and environment" was incorporated for the first

[19] Md Zafar, M. N., (2000). Enviro-constitutional ethos in right duty discourse: Towards the creation of an equitable and sustainable socio-legal order. *International Journal of English and Literature, 1*, 61.

[20] Constitution of India, art. 47.

[21] Constitution of India, art. 48.

[22] Constitution of India, art. 49.

time through Articles 48A and 51A (g).[23] Thus, India took a bold step to include environmental protection rights and duties in its Constitution. By incorporating Article 48A in part IV of the Constitution, which contains the directive principles of state policy, the state has been given the constitutional mandate to protect and improve the environment and to safeguard the forest and wildlife of the country. Since the principles laid down in the part IV of the Constitution are fundamental in the governance of the country, therefore it has been now the constitutional duty of state to deal with the matters relating to environment, forest and wildlife of the country.[24]

5.7. LEGISLATIVE MANDATES FOR PROTECTION OF ENVIRONMENT AND CLIMATE CHANGE IN INDIA

The republic of India, being the world's second most populous country has strong influence on environment pollution and global energy consumption, thereby, become fourth largest GHG emitters in the world. India occupies 2.4% of the world Geographic areas, having nearly 17% of world population, which emits less than 7% of GHG emissions. Though its rank 4th in total emissions of greenhouse gases but it has been predicted to become third largest emitters in coming years. In view of this, India has great responsibility on shoulder to take urgent step to mitigate the impact of climate change at National and international level as well.

The Constitution of India under fundamental rights provides a regime of protection for privileged set of rights including right to life and liberty. The rights has over the period has been extended through judicial creativity to right to live with human dignity, the right to livelihood, and most importantly for current purposes, the right to enjoyment of pollution free water and air. However, no climate change related principles have been evolved by the apex courts. But the courts Cherished out the environmental right by integrating certain *Principles Like Polluter Pays Principles, Precautionary Principle, Intergenerational Equity Principles, Principle of Sustainable Development* and notion of the state as a Trustee of earth and its natural

[23] Constitution of India, art. 48A – "The State shall endeavor to protect and improve the environment and to safeguard the forests and wildlife of the country. Art. 51A (g) – It shall be the duty of every citizen of India- to protect and improve the natural environment including forests, lakes, rivers and wild life, and to have compassion for living creatures."

[24] Bhaskar, K. C., (2006). Environmentalism: Indian constitution and judiciary. 48 *Journal of Indian Law Institute*, *48*, 100.

resources.[25] Hence, the issue of Climate Change could well be related to the duty of a state as a Trustee to protect the atmosphere from greenhouse gases emission and to follow these principles in making the law and policies for mitigating the climate change. Besides, the citizen of this country has been assigned fundamental duty[26] to protect environment wildlife and natural resources which could also help in mitigating climate change.

In addition, India has wide array of laws, rules, regulations, plans and policies with administrative authorities to address the environmental pollution and energy conservation, but there are no specific laws on the climate change yet. Then, all such norms and principles could well utilize to address the reduction of greenhouse gases emission and other mitigation measures. Like EPA, 1986[27] empowers the central government to take all necessary measures for protecting and improving environment and preventing controlling and abetting environmental pollution. Section 3(1), of the Act empowers the Centre "to take all such measures as it deems necessary or expedient for the purpose of protecting and improving the quality of the environment and preventing, controlling, and abating environmental pollution." Section 7 of the EPA prohibits the discharge or emission of environmental pollutants in excess of the prescribed standards. To implement this mandate, the government has framed the Environment (Protection) Rules of 1986 (EPR). The standards are set out in the schedule appended to the EPR.[28] The Air (Prevention and Control of Pollution) Act, 1981[29] which defines 'air pollutant' any solid, liquid or gaseous substance including noise present in the atmosphere in such concentration as may be or tend to be injurious to human being or other living creatures or plants for environment. Such definition could also arguably cover for greenhouse gases for regulation and mitigation of climate change. The Forest Conservation

[25] Basically, the ancient Roman Empire developed this legal theory, i.e., Doctrine of the Public Trust. The Public Trust Doctrine primarily rests on the principle that certain resources like air, sea, waters and the forests have such a great importance to the people as a whole that it would be wholly unjustified to make them a subject of private ownership. The said resources being a gift of nature, they should be made freely available to everyone irrespective of the status in life. The doctrine enjoins upon the Government to protect the resources for the enjoyment of the general public rather than to permit their use for private ownership or commercial purposes.

[26] Article 51A(g) in The Constitution of India "(g) It shall be the duty of every citizen of India- to protect and improve the natural environment including forests, lakes, rivers and wild life, and to have compassion for living creatures."

[27] The Environment (Protection) Act, 1986 (Act 29 of 1986).

[28] See Rule 3 of EP Rules 1986.

[29] The Air (Prevention and Control of Pollution) Act, 1981 (Act 14 of 1981).

Act, 1980[30] restrict the conservation of forest land to non-forest use, which could be useful for reforestation. The Wildlife Protection Act, 1972[31] specifically provides the provision for protection and conservation of wildlife, birds and plants. It prohibits any person from picking, uprooting, damaging, destroying, acquiring for collecting any is specified plant from any forest land and area specified by notification by the central government. The Biological Diversity Act, 2002[32] and its rule addresses the issues concerning conservation, sustainable use and the access to genetic resources and associated traditional knowledge and equitable sharing of benefits arising out of the use of these resources and associated knowledge which support the purpose of mitigation of climate change. The National Green Tribunal Act, 2010 has also formed and empowered is specified Tribunal to hear the cases on environment degradation and climate change as well.

5.8. RELEVANT LEGISLATIONS

The Wildlife (Protection) Act, 1972:

First ever Act providing for the protection to wild animals, birds and plants introduced and enforced at the time when there was no discussion on the climate change and conservation of biodiversity drawing worldwide focus until 1992.[33]Hunting of wild animals specified in schedules made prohibited[34]unless the animal become dangerous to human life or, is as disabled or diseased as to be beyond recovery or is required for educational or research purpose. Further, this Act prohibits picking, uprooting, damaging, acquiring or collecting any specified plant[35] from any forest land unless it is permitted by the concerned authority for certain definite purposes. The Centre and the State government has been empowered under act that they may, by notification, declare any area as sanctuary[36] or park[37] if the area is of adequate ecological, faunal, floral, geo-morphological or of natural significance, for the purpose of protecting, propagating or developing

[30] The Forest (Conservation) Act, 1980 (Act 69 of 1980).

[31] The Wild Life (Protection) Act, 1972 (Act 53 of 1972).

[32] The Biological Diversity Act, 2002 (Act 18 of 2003).

[33] Ali, M., (2010). Climate Change and biodiversity: India's perspective and legal framework. *Journal of the Indian Law Institute,* 52(3, 4), 343–365.

[34] The Wild Life (Protection) Act, 1972 (Act 53 of 1972), s 9.

[35] The Wild Life (Protection) Act, 1972 (Act 53 of 1972), s 17 A.

[36] The Wild Life (Protection) Act, 1972 (Act 53 of 1972), s 18.

[37] The Wild Life (Protection) Act, 1972 (Act 53 of 1972), s 35.

wildlife or environment. There for the Act lays down for in-situ conversation as well as ex-situ conservation.

The Air (Prevention and Control of Pollution) Act, 1981:

The Air Act may be considered to be important for addressing climate change, as we know there is a direct link between air pollution and the climate change. It is in fact due to the emission of six greenhouses gases namely, nitrous oxides, carbon dioxide, CFCs, PFCs, methane and ground level ozone, that the infrared radiation emitted by the earth's surface is trapped resulting into raising earth temperature, and leading to global warming, hence climate change.[38]The only means to prevent climate change is to reduce emissions dramatically. The Act provides for prevention, control and abatement of air pollution with a broader approach including preservation of natural resources of the earth including the preservation of quality of air and establishment of air pollution control boards to carry out the objects of the Act. It says that "emission means any solid or liquid or gaseous substance coming out of any chimney, duct or flue or any other outlet." The Act also provide for the institutional and regulatory framework which would restrict the emission of gases from automobiles and that of industries.[39]This restriction of emissions would have impact on reduction of the greenhouse gas emissions as we know transportation and industrial emissions are two important sectors that have substantial contribution towards the greenhouse gases. The pollution control boards are authorized to lay down standards for emission of air pollutants into the atmosphere from industrial plants, automobiles or from any other source not being ship or an aircraft. The boards are also empowered to issue directions to the persons violating the law and also lodge complaints against such persons in a court of law.

The Forest (Conservation) Act, 1980:

Forests have been identified as carbon sink as well as richest source, amongst the natural resources, to be exploited for commercial gain. The Britishers in order to monopolize the control over forest, introduced the Forest Act, 1865, having right of ownership denying others and the power to protect the same. Such approach of protection failed because the local people lost

---

[38] Ministry of Environment & Forests, Ministry of Power, Bureau of Energy Efficiency, Government of India, India: Addressing Energy Security and Climate Change, envfor.nic.in/divisions/ccd/Addressing_CC_091007.pdf (accessed on 07 July 2023).

[39] The Air (Prevention and Control of Pollution) Act, 1981 (Act 14 of 1981), ss.20, 22.

interest in taking care of the forests as they were not the beneficiaries. Post-independence government continued the practice[40] until it was realized to have a law in the form of Forest (Conservation) Act in 1980 providing for the protection of forests. The underlining object of the law is to impose restrictions on use of forest land for non-forest purpose and to protect and conserve the green cover essential for the trapping of carbon dioxide as carbon sink.

The Environment Protection Act, 1986:

After the special legislations, i.e., aforesaid the Air Act, 1981 and the Water Act, 1974 dealing with the air and water pollution problems, it was found that there were still some areas left out of their contours and the environment cannot be protected due to activities which is not directly connected to water and air pollution, there were lack of provisions demanding some affirmative actions to improve the quality of the environment. The EPA 1986 is a comprehensive legislation which empower the central government to take measures to protect and improve the quality of environment by constituting authority/authorities[41] that was not possible under the previous legislations and this feature distinguishes it from other environmental laws. Further, the Act empowers[42] the central government for making rules to carry out the purposes of the Act. By invoking these provisions, the central government has constituted several authorities[43] with different term of references for lessening the pollution load and also suggest viable options to avoid the impact of restrictions imposed to protect and improve the environment. One of such authorities is the central empowered committee which was constituted in 2002 to recommend the clearance of big projects involving exploitation of natural resources like mining and forests. Under the rule making power, the central government, has made several rules and regulations including concern of biological diversity, which gives the Act a status of umbrella legislation. The Act, therefore, is highly significant in order to prevent and control the climate change.

---

[40] National Forest Policy, 1952, the first forest policy in independent India.

[41] The Environment (Protection) Act, 1986 (Act 29 of 1986), s. 3 (3).

[42] The Environment (Protection) Act, 1986 (Act 29 of 1986), s. 25.

[43] The Environment Pollution (Prevention and Control) Authority, Loss of Ecology (Prevention and Payment of Compensation) Authority, Environment Impact Assessment Authority, etc., (2010)

The Motor Vehicles Act, 1988:

This Act doesn't come under environmental law but also comes into play in reduction of emissions from automobiles as central government is provided with the power to makes rules for regulating emissions. Pursuant to this power the emission norms such as Euro-I, Euro-II have been introduced.[44] It says emission 'means any solid or liquid or gaseous substance coming out of any chimney, duct or flue or any other outlet.' The boards under this Act, are authorized to lay down standards for emission of air pollution into the atmosphere from industrial plants, automobiles or from any other source not being ship or an aircraft.

5.8.1. Energy Conservation Act, 2001

The Energy Conservation Act of 2001[45] laid the foundation for promoting energy efficiency by establishing the Bureau of Energy Efficiency (BEE) with the primary objective of reducing energy intensity. The Act empowered the government to set energy consumption standards for equipment and appliances, specify energy conservation building codes (ECBCs) for commercial buildings and setting up energy consumption norms and standards for consumers. It also establishes a compliance mechanism through energy audits to be conducted by accredited auditors. This was made mandatory for large consuming units – the units falling under the ambit of the Energy Conservation Act 2001, are 1. Thermal Power Stations, 2. Fertilizers, 3. Cement, 4. Iron and Steel, 5. Chlor- alkali, 6. Almninium.7. Ralways.8. Textiles and 9. Pulp and Paper. These units have to employ certified 'energy managers' and report energy consumption and energy conservation data annually.

5.8.2. The Electricity Act, 2003

The Electricity Act, 2003[46] was a landmark legislation for power sector reforms in India. It encouraged the development of renewable energy by mandating that State Electricity Regulatory Commissions (SERCs) promote renewable energy by allowing connectivity and sale of electricity to any interested person and specified that a certain percentage of the electricity consumption should be from renewables. Electricity Act 2003 requires

[44] Motor Vehicles Act, 1988 (Act 59 of 1988), s.110.

[45] The Energy Conservation Act, 2001 (Act 52 of 2001).

[46] The Electricity Act, 2003 (Act 36 of 2003).

SERCs to specify certain percentage of electricity that the Distribution Cost must procure from renewable sources preferential prices from renewables had been notified. Renewable energy will form part or our new energy recipe soon.

5.8.3. Establishment of National Green Tribunal (NGT)

The National Green Tribunal Act, 2010[47] was passed by our Parliament and the same day it was notified by the Government of India on 18th October 2010. The National Green Tribunal Act, 2010 provides that the central government shall by notification establish a Tribunal to be known as the National Green Tribunal.[48] In exercise of the powers conferred by section 3 of the Act, the central government established "The National Green Tribunal" to exercise the jurisdiction, powers and authority conferred on it by or under the Act. Thus, the National Green Tribunal came into existence and previous Acts providing tribunal and appellate authority were repealed.[49] The first information about the abolishment of the NEAA[50] was given by an innocuous little notice outside the authority's office on 18th October 2010.

The National Green Tribunal has been enacted to fulfill the long felt need for alternative forum to deliver speedy and inexpensive justice. The philosophy of PIL is echoed in this legislation because now a class of victims can knock the doors of this forum for redressal. The NGT Act provides one more innovation by providing strict penalty for non-observation of the orders of the tribunal. The tribunal is playing a crucial and vital role in combating the climate change.[51]

5.8.4. The Companies Act, 2013

The new Companies Act, 2013 has introduced concept of corporate social responsibility (CSR)[52]. This act has made mandatory for companies to spend

[47] The National Green Tribunal Act, 2010 (Act 19 of 2010).

[48] The National Green Tribunal Act, 2010 (Act 19 of 2010), s.3.

[49] Both the Acts, i.e., the NETA,1995 and the NEAA, 1997 were repealed by the National Green Tribunal Act, 2010

[50] The National Environment Appellate Authority Act, 1997 (Act 22 of 1997).

[51] Vinod, S. M., (2010). National green tribunal: Alternative environment dispute resolution mechanism. *Journal of the Indian Law Institute (Special Issue on Climate Change and Environmental Law), 52*(3, 4), 522–552.

[52] Corporate social responsibility is a way of describing how companies try to measure and control their impact on society. This includes a company's contributions -both negative and positive – to the economy, environment and greater community. In other words, it's a corporate approach to ensuring a company does the right thing more often than not.

2% of their profit on CSR. It has set stricter compliance and created new opportunities for business spending on environmental issues and clean energy. Companies has been advised to give preference to CSR activities that best align with the business. 'Environment Sustainability' is listed as one of the eligible CSR activities.[53] Sub Section 1 of Section 135 to comply with the provisions relevant to CSR.[54] This provision requires that companies having a net worth of Rs. 500 crore or more or a turnover of Rs. 1,000 crore or a net profit of 5 crore must spend 2% of net profit on CSR activities. A schedule in the Corporate Social Responsibility Policy) Rules, 2014 lists various activities companies can undertake as CSR. Companies should take measures to check and prevent pollution; recycle, manage and reduce waste, manage natural resources in a sustainable manner and ensure optimal use of resources like land and water, proactively respond to the challenges of climate change by adopting cleaner production methods, promoting efficient use of energy and environment friendly technologies.[55]

Thus, it can be said that this move makes India one of the countries in the world to mandate spending on social welfare as part of company activity by law. The Climate Group is of the view that, this new provision is facilitating an enabling environment to develop longer term strategies to address some of the most persistent economic, social and environmental issues facing Indian society.

5.8.5. The Climate Change Bill, 2015

The Climate Change Bill, 2015[56] was presented by Kalikesh Narayan Singh Deo on November 27, 2014 in New Delhi, section 2 of the bill defines various terms like annual range for the net carbon account, carbon budget, and carbon unit, net carbon account targeted greenhouse gases. Chapter II of the bill deals with carbon target setting. Chapter III and chapter IV of the bill deals with carbon budgeting, national committee on climate change

---

[53] http://www.theclimategroup.org/what-we-do/news-and-blogs/indian-businesses-must-spend-2-oftheir-earnings-on-csr/a (accessed on 07 July 2023).

[54] Megha, K., *India: Corporate Social Responsibility: Mandating Companies to Contribute Towards Society*. http://www.mondaq.com/india/x/305620/Corporate+Commercial+Law/Corporate+ Social+Responsibility+Mandating+Companies+To+Contribute+Towards+Society (accessed on 07 July 2023).

[55] Lok Sabha Secretariat, Parliament Library and Reference, Research, Documentation and Information Service (LARRDIS) Corporate Social Responsibility, http://164.100.47.134/intranet/CorporateSociaResponsbility.pdf (accessed on 07 July 2023).

[56] The Climate Change Bill, 2015 (Bill 23 of 2015).

respectively. Chapter V discusses carbon trading scheme. This bill contains 19 section. Till date this bill has not been passed by parliament.[57]

5.9. INDIA'S CLIMATE CHANGE STRATEGY

As a non-annexe one party to the UNFCCC and the Kyoto protocol, earlier India had no legal obligation to reduce it GHG emissions. Instead, India has long argued that the historical responsibilities to deal with climate change lies only with the developed countries as they were the creators of the problem and that they must take on a leadership role in addressing climate change. The overriding priority of India (and other developing countries) is to address development and poverty eradicates and any action to reduce emissions should only be voluntary not mandatory. This principle, commonly known as the principle of *Common but Differentiated Responsibilities*, is one that India has long been an advocate of. The firewall created the principle between developed and developing countries shielding India from undertaking any significant action to reduce emissions.[58]

India's domestic approach to climate change was largely mirrored its multilateral stance. Domestic policies per se did not exist and climate change was address only in a piecemeal manner, mostly through energy and forestry policies in which climate was never the central focus. The main thrust of these policies for social and economic development. Climate change was mentioned, if at all, only as a "co benefit" of policy.

India has also record of plans and policies for addressing the mitigation mechanism for climate change. The National Climate Change Council issued India's comprehensive National Action Plan in June, 2008 which called as National Action Plan on Climate Change. There have been 8 missions announced under this action plan for combatting the climate change. National Solar Mission seek to deploy 20,000 megawatt of solar electricity capacity in India by 2020. National Mission for Enhanced Energy and Efficiency creates new institutional mechanism to enable the development and strengthening of Energy Efficiency markets for lowering the electricity consumption. National Mission on Sustainable Habitat under NAPCC promotes the introduction of sustainable transport, Energy efficient building and sustainable waste management in cities. National Water Mission

[57] *The Climate Change Bill, 2015* (Bill 23 of 2015), http://164.100.47.4/billstexts/lsbilltexts/asintroduced/4367LS.pdf (accessed on 07 July 2023).

[58] Raghunandan, D., (2019). India in international climate negotiations: Chequerd trajectory. In: Navroz, K. D., (ed.), *India in a Warming World: Integrating Climate Change and Development* (pp. 187-189). New Delhi: Oxford University Press.

promotes the integrated management of water resources. National Mission for Green India (NMGI) seek to afforestation in additional 10 million hectare of forest land in India, water land and community lands. National Mission on Sustainable Agriculture focuses on enhancing productivity and resistance of agriculture so as to reduce vulnerability to extremes of weather condition, long day spells, flooding and variable moisture availability. National Mission on Strategic Knowledge for Climate Change identifies challenges arising from climate change promotes the development and diffusion of knowledge on responses to the challenges in the area of health, demography, migration and livelihood of coastal communities. Thus, these different missions, in addition with 'Seven Policy Thrust Areas' together encompasses action for greenhouse gases mitigation as well as adaptation as well.

Besides NAPCC the National Forest policy (1988) has been pioneer for stability of environment and ecological balance to boost the climate mitigation. There is also Joint Forest Management (1996) which enables the local communities for forest conservation and management. The National Afforestation and Eco-Development Board (1992) was set up for focusing on afforestation in forest and adjoining lands. The National Environment Policy (2006) also promotes the betterment of the environment. The National Biodiversity Action Plan (2008) supports the conservation of biological resources in the country to enhance the environment balance and stability. Apart from this, there are different supportive policies like New and Renewable Energy Policy (2005) which promotes dependence on sustainable renewable energy sources, accelerated deployment of renewable energy instrument and technology. The Rural Electrification Policy (2006) promotes renewable energy with great connectivity and cost-effective. The Electricity Tariff Policy (2003) promotes tax Reform to promote optimal fuel choices. The Energy Conservation Building Code (2007) was initiated to optimize new building energy demand these are not least, but major legislation and policies in India to combat the climate changes are yet to come in the coming days.[59]

5.9.1. Five Year Plans

Many policies developed during this period that had a climate "co-benefit "originated in India's five-year plans. Five-year plans, must like in other countries, lay the foundation for economic planning and development for the country. In India's case, its five-year plans are developed by its planning

[59] Usha, T., (2016). Mitigation of climate change. In: Usha, T., (ed.), *Climate Change Law Policy and Governance* (p. 91). Lucknow, India: Eastern Book Company, Lucknow.

commission. Environmental concerns were addressed for the first time in the sixth five-year plan (1980 to 1985)[60] and one of the primary objectives of the ninth five-year plan (1997 to 2002) was to "ensure environmental sustainability of the development process through social mobilization and participation of people."[61] Climate change was given due important for the first time as late as eleventh five-year plan (2007 to 2012). Most of the objectives in the eleventh five-year plan was driven by the need to build energy resources to meet the growing energy demand and ensuring India's economic growth. Objectives included reducing energy intensity by 20% by 2016 to 2017, building additional capacity for nuclear power and renewable energy, and the plan also set the specific objective of increasing forest cover by 5%.[62] Twelfth five-year plan aimed to Promotion and adoption of cleaner technology, strengthening and initiation of reforms in regulations, policy making and enforcement institutions for environmental governance. And also, to reduce emission intensity of our GDP in line with the target of 20 to 25% reduction over 2005 levels by 2020.[63]

5.9.2. Energy Audit of Large Industrial Consumers

The National Electricity Policy, 2005 called for increased participation by the private sector to exploit non-conventional energy resources and allowed for renewable energy procurement through a competitive bidding process. It also called for a more level playing field between non-conventional technologies and conventional ones by allowing for differential tariffs. To further encourage renewable energy development, the National Tariff Policy, 2006 stipulated the SERCs to purchase a minimum percentage of power from renewable sources.

5.9.3. National Environment Policy 2006

India is committed to the principle of common but differentiated responsibility of countries in relation to climate change. This is clear

[60] *8th Five Year Plan,* (Vol. 2). https://niti.gov.in/planningcommission.gov.in/docs/plans/planrel/fiveyr/8th/vol2/8vol2.html (accessed on 07 July 2023).

[61] *9th Five Year Plan,* (Vol. 1). https://niti.gov.in/planningcommission.gov.in/docs/plans/planrel/fiveyr/9th/vol1/v1c1-2.htm (accessed on 07 July 2023).

[62] *Eleventh Five Year Plan,* (2007–12), https://niti.gov.in/planningcommission.gov.in/docs/plans/planrel/fiveyr/11th/11_v1/11th_vol1.pdf (accessed on 07 July 2023).

[63] *Twelfth Five Year Plan (2012–2017) Faster, More Inclusive and Sustainable Growth Volume I,* https://niti.gov.in/planningcommission.gov.in/docs/plans/planrel/12thplan/pdf/12fyp_vol1.pdf (accessed on 07 July 2023).

from the Draft Environmental Policy statement issued by the Ministry of Environment and forests (MoEF) in 2004. "Another major set of challenges arise from emerging global environmental concerns such as climate change, stratospheric ozone depletion, and biodiversity loss. The key is to operationalize the principle of common but differentiated responsibility of countries in relation to these problems. Multilateral regimes and programs responding to these global environmental issues must not adversely impact the development opportunities of developing countries. Further, the sharing of global natural resources must proceed only on the basis **of** equal sharing per-capita across all countries"[64] NEP INCLUDES identification of key vulnerabilities of India to Climate Change, in particular impacts on water resources, forests, coastal areas, agriculture and health, assessment of the need for adaptation to Climate Change and encouragement to the Indian Industry to participate in the Clean Development Mechanism (CDM).[65]

5.9.4. Energy Conservation Building Code, 2006 (ECBC)

The ECBC 2006 had been developed by the International Institute for Energy Conservation (IIEC) under contract with the United States Agency for International Development (USAID) as a part of the Energy Conservation and Commercialization (ECO) Project providing support to Energy Efficiency Action Plan. The ECBC prescribes standards to be followed in the construction of buildings especially dealing with heating, air conditioning, lighting and ventilation. This is in an effort to conserve energy. This code came into effect from March 2007 which addresses the design of new, large commercial buildings to optimize their energy demand. The compliance with this code has also been incorporated into environmental impact assessment requirement for large buildings.

5.9.5. Introduction of Labeling Program for Electrical Appliances

This program was launched in 2006 and comparative star-based labeling introduced for fluorescent tube- lights, Air conditioners, Distribution Transformers and Refrigerators.

[64] "The Draft Environmental Policy Statement," Ministry of Environment and Forest, Government of India (2010).

[65] http://164.100.47.134/intranet/CLIMATE_CHANGE-INDIA's_PERSPECTIVE.pdf (accessed on 07 July 2023).

5.9.6. Integrated Energy Policy (2006)

The integrated energy policy 2006[66] was an effort of the Planning Commission to address the country's growing energy demands. The Integrated Energy Policy provides a comprehensive framework for energy policy in India. The policy's vision is to "to reliably meet the demand for energy service of all sectors including the lifeline energy needs of vulnerable households in all parts of the country with safe, clean and convenient energy at the least –cost." The policy states that India should pursue all available fuel options and forms of energy and that coal will remain the mainstay of the Indian economy till 2031 – 2032. In terms of addressing climate change, the integrated energy policy makes recommendations that include power sector reforms, ramping up mass transit, increasing nuclear power and renewables, and highlighting energy efficiency in all sectors.[67]

5.9.7. Prime Minister's Council on Climate Change

A high-Level advisory group on climate change was constituted in June 2007 and reconstituted in November 2014 with the following objectives:

i. To coordinate national action plans for assessment, adaptation and mitigation of climate change.

ii. To advise government on pro-active measures that can be taken by India to deal with the challenge of climate change.

iii. To facilitate inter-ministerial coordination and guide policy in relevant areas.

Most of the legislative and policy initiatives described in the section were driven primarily by India's development imperatives- a need to exploit all potential energy resources to ensure that there is an adequate energy supply and to continue to spur economic growth. It was only in the late 2000s, due to various factors most notably the growing international pressure to address climate that led India to develop a climate – specific policy framework.

5.9.8. Late 2000s Onwards: India's Growing Status as a Major Emerging Economy

The 21st century witnessed the arrival of India on the world stage as a global power. With the Indian economy growing at a phenomenal rate of 8% a year, India was no longer viewed solely as a "developing country" in the world. In

[66] Planning Commission, Integrated Energy Policy, (2006).

[67] Deepa, B. *India's Integrated Energy Policy: A Source of Economic Nirvana or Environmental Disaster?* https://elr.info/sites/default/files/articles/40.10706.pdf (accessed on 07 July 2023).

fact, during the 2007 financial crisis[68] both India and China were considered as critical partners in the G20 talks that helped avert a global meltdown due to their growing economies. This acknowledgment of India's status as a major emerging economy conveyed with it more responsibility. To be recognized as a legitimate global player the world wanted to see India as a country willing to shoulder responsibilities. This expectation was reflected within the multilateral climate negotiations where India started to feel international pressure to be more proactive and take action to reduce its own GHG emissions. The shift in India's stance on taking action domestically and launching the National Action Plan on Climate Change (NAPCC) can be attributed to a large extent to this growing pressure internationally. Facing the fear of isolation and continued international pressure, prior to the much-hyped Copenhagen Summit in 2009, India further announced that it would voluntarily reduce its emissions intensity between 20 and 25% below 2005 levels by 2020.[69]

While international diplomatic pressure played a very critical role that led to India as more proactive stance, one cannot displace the importance that domestic priorities played in this context as well. A realization by the Indian government that India is extremely vulnerable to climate change impacts meant that it was in India's personal interest to take action both domestically and internationally. With its accelerating energy demands and India importing more than 70% of its oil, it made sense for India to privatize energy security and focus on Energy efficiency and renewable energy. This is reflected in specific missions of the NAPCC. Energy access, another domestic concern, could put India's development at risk. The Indian governments focused on developing renewable energy helps it to achieve this goal.[70]

5.9.9. National Action Plan on Climate Change

The National Action Plan on Climate Change (NAPCC) released by Prime Minister Manmohan Singh in 2008 marked a watershed moment in India's

---

[68] Financial crisis of 2007-08, also called subprime mortgage crisis, severe contraction of liquidity in global financial markets that originated in the United States as a result of the collapse of the U.S. housing market.

[69] See Aditya, V. P., & Navroz, K. D., (2021). The limits of opportunism: The uneven emergence of climate institutions In India. *Environmental Politics*. https://www.tandfonline.com/doi/epub/10.1080/09644016.2021.1933800?needAccess=true (accessed on 07 July 2023).

[70] Patodia, R., (2013). India's evolving climate change strategy. In: Erkki, J. H., et al., (eds.), *Climate Change and the Law* (pp. 610, 611). Dordrecht: Springer.

domestic engagement on the climate issues. Consisting of eight national missions, the NAPCC is a package of existing and planned initiatives, policies and programs focused both on adaptation and mitigation. While the Plan emphasized India's development objectives, it established for the first time, a concrete framework to address climate specifically in the domestic context.

The eight Mission comprehensively address mitigation, adaptation as well as research and development (R&D). The Mission are the National Solar Mission, National Mission on Sustainable Habitat, National Mission for Sustaining the Himalayan Ecosystem, National Water Mission, the National Mission on Enhanced Energy Efficiency, National Mission for a green India, National Mission for sustainable Agriculture and National Mission on Strategic Knowledge for Climate Change. In early 2012, the Government of India announced its plans to establish an additional mission on clean coal technologies. These Mission are to be institutionalized through inter-sectoral groups, consisting of members from the relevant ministries, civil society, industry and academia. Detailed plans that include targets, timelines and objectives are to be developed by these groups and submitted to the Prime Minister's Council on Climate Change.

Some of the missions build on existing legislation or policies, like the National Mission on Sustainable Habitat or the GIM, while other like the Solar Mission, chart a new and ambitious path by the government to access a previously untapped energy source. Considering the significant progress made by the National Mission on Enhanced Energy Efficiency (NMEEE) and the National Solar Mission, this section provides an overview of the achievements specifically of these two Missions.

5.9.10. National Clean Energy Fund (NCEF)

The Government of India created has the National Clean Energy Fund (NCEF) in 2010 for the purpose of financing and promoting clean energy initiatives and funding research in the area of clean energy in the country. The corpus of the fund has been built by levying a cess of INR 50 (subsequently increased to INR 100 in 2014, it further increased to Rs.400 per ton in 2016) per ton of coal produced domestically or imported. Now this fund has been renamed as National Clean Energy and Environment Fund (NCEEF).[71]

[71] *National Clean Energy and Environment Fund (NCEEF).* https://www.indiafilings.com/learn/national-clean-energy-and-environment-fund-nceef/ (accessed on 07 July 2023).

In 2014, the Government of India expanded the scope of the NCEF to include financing and promoting clean environment initiatives and funding research in the area of clean environment. To finance these additional initiatives, the Clean Energy Cess has been increased from INR 50 per ton to INR 100 per ton of coal.

5.9.11. National Adaptation Fund (NAF)

On July 10, 2014 the Honorable Finance Minister of India announced an allocation of INR 100 crore towards a newly established NAF. This fund was created to assist national and state level activities to meet the cost of adaptation measures in areas that are particularly vulnerable to the adverse effects of climate change.

5.9.12. National Mission on Enhanced Energy Efficiency – Perform Achieve and Trade Scheme

The National Mission on Enhanced Energy Efficiency was mandated to implement four new initiatives Under the NAPCC; the Perform, Achieve and trade (PAT) Scheme; Market Transformation for Energy Efficiency; Energy Efficiency Financing Platform; Framework for Energy Efficiency Economic Development.[72] While the latter three are still in the early stages of implementations, the Perform, Achieve and Trade scheme (PAT Scheme) was recently launched by the Government of India.[73] The PAT scheme is an innovative market mechanism for trading energy efficiency certificates in energy-intensive sector. After many delays mostly due to industry pushback and legal holdups, the PAT scheme was finally rolled-out in April 2012 making India the first developing countries country to implement a market-based mechanism.

The PAT scheme is being implemented in three phases – the first phase covers 478 facilities from eight energy-intensive sectors, namely aluminum, cement, chlor-alkali, fertilizer, iron and steel, pulp and paper, textiles and thermal power plants. The governments expect the scheme to deliver reductions of about 100 million tons of CO_2 annually by the end of its first phase. The Energy Conservation Act 2001, the legal framework on which the PAT scheme is based, was amended by the Parliament in 2010 to allow

---

[72] *National Mission for Enhanced Energy Efficiency*. http://moef.gov.in/wp-content/uploads/2018/01/Mission-SAPCC-NMEEE.pdf (accessed on 07 July 2023).

[73] *Energy Conservation Rules, 2012 (PAT Rules)*. https://beeindia.gov.in/sites/default/files/PAT%20Rules%202012.pdf (accessed on 07 July 2023).

for the establishment of the scheme. The PAT scheme is a baseline-credit scheme that allows facilities to trade certificates to meet their compliance requirements and simultaneously reduce costs. Each facility has a specific energy consumption target (a reduction in energy consumption from the facility's baseline) with less energy efficient facilities having a greater reduction target compared to more efficient ones. A facility's baseline is based on its historic specific energy consumption over the period 2007–2010. Facilities that make greater reductions than their target will be receive "Energy Saving Certificates" (EsCerts) which can be traded with other facilities that have difficulty meeting their target or bank them for use in a subsequent phase. Those facilities that are unable to meet their targets must buy EsCerts or pay a penalty. One Energy Saving Certificate is equivalent to 1 tons of oil equivalent, an energy consumption measure rather than a carbon reduction measure.

The first phase extends over a 3-year period (2012–2015) and covered facilities are expected to meet their target by the end of the first phase. Monitoring and verification will be conducted by auditors at the end of the first phase and Energy Saving Certificates will be issued ex-post.

While details of the subsequent phases of the PAT scheme are still being fleshed out, early signs have hinted at the possibility of broadening the scheme to include other energy-intensive sectors like petroleum refineries, petrochemicals, chemicals etc., and further tightening the targets.

5.9.13. Solar Mission

The then Prime Minister Manmohan Singh had emphasized the importance of the Indian economy to gradually shift away from a fossil fuels-based economy to ones more dependent on non-fossil fuels and renewable sources of energy. This principle is reflected in the Solar Mission,[74] one of the most ambitions Mission launched by the government. The mission sets a goal of generating 20 GW grid-connected solar power plants by 2022, a several thousand-fold increase from current levels. While extremely ambitions, the Government of India has already demonstrated some early success in getting this Mission off the ground.

74 *Renewable Policy Framework and Wind Energy Program in India.* https://mnre.gov.in/img/documents/uploads/94e402c36ee44fe29e2b96a6b1b69a30.pdf (accessed on 07 July 2023).

5.9.14. Expert Group on Low Carbon Strategies

Recognizing the importance of transitioning to a low carbon economy, the Indian government set up an "Expert Group on Low Carbon Strategies for Inclusive Growth" consisting of experts from various ministries, civil society and the private sector. The Expert Group was based on the Planning Commission Recommendations from the Expert Group's work are expected to feed into India's Twelfth Five Year Plan (2013–2017). The group submitted its interim report in May 2011. The report, while interim, gives a flavor of the range of actions to increase investment in renewable technologies, reduce losses from transmission and distribution from the power sector; adopt super-critical technologies in coal based thermal power generation; get iron, steel and cement sectors to adopt best-available technology and; to increase the share of rail overall freight transport. The group submitted its final report in April 2014.[75] The Expert Group on Low Carbon Strategies for Inclusive Growth evolved a macro-model to fully elucidate the inter-sectoral implications of different mitigation measures and also ensured that the low carbon strategies being recommended are mutually consistent with each other. The model output was summarized through two endpoint scenarios: BIG (Baseline, Inclusive Growth) and LCIG (Low Carbon, Inclusive Growth). The report described these models and the energy efficiencies of various sectors in detail.

5.9.15. Engagement at the State Level

Soon after the launch of the NAPCC in August 2009, Prime Minister Manmohan Singh addressed state environment ministers and urged them to develop State Action Plans on Climate Change (SAPCC). Following this, the MoEF provided additional guidance by providing states with a common framework for developing these plans. The states Plans are to be prepared under the auspices of the NAPCC and the intention is to have a top-down approach where national and state actions are "harmonized."

Responses by states have been mixed with some states making climate change a priority while others have yet to submit their action plans. As can be expected, most of these sub-national plans are driven by each state's assessment of their vulnerabilities and opportunities, as well as their own development agenda.

[75] Planning Commission Government of India, *The Final Report of the Expert Group on Low Carbon Strategies for Inclusive Growth*. https://cstep.in/drupal/sites/default/files/2019-02/CSTEP_Low_Carbon_Strategies_for_Inclusive_Growth_Report_2014.pdf (accessed on 07 July 2023).

5.10. SOCIAL PROTECTION PROGRAMS AND CLIMATE CHANGE IN INDIA

Country experiences of a diverse range of social protection instruments weather-indexed crop insurance, employment guarantee schemes, asset transfers and cash transfers reveal how measures can enhance the resilience of vulnerable communities. These point to ways in which the design of social protection measures could he strengthened taking into account current and potential future climate related shocks.

Finding work in the urban areas is particularly challenging for excluded People who may face discrimination from employees or lack supportive social networks in their destination community. There are also few opportunities for off-farm employment in rural areas. A legislative guarantee of employment can help build the resilience of these people to the impacts of climate change. The National Rural Employment Guarantee Act (NAREGA) of India was enacted on 2005 in two hundred of the poorest rural districts. In section 3(1), it guarantees 100 days of employment to the rural poor. Wages are fixed at the State minimum wage[76]. Program can support climate-smart development. In Schedule (I) it considers significant labor toward appropriate adaptive works, including water conservation, catchment protection, and plantations. It provides funds for tools and other items necessary to complete activities and technical Support for designing and implementing the projects[77]. It can thus become a core part of village development through productive, climate-resilient asset creation and maintenance.

Employment guarantee schemes can have gender equity objectives, and target women and female headed households, however, critics have questioned the implications for women's workloads. Where the work involves heavy manual labor, this can exclude highly vulnerable people such as the elderly and disabled. Other forms of social protection would be needed for the labor constrained poor and further evaluation of the impact of employment guarantee schemes on vulnerable and excluded people is required. There is also evidence that employment guarantee schemes may not be efficient relative to cash transfers. NREGA, by encouraging works on water harvesting, flood protection, afforestation and plantation, helps to insulate local communities from adverse effects of climate change. There is an increase in groundwater level and in water percolation, and improvement

[76] The National Rural Employment Guarantee Act (NAREGA), 2005 (Act 42of 2005) s.6.

[77] The National Rural Employment Guarantee Act (NAREGA), 2005 (Act 42 of 2005), s.20-24.

in soil fertility leading to improved land productivity[78]. This Act aims to improve the livelihood security of rural households by providing at least one hundred days of guaranteed wage employment in every financial year to a household whose adult members volunteer to do unskilled labor. Components of the work program aim to reduce vulnerability to climate risks (water conservation, drought proofing and flood protection) as well as increasing agricultural productivity and food security levels.

Some other major schemes/policies significantly addressing adaptation objectives are as follows:

a. Swarnajayanti Gram Swarozgar Yojana (Rural Self-Employment Program)[79]
b. Sampoorna Grameen Rozgar Yojana (comprehensive rural employment scheme)
c. Pradhan Mantri Gram Sadak Yojana (Prime ministers rural roads program)
d. National Rural Health Mission.
e. National Rural Drinking Water Program.
f. Desert Development Program.
g. Major and Medium Irrigation.
h. Sustainability of Dryland/Rainfed Farming System.
i. Disaster Management.

NREGS has become one of the largest rights-based social protection initiatives in the world, reaching around 52.5 million households[80]. Additionally, India recently launched the social insurance scheme Swasthya Bima Yojana (RSBY) RSBY uses Smart card technology to enroll beneficiary. A biometric smart card is given to each family below the poverty line in the informal economy, which enables the members to avail themselves of benefit under the scheme throughout India.

---

[78] Amita, S., (2010). *Right Based Legal Guarantee as A Social Protection Framework: The Mahatma Gandhi National Rural Employment Guarantee Act*, 5. Asian Development Bank. https://www.scribd.com/document/29955915/Rights-based-Legal-Guarantee-as-Social-Protection-Framework-A-Discussion-on-the-Mahatma-Gandhi-National-Rural-Employment-Guarantee-Act-MGNREGA-India (accessed on 07 July 2023).

[79] The objective of SGSY is to provide sustainable income to the rural poor. The program aims at establishing a large number of micro-enterprises in the rural areas, based upon the potential of the rural poor.

[80] Report of the Advisory Group chaired by Michelle Bachelet Convened by the ILO with the collaboration of the WHO "Social Protection Floor for a Fair and Inclusive Globalization.15 International Labor Organization (2011).

By exploring the relationship between climate change adaptation, disaster risk reduction (DRR) and social protection, the Institute of Development Studies (IDS) researchers have developed the concept of "adaptive social protection (ASP)."[81] In Indian framework, this study uses the ASP framework to identify the opportunities that MGNREGA as a development program holds for integrating climate change adaptation and the challenges it is confronted with. The opportunities and challenges will be discussed simultaneously as they are intricately linked. The ASP framework combines the components of social protection, DRR and climate change adaptation and considers four different dimensions – protective, preventive, promotive and transformative. The first two measures denote coping strategies whilst the remaining two signify building adaptive capacity.[82]

5.11. JUDICIAL APPROACH TO CLIMATE CHANGE IN INDIA

Although climate change concerns have still to form the core subject matter of a dispute before the Indian Courts, there is potential, in particular, given the filtering through of climate concerns to the courts, for the increasing use of litigation to further climate goals. Climate litigation, as is apparent from jurisdictions such as the US and Australia where climate litigation is pervasive, can take many forms.[83] In India too, many hooks exist for climate litigation in public and private law. The greatest potential for climate litigation in Indian court lies in rights-based climate claims. This is not only because there is a rich culture of judicial activism and PIL in India but also because this is complemented by an expansive indigenously-developed rights jurisprudence.

The Indian judiciary has been very conscious of the need to protect and preserve our environment. The Apex court has held that the right to a clean and healthy environment is a fleet of the right to life enshrined under Article

[81] See Davies, M., Bruce, G., et al., (2008). Adaptive social protection: Synergies for poverty reduction. *Institute of Development Studies, 39*(4), 105–111. https://citeseerx.ist.psu.edu/viewdoc/download?doi=10.1.1.847.5969&rep=rep1&type=pdf (accessed on 07 July 2023).

[82] Younes, C., (2016). The role of social protection mechanism in climate change mitigation in India. In: Usha, T., (ed.), *Climate Change Law Policy and Governance* (p. 24). Lucknow, India: Eastern Book Company.

[83] See Michael, B. G., & Cullen, H. J., (2012). *Climate Change Litigation in the U.S.* http://www.climatecasechart.com/ (accessed on 07 July 2023); Michael B. G., & Cullen, H. J., (2010). *Non U.S. Climate Change Litigation Chart.* http://www.law.columbia.edu/null/download?&exclusive=filemgr.download&file_id=163021 (accessed on 07 July 2023).

21 of' the Constitution On various occasions. The Court has taken cognizance of the hazards of illegal and indiscriminate mining of the iron ore, limestone, sand, etc.[84] discharge of toxic effluents into water bodies[85]. The deteriorating quality of air, damage done to the ecology on account of the industrial construction activity etc. In all these cases, the Hon'ble Supreme court emphasized that development of industry is essential for the economy of the country, but at the same time, the environment and ecosystem have to be protected and that '*sustainable development*' is the answer. These decisions are quite well known and therefore, the authors propose to highlight only those decisions where the dangers of overall climate change, in the context of global warming, are taken note of.

5.11.1. Judicial Activism and Public Interest Litigation

The Indian judiciary, unlike other judicial system is a dynamic actor and plays important role in shaping law, evolves policy, and plays a central determinative role in the governance of modern India. The Court plays this role mainly through the exercise of its self-fashioned public interest jurisdiction. Under PIL the Court is most likely to exercise in a rights-based climate change claim.[86]

The origins of public interest jurisdiction in India were in late 1970s, early 1980s, and in particular the case of *S. P . Gupta v. Union of India*[87] in this case Justice Bhagwati[88] relaxed the rule of *locus standi*, and opened up the doors of the Supreme Court for public-spirited citizens.[89] Those who were wishing to espouse the cause of the poor and oppressed (representative standing) and those wishing to enforce performance of public duties (citizen

---

[84] See decisions of the Hon'ble Supreme Court in *Goa foundation v. union of India* (2014) 6 SCC 590: *R.L. & E Kendra, Deharadun V. State of U.P* (1989) sup 1 SCC537: *M.C Mehta v. Union of India* (2004) 12 SCC 118.

[85] See *Vellore Citizen's Welfare Forum* v. *Union of India,* (1996). 5 SCC 647 wherein the Hon'ble Supreme court imposed a find on tanneries discharging untreated effluent in agricultural fields, road-sides, Waterways and open lands.

[86] Pritam, K. G., (2013). Judicial activism and public interest litigation In India. *Galgotias Journal of Legal Studies, 1*, 77–96.

[87] AIR 1982 SC 149 popularly known as "Judges Transfer Case."

[88] Prafullachandra Natwarlal Bhagwati (21 December 1921 – 15 June 2017) was the 17th Chief Justice of India, serving from 12 July 1985 until his retirement on 20 December 1986. He introduced the concepts of public interest litigation and absolute liability in India, and for this reason is held, along with Justice V. R. Krishna Iyer, to be a pioneer of judicial activism in the country.

[89] Public spirited individuals like M. C. Mehta, T. N. Godaverman, Almitra Patel have filed a number of petitions to protect and preserve wholesome environment.

standing).[90] PIL in India can be filed either in the High Court or Supreme Court. If the grievance is of a relating to legal wrong, Article 226 of the Constitution permits recourse to the High Court of the State. If the complaint is relating to a violation of fundamental rights, Article 32 of the Constitution permits direct recourse to the Supreme Court. For violations of fundamental rights, the Supreme Court may issue an order, direction or writ, including a writ in the nature of *habeas corpus, quo warranto, mandamus, prohibition* or *certiorari*.[91] The High Courts of any state can pass similar orders for enforcement of fundamental rights as well as other legal rights.[92]

At the request of public-spirited citizens, the Courts have passed and continue to pass orders in a range of cases. In the environmental field the Supreme Court, for instance, has passed hundreds of orders inter alia to protect the Taj Mahal from corrosive air pollution,[93] *M. C Mehta v. Union of India*, (popularly known as Ganga Water Pollution Case or Kanpur Tanneries Case) a PIL was filed, for the issuance of directions restraining the tanneries from discharging trade effluent into the river Ganga till such time they put up necessary treatment plant for treating the trade effluents in order arrest the pollution of water in the said river. The Supreme Court observed:

"*We are therefore issuing the direction for the closure of those tanneries which have failed to take minimum step required for the primary treatment of industrial effluent. We are conscious that closure of tanneries brings unemployment, loss of revenue, but life health and ecology have greater importance to people.*"

In *T. N Godaverman Thirumalkapad v. Union of India*[94], (popularly known as forest conservation case), the Supreme Court issued interim direction that all the on-going activities within any forest in any State throughout the country without the permission of central government must be stopped forthwith. Running of saw mills including veneer or ply wood mills within the forest was also stopped. Felling of trees in other State of Arunachal Pradesh was totally banned in certain forest whereas in other forest it was suspended in accordance with the working plan of the state government. Movement of cut trees and timber from any of the seven North-

[90] *S. P. Gupta* v. *Union of India*, (1981). Supp SCC 87, at 233.

[91] The Constitution of India, 1950, art. 32.

[92] See Sarbani, S., (2012). *Public Interest Litigation in India: Implications for Law and Development* (p. 7–9). Kolkata: Maharaniban Culcutta Research Group.

[93] *M. C. Mehta v. Union of India* (Ganga Pollution Case), Writ Petition Number 3727 of 1985, AIR 1988 SC 1115.

[94] (1997) 2 SCC 267.

Eastern State to any other state was completely banned. The court issued direction to stop felling of trees in other states such as the state of J&K, Himachal Pradesh and Tamil Nadu, with a view to protect and preserve the forest.

In *T. N Godaverman Thirumalkapad v. Union of India*[95], the Supreme Court has once again emphasized on the principle of sustainable development. The court held that forest policy, 1998, which has a statutory flavor, dictates that deprivation of economic benefit must be subordinate to ensuring environmental stability and maintenance of ecological balance. Non-fulfillment of this principle would be violative of Article 14 and 21 of the Indian constitution.

*Indian Council for Enviro-Legal Action v. Union of Ind*ia[96], is a monumental judgment on environment protection and sustainable development. In this case a PIL was filed alleging environmental pollution caused by private industrial unit. The PIL was filed not for the issuance of writ order direction but against the union of India, State Government and state pollution board concerned to compel them to perform their statutory duties on ground and that their failure to carry out their duties violated the right to life of citizens under Article 21 of the constitution.

In this case the industrial units were located in Bichhri village in Udaipur. They were producing certain chemical like oleum and H-Acid etc. They had not obtained the necessary clearances. Nor did they install any equipment for treatment of highly toxic effluents discharged by them. The highly toxic effluent of these industries percolated deep in to the bowls of earth polluting the ground water and making it unfit for drinking by human beings and cattle and for irrigating the land. The soil became unfit for cultivation. In fact, it spread diseases, death and disaster in the village and surrounding areas.

The Supreme Court issued various directions for implementation and enforcement of law to protect the ecology and the fundamental right to life of people. The court pointed out that enactment of law, but tolerating its infringement, is worse than not enacting a law at all. Violation of anti-pollution laws not only adversely affects the existing quality of life but the non-enforcement of legal provision often results in ecological imbalance and degradation of environment, the adverse effect of which will have to be borne by the future generations. While emphasizing on sustainable development court observed:

---

[95] (2006) 1 SCC 1.

[96] (1996) 3 SCC 212. (This case is popularly known as H-Acid case).

"Both development and environment go hand in hand, in other words, there should not be development at the cost of environment and vice-versa, but there should be development while taking due care and ensuring the protection of environment."

R.L. and E Kendra, Dehradun V. State of U.P.[97],(popularly known as Doon Valley Case) was the first case of its kind in country involving issues relating to environment and ecological balance which brought into sharp focus the conflict between development and conservation and the court emphasized the need for reconciling the two in the larger interest of the country, mining which denuded the Mussoorie Hill of trees and forest cover and accelerated soil erosion resulting in landslides and blockage of underground water which fed many rivers and springs in the river valley. The court appointed an expert committee to advice the Bench on the technical issues and on the basis of report of committee, the court ordered the closure of number of lime stone quarries.

The court observed that;

"This would undoubtedly cause hardship to them, but it is a price that has to be paid for protecting and safeguarding the right of the people to live in healthy environment with minimal disturbance of ecological balance and without avoidable hazard to them and to their cattle, homes and agricultural land and undue affectation of air water and environment."[98]

It is submitted that the decision the decision of court has rightly reaffirmed that development is not antithetical to environment. However, thoughtless development can cause avoidable harm to the environment. The court also recorded a sounding note that preservation of the environment and keeping the ecological balance unaffected is a task which not only government but also every citizen must undertake.

In *Kinkry Devi v. State*[99], a PIL was filed in which it was alleged that the unscientific and uncontrolled quarrying of the lime stone has caused damage to the Shivalik Hills and was imposing danger to the ecology, and inhabitants of the area. The Himachal Pradesh High court relied on Doon Valley case and pointed out that if a just balance is not struck between development and environment by proper tapping of natural resources, there will be violation of Articles 14, 21, 48-Aand 51 A (g) of the Constitution. The Court rightly observed that natural resources have got to be tapped for the purpose of

---

[97] AIR 1985 SC 652.

[98] *Id* at 656 (emphasis supplied).

[99] AIR 1988 HP 4.

social development but the tapping has to be done with care so that ecology and environment may not be affected in any serious way.

In *A.R.C Cement Ltd v. State of U. P.*[100]., the Supreme Court did not permit the cement factory to run in Doon Valley area where the mining operation had been stopped and in order to restore the Doon Valley to its original character it was directed to be declared as non-industrial. However, the government was asked to provide an alternate site for shifting the cement factory of the petitioner.

A monument judgment was delivered by the Supreme Court in *M. C Mehta v. Union of India*[101]. Bhopal catastrophe is only a manifestation of the potential hazards of all chemical industries in India, none of which are amenable to effective regulation. Hardly had the people got out of the shock of Bhopal disaster when major leakage of oleum gas took place from one of the units of Shriram Chemicals in Delhi and this leakage affected large number of persons. In this case the court has evolved many principles which are new to the Indian "environmental jurisprudence."

In this case the approach of the Supreme Court was in consonance with environment protection and sustainable development.

The court also suggested the government to set up "Ecological Science Group" for the dissemination of information and the need to set up "environmental courts" to deal with environmental cases. Regarding the liability of an enterprise which is engaged in hazardous and inherently dangerous industry, the court observed:

"*We are of the view that an enterprise which is engaged in a hazardous or inherently dangerous industry which possess a potential threat to the health and safety of the persons working in the factory and residing in the surrounding areas owes an absolute and non-delegable duty to the community to ensure that no harm result to any one on account of hazardous or inherently dangerous nature of the activity which it has undertaken.*"

In *Vellore Citizen's Welfare Forum v. Union of India* (popularly known as T. N. Tanneries case)[102], is a landmark judgment of the Supreme Court where the principle of sustainable development has been adopted by the court as balancing concept.

It was held that "the Precautionary Principle" and "the Polluter's Pays Principle are essential features of "Sustainable Development." "Polluter

[100] (1993) supp. (1) SCC 57.

[101] AIR 1987 SC 965.

[102] (1996) 5 SCC 647.

Pays Principle" as interpreted by the Supreme Court means that the absolute liability for harm to the environment extends not only to compensate the victims of pollution but also the cost of restoring the environmental degradation.

In *Charanlal Sahu v. Union of India*[103], the Supreme Court while deciding the question of constitutional validity of Bhopal gas leak disaster act, 1985 held that the said act was valid but directed the interim compensation or maintenance to be paid and further suggested for taking "precautionary measures" and for effectively dealing with such industrial disaster in future. In Ishwar Singh v. State of Haryana [104] the high court issued a direction for closing down the stone crushing business of those which were not situated within the identified zone. The Court further directed that those who wanted to carry on their business of stone crushing should shift to identified zone. The Court directed that stone crusher owner were liable to pay compensation to citizens of the area, who have suffered, within a period of two months failing which their license to carry to carry the business of stone-crushing was to be canceled.

In *Tarun Bharat Sangh v. Union of India*[105], the petitioner through PIL brought to the notice of Court that State Government of the Rajasthan, though professing to protect the environment by means of notification and declarations, was itself permitting the degradation of the environment by authorizing mining operation in the area declared as reserved forest. In order to protect environment and wildlife within the protected area, the Supreme Court issued direction that no mining operation of whatever nature shall be carried on within the protected area.

In *M. C Mehta v. Union of India*[106]a PIL was filed seeking direction from the court to stop the mining activities in the vicinity of touring resort of Badkal lake and Suraj Kund in Haryana. The Haryana pollution control board recommended that mining activities within the radius of 5 Km. from tourist resort should be stopped. Similar recommendation was also made by NEERI. Having regard to the opinion of two expert bodies Court held that mining activities should be stopped within 3 Kms. of such tourist resort. In State of Himachal Pradesh v. Ganesh Wood Products[107], the following question was raised before the Hon'ble Supreme Court: *Is it possible to*

[103] (1990) 1 SCC 613 at 707-708.

[104] AIR 1996 P & H 30.

[105] (1992) Supp 2 SCC 448.

[106] (1996) 8 SCC 462.

[107] AIR 1996 SC 149.

derive a right to Intergenerational Equity from Article 14 and 21? The Apex Court observed that the central tenet of this theory is the right of each generation of human beings to benefit from the cultural and natural inheritance from the past generations as well as the obligation to preserve such heritage for future generations. This requires conserving diversity and equality of biological resources and of renewable resources such as forests, water and soil.

The same question raised above can also be posed in the following manner: *Is there any temporal dimension to Article 21 and the right to equality under Article 14 which recognizes "intergenerational equity"?* The answer to this question lies in the following notion: All members of each generation of human beings, as a species, inherit a natural and cultural patrimony from their previous generations, both as beneficiaries and as custodians who are under obligation to pass on this heritage to future generations. Therefore, the right of each generation to benefit from and develop this cultural and natural heritage is inseparably coupled with the obligation to use this heritage in such a manner that it can be transmitted to future generations in no worse condition than it was received from the previous generations[108].

- How do we know what future generations will want or need?[109]
- How do we know what is just for people who do not yet exist?[110]
- How do we prioritize justice for future generations when there is an abundance of injustice within the current generation?[111]
- Who within the present generation is responsible for future generations?
- Which aspects of health, diversity and productivity of the environment should be prioritized?
- What process should be put into place in order to ensure that this occurs?

---

[108] See Edith, B. W., (1992). "In Fairness to Future Generations: International Law, Common Patrimony, and Intergenerational Equity," 16, *The Fletcher Forum of World Affairs*, (Winter 1992): 172-175; Edith, B. W., (1990). "Our Rights and Obligations to Future Generations for Environment," 84 *Am. J. Int'l L.* 198 (1990):198; and Edith Brown Weiss "Intergenerational Rights and Rights of Future Generations."

[109] Edith Brown Weiss argues that this problem should be addressed by ensuring that the options are conserved for future generations. Edith Brown Weiss "In Fairness to Future Generation: International Law, Common Patrimony and Intergenerational Equity," 16, *The Fletcher Forum of World Affairs*, (Winter 1992): 172-175.

[110] Derek, P., (1986). *Reasons and Persons*. London: Oxford Scholarship.

[111] Frederic, G., & Alex, G., (2007). Are Generational Savings Unjust? *Politics, Philosophy & Economics, 6*, 193–217.

- And how should the interests of future generations be balanced against those of the current generations?

These uncertainties have significant implications for the content of the concept of Intergenerational Equity. In effect, it means, in the absence of further legislations, that how the principle is applied in a particular setting remains largely up to judges.

5.11.2. Climate Concerns Raised in the Indian Courts

Climate change concerns have, however, begun to filter through to courts and tribunals at various levels in India, as the cases that follow will demonstrate. The existing case law relating to climate concerns, can be divided into three broad categories:

I^st group of cases in which petitioners raises climate concerns, among others, to challenge what they perceive as ill-informed decision making on environmental matters.

2nd group of cases in which respondents raises climate concerns, among others, to justify, in the face of a challenge, what they perceive as environmentally sound decision making.

3rd group of cases in which judges appear of their own volition to refer to climate concerns, albeit in passing and as obiter, as one among the reasons for their decision.[112]

If we see first the first category of cases, petitioners raise climate concerns, among others, to argue for more environmentally friendly decision-making. In such type of cases petitioners appear to be using 'climate concerns' as a sword to encourage better-informed decisions and actions relating to the environment. For example, in *Manushi Sangthan, Delhi v. Govt. of Delhi & Ors*[113] in this case the petitioners challenged the limit on the issuance of cycle rickshaw licenses set by the Delhi Municipal Corporation, arguing, inter alia, that the IPCC's Fourth Assessment Report, 2007, had emphasized the need for policies that encourage use of more fuel-efficient vehicles, hybrid vehicles, NMT, (such as cycling and walking), and better land-use and transport planning. Although not directly with reference to this argument, the Court held the limit imposed by the Delhi Municipal Corporation to be arbitrary, and ordered a more detailed study on urban transportation options.

[112]Lavanya, R., (2013). Rights based climate litigation in the Indian courts: Potential, prospects & potential problems. *Centre For Policy Research Climate Initiative* (Working Paper 2013/1 May).
[113] (2010) DLC 168.

In case of *We the People v. Union of India*[114] the petitioners argued that State authorities were cutting down old growth trees in the implementation of development projects such as road expansions, thus contributing to global warming, without planting oxygen-generating trees to compensate for the loss of such old growth trees. The Allahabad High Court found merit in his argument and ordered the Principal Chief Conservator of Forest to appear before the Court and make available all the details of compensatory tree planting measures. The Court heard details of trees, including varieties of trees, felled and planted in the State of Uttar Pradesh in the context of road construction and expansion projects, and also directed the government to make provision for sufficient space to plant trees, the majority of which should be old growth trees, while constructing roads.[115]

If we see the second category of cases, in which respondents raise climate concerns to justify their actions. In such type of cases respondents appear to be using climate concerns as a shield to defend their actions. In case of *Nar Bahadur Bhandari and Ors v. the State of Sikkim and Ors*[116] the petitioners challenged the construction of a hydroelectric power project on the Teesta River. The Court referred to a Ministry of Environment and Forests affidavit that had been filed in the Supreme Court in a related case before the T. N. Godavarman bench. The Ministry of Environment and Forests justified their decision to permit the construction of this hydroelectric power project in this affidavit. They contended inter alia that India suffered from a severe peak power shortage, and "this position needs to be corrected through execution of more and more hydro power projects to generate environment friendly and peak power and reduce dependence on power generation based on fossil fuels which are contributing enormously towards atmospheric pollution and global warming and climate change."

In the third category of cases in climate litigation, judges, in so far as this can be inferred from the judgments, appear to refer *suo moto* to climate concerns, among others, in their judgments. The Supreme Court's decision in *Narmada Bachao Andolan v. Union of India.*[117] The Supreme Court while allowing the continued construction of the controversial Narmada Dam, noted as part of the rationale for favoring hydroelectric power projects that

---

[114] Order of the Allahabad High Court in Misc. Bench No. 5750 of 2010, 16 June 2010, http://www.indiankanoon.org/doc/1558452/ (accessed on 07 July 2023).

[115] *Ibid.*, 18th June 2010, http://www.indiankanoon.org/doc/1172427/MANU/ (accessed on 07 July 2023).

[116] MANU/SI/0023/2010.

[117] AIR 2000 SC 3751.

"...thermal power projects use fossil fuels, which are not only exhausting fast but also contributing towards environmental pollution. Global warming due to the greenhouse effect has become a most important cause of concern. One of the various factors responsible for this is the burning of fossil fuel in thermal power plants. ...On the other hand, the hydel power's contribution in the greenhouse effect is insignificant and it can be termed ecology friendly." A more recent case in a similar way is *Tamil Nadu Newsprint and Papers Ltd. v. Tamil Nadu Electricity Regulatory Commission.*[118] In this case the Appellate Tribunal for Electricity upheld an order fixing a tariff for the purchase of power from non-conventional energy sources. The Tribunal observed in its order that, "the danger of climate change needs to be averted by undertaking measures to curtail emission of greenhouse gases. Though largely it is the developed countries which are major contributories of greenhouse gases, we also need to regulate electricity sector for protection of environment in accordance with the spirit of the Constitution and the Electricity Act, 2003. Small steps in the first instance, to reduce dependence on fossil fuel to the extent possible, which does not impact the progress of electricity sector, can ultimately lead to generation of momentum for a giant leap in the development of technology for production of clean energy."

Climate change concerns have been taking in note as one among many pressing environmental concerns, by the Supreme Court of India and various High Courts. In *Karnataka Industrial Areas Development Board v. Sri C. Kenchappa & Ors*[119]the Supreme Court, in this case ordered authorities to properly consider the adverse environmental impacts of development before acquisition of any lands for development, also referred to the devastating impacts of human intervention on the planet, including the impacts of climate change and ozone layer depletion. Similarly, in the case of *Bombay Dyeing and Mfg. Co. Ltd. v. Bombay Environmental Action Group and Ors*,[120] the Supreme Court observed the need of prioritizing environmental issues, inter alia, due to climate change concerns. The Court has also acknowledged climate concerns from the perspective of considering the merits of different sources of energy. In *Reliance Natural Resources Ltd. v. Reliance Industries Ltd*,[121] in this case Supreme Court while describing the benefits of natural gas, the Court observed that "Its low carbon content, relative to other fossil

[118] 2007 ELR (APTEL) 157.

[119] (2006) 6 SCC 371.

[120] (2006) 3 SCC 434.

[121] (2010) 7 SCC 1.

fuels, implies that its use may help in combating global warming problems." The Allahabad High Court in *Swami Parmanand Bhatta Company v. Union of India,*[122] ruled that the government could regulate in the interests of the protection of environment the petitioner's exercise of his right to operate his brick kiln. The Court noted in this context that the "adverse effect of environmental pollution are now felt, as evidenced, like global warming, recurring natural calamities and on health of people." The Delhi High Court in *Outdoors Communication v. PWD and Municipal Corporation of Delhi,*[123] a case relating to the tendering of outdoor advertising space, noted that "the warnings of global warming have deserved scant attention." The Bombay High Court in *Goa Foundation v. Goa State Coastal Zone Management*[124] declined to review a decision which allowed the construction of a resort on Bagga beach on the grounds that their power of judicial review is limited. However, the Court observed that, "as far as the State of Goa is concerned, the entire coastline is filled with sand dunes. Sand dunes do play a key role in protection of the hinterland, in as much as the sand dunes as sentinel against any destructive cyclones, rising water level of the sea due to global increase in temperature. ...Their protection is, therefore, absolutely necessary and they are rightly placed in CRZ I (Coastal Regulation Zone) category."

In *Bombay Dyeing and Manufacture Co Ltd v. Bombay Environmental Action Group*[125], the Hon'ble Court stressed that "with the major threats to the environment such as climate change, depletion of natural resources, the eutrophication of water system and biodiversity and global warming the need to protect environment has become priority.

All the international convention mentioned in chapter IV find a mention in the decision of the Hon'ble Supreme Court in *G Sunderrajan v. Union of India*[126]. Holding that the setting up of a nuclear power plant Kudankulam, Tamil Nadu was not contrary to public policy, the Hon'ble Supreme Court referred to a report of the International Atomic Energy Agency which highlighted that the world is still dependent on burning coal, oil and natural gas for 80% of its energy supply surging energy use cause surging emissions of greenhouse gases disrupting the climate with potentially catastrophic results. The report further stated that nuclear energy is a readily expandable source of low-carbon baseload electricity and in the future might also

[122] MANU/UP/2765/2010.
[123] MANU/DE/2913/2007.
[124] MANU/MH/0964/2000.
[125] (2006) 3 SCC 434.
[126] (2013) 6 SCC 620.

help to meet other energy need. Such as hydrogen production and water desalination. In the case of *Karnataka Industrial Area Development Board v. C Kenchppa*,[127] the Hon'ble Supreme court directed that adverse impact on the environment must be properly comprehended by the Appellant before acquiring any lands for industrial purposes, in doing so, the Hon'ble court stressed on the importance of sustainable development while noting that:

"*The entire world is facing a serious problem of environmental degradation due to indiscriminate development. Industrialization, burning of fossil fuel and massive deforestation are leading to degradation of environment. Today the atmospheric level of carbon dioxide, the principal source of global warming, is 26% higher than pre-industrialized concentration. The earth's surface reached its record level of warming in 1990. In fact, six of the seven warmest years on record have occurred since 1980, according to the World Watch Institute's 1992 report. The rise in global temperature has also been confirmed by the Intergovernmental Panel on Climate Change set up by the United Nations and its final report published in 1990. The global warming has led to unprecedented rise in sea level. Apart from melting of the polar ice it led to inundation of low-lying coastal region Global warming is expected to profoundly affect specie, and ecosystem... Melting polar ice, glaciers thermal expansion of seas would cause worldwide flooding and unprecedented rise in sea level lf gas emissions continue at the present rate. Enormous amount of gases and chemicals emitted by the industrial plants and automobiles have led to depletion of ozone layers winch serve as a shield to protect life on the earth, from the ultra-violet rays of the sun.*"

A similar observation was made by the Hon'ble Court in the case of *Narmada Bachao Andolan v. Union of. India*[128], approving the decision to construct a high dam having a "Full Reservoir Level" of 455 ft with the capability of developing hydel power the Court favored an environment-friendly option for generation of electricity over a pollution-causing thermal power project, even though the former would mean displacement of nearby inhabitants, and observed as follows:

"The world over, countries having rich water and river systems have effectively exploited these for hydel power generation. In India, the share of hydel power in the total power generated was as high as 50% in the, year 1962–63 but the share of the hydel power started declining rapidly after 1980. There is more reliance now on thermal power projects. Bu t these

[127] (2006) 6 SC C 371.

[128] (2000) 10 SCC 664.

thermal power projects use Fossil fuels which ore not only depleting fast but also contributing towards environmental pollution. Global warming due to the greenhouse effect has become major cause of concern. One of the various Factors responsible. for this is the burning of Fossil fuel in thermal power plant· there is there for international concern for reduction of greenhouse gases which is shared by the World Bank resulting in the restriction of funds for the thermal power projects. On the other hand, the hydel power contributions in the greenhouse effect are negligible and it can be termed ecology friendly. Not only is this but the cost of generation of electricity in hydel projects significantly less.... Giving the option of generating eco-friendly electricity and substituting it by thermal power may not, therefore, be the best option. Perhaps the setting up of a thermal power plant may not displace as many families as hydel projects may but at the same time the pollution caused by the thermal power plant and the adverse effect on the neighborhood could be far greater than the inconvenience caused in shifting and rehabilitating the outees of a reservoir."

The High Court and national green tribunal have also taken cognizance of this phenomenon in several judgments. In the case of *State of Maharashtra v. Suraj Pal,* [129]while dealing with the poaching of endangered wild animals particularly tigers, the Hon'ble Bombay High Court noted that:

"*Carbon sequestration is extreme, important from the viewpoint of reducing carbon Footprint of a nation and is considered a major contribution towards lowering down global warming brought about by green house or carbon dioxide emitting gases. It also helps in strengthening the economy of a nation Apart from increasing world image by reducing ifs carbon footprint as it would show that the nation does not leave behind a big bad footprint created by burning fossil fuels like coat, oil and natural gas which emit huge amount of carbon dioxide and other toxic gases into the atmosphere... a healthy forest ecosystem determines sustainability of life on earth itself and since tigers are so sustainability of life on earth itself and since tigers are so important a bench mark of a healthy forest ecosystem, the habitant where wild tigers abound, are considered to be extremely high value ecosystem.*"

In *Abhimanyu Rathor v. State of H. P*[130]*, the Hon'ble Himachal Pradesh High* court lamented:

"We have already caused immense damage to the environment and ecology resulting in global warming. The global warming has even reduced

[129] (2015) 1 Bom CR (Cri) 576.

[130] (2014) SCC OnLine HP 5202.

the crop yield and has changed the weather pattern. The Kedarnath and Srinagar tragedies are the outcome of reckless and irreparable loss caused to the environment and ecology. Every citizen has a right to breathe fresh air and access to potable water. The M. C. Shimla, could not be oblivious to these grim facts."

Taking cognizance of unregulated and heavy tourism, overcrowding, mis-use of natural resources, construction of buildings and infrastructure, littering of waste and other activities associated with tourism. The National Green Tribunal in *Court of Its Own Motion v. State of H. P.*[131], cautioned against the dangers of global warming in the following manner:

"Global warming has a direct impact on environment and ecology of any zone. Global warming, a succession of hot summer through the 1980s seemed to bear out the conservationist's warming that the earth's climate was warming up. This global warming, if it is indeed the case, is a consequence of the large quantities of carbon dioxide released into the atmosphere by industries power stations. And motor vehicle. The earth's temperature will rise, the polar cap will melt, raising sea levels and causing catastrophic flooding in low lying areas and large areas of productive land will become desert. This global warming-heat-has affected various countries in the world. The glaciers on the Mount Blanc were melting causing slush: the impact of what keeps on happening in the solar atmosphere and the sun directly affect our planet, hence, there is a need to tackle global warming. Environmentalists present a grim picture of the effect of global warming with the mean temperature of the earth increased by about 1.6°C. This is a declaration made at a conference of parties to the Framework Convention on Climatic Changes. Global warming represents the increase in the average temperature of earth's near-surface air and oceans (ref: "according and pollution laws in India" second edition)

According to a statistical report from the World Meteorological Organization, global temperature has registered an increase of more than 0.6°C in the past 100 years. In the near future contamination of underground and surface fresh water supplies is likely near coastal areas (Ref: Law relating to Environmental Pollution & Protection, 5th Edition. Vol 1, by Dr. N. Maheshwara Swamy).

Global warming has its impacts in other parts of the World, as in the Indian sub-continent. It is likely to affect the glaciers. There will be early and untimely melting of ice resulting in various environmental issues Rohtang

[131] (2014) SCC On Line NGT.

Pass being one of the eco-sensitive and fragile areas of the glacier is likely to get affected more than other areas. Thus, there is a need for evolving schemes and mechanism to take greater care of the glacier in the interest of environmental and ecological balance."

The National Green Tribunal has also banned petrol vehicle older than 15 years and has recently banned diesel vehicles older than 10 years from plying on the roads on Delhi, noting that they were responsible for rising pollution levels.[132] It is indeed comforting that our Judiciary appears to have adopted a Hand-on approach in acknowledging the challenges that climate change poses and placing environmental interest at a higher pedestal. It is only if such an approach is adopted, the solutions to these challenges can be identified and implemented promptly.

In Om Dutt Singh case, the applicants challenged the construction of a massive irrigation project on the ground that submergence of large tracts of forest land would lead to methane emissions.[133] The NGT did not address this issue in its final order and it rejected the challenge, the primary reason seems to be from the judgment that because shutting down the project at such a late stage would have meant a massive waste of public money.

The NGT in Sukhdev Vihar case had had to consider whether a waste-to-energy plant should be permitted to operate in a densely populated residential area.[134]One of the applicants of the case contented that incineration of unsegregated waste leads to GHG emissions but The Tribunal rejected the applicants' case on the ground that the plant met all of the prescribed emission standards. Significantly, the NGT noted that the project was a CDM project, and was therefore required to meet international standards, including stoppage of GHG emissions in the atmosphere. In Society for Protection of Environment and Biodiversity, the NGT found some part of a government notification which were exempting the construction industry (responsible for 22% of India's annual GHG emission) from an environmental regulatory approval process to be illegal.[135] The Tribunal held that the exemption was made completely in derogation of India's commitment under the Paris

---

[132] Shirin, B., (2015). *NGT Bans Diesel Vehicles Older Than 10 Years in Delhi*. Down To Earth https://www.downtoearth.org.in/news/ngt-bans-diesel-vehicles-older-than-10-years-in-delhi-49325 (accessed on 07 July 2023).

[133] Nat'l Green T [NGT], *Om Dutt Singh v. State of Uttar Pradesh,* Application No. 521/2014 (May 7, 2015).

[134] NGT, *Sukhdev Vihar Residents' Welfare Ass'n v. Union of India [UoI]*, Application No. 22 (THC)/2013 (February 2, 2017).

[135] NGT, *Society for Protection of Environment & Biodiversity v. UoI*, Application No. 677/2016 (Dec. 8, 2017).

Agreement and the Rio Declaration. The NGT, in *Ratandeep Rangari v. State of Maharashtra & Others,*[136] directed the MoEFCC to come up with a monitoring and compliance protocol for ensuring that all thermal power plants and coal mines effectively implement a notification requiring coal-based thermal power plants to use coal with ash content not exceeding 34%.

In case of *Manu Anand v. State of Kerala*[137], while directing the government to formulate policy on the use of agricultural land for mining operations, the Kerala High Court specified that the Kyoto Protocol "reminds the nation to strive for the policies and measures to minimize adverse effects on climate change and to promote sustainable forms of agriculture in the light of climate change conditions."[138]

Ridhima Pandey v. Union of India & others.[139] A nine-year-old girl Ridhima Pandey, from the Uttarakhand region, filed a climate change case in March 2017 with the National Green Tribunal of India. In this case Plaintiff argued that the Public Trust Doctrine, India's commitments under the Paris Agreement, and India's existing environmental laws and climate-related policies oblige greater action to mitigate climate change. It also argued that the term "environment," as used in the Environment (Protection) Act 1986, necessarily encompasses the climate. The case was brought in pursuant to section 2(m) of the National Green Tribunal Act 2010, which authorizes claims that raise "a substantial question relating to the environment." In addition to those legal provisions, the petitioner cites the principles of sustainable development, precautionary principal, and intergenerational equity, as well as judicial decisions based on similar legal principles in the Netherlands (Urgenda Foundation v. Kingdom of the Netherlands), Pakistan, (Leghari v. Pakistan), and the U.S. (Juliana v. United States).

The petitioner also stated that India is the third-largest national emitter of greenhouse gases (behind China and the U.S.) and India is among those countries that are most susceptible to adverse impact of climate change.

To remedy the alleged injury to the present and future climate, the petitioner asked the court to order the national government to undertake a variety of measures, including but not limited to inclusion of climate change

[136] Application No. 19/2014 (WZ) (Oct. 15, 2015).

[137] MANU/KE/0461/2016.

[138] Shibani, G., (2020). Litigating climate claims in India. *American Journal of International Law, 114*, 45–49.

[139] Application No. 187/2017 (January 15, 2019). In this case, a minor sought directions from the NGT to the government and environmental regulatory agencies to consider climate-related issues in their decision-making.

in the issues considered by environmental impact assessments, preparation of a national greenhouse gas emissions inventory, and preparation of a national carbon budget against which particular projects' emissions impacts can be assessed.

On January 15, 2019, the National Green Tribunal dismissed the case, reasoning that the climate change is already covered in the process of impact assessments under the EPA of 1986, and therefore, "There is no reason to presume that Paris Agreement and other international protocols are not reflected in the policies of the Government of India or are not taken into consideration in granting environment clearances."

In the *Mahendra Pandey v. UoI,* [140] case, climate change was the core issue, the applicant moved toward the Tribunal with regard to the formulation of the action plan by the State of Delhi. The matter before tribunal was disposed of when the action plan was submitted by Delhi to the central government for final approval. From an understanding of the Tribunal's orders, it does not appear that the Tribunal played any role in formulating the plan, but can be said judicial attention may have expedited the plan's submission, which was long overdue.

5.12. EVOLUTION OF LEGAL PRINCIPLES AND DOCTRINE IN ENVIRONMENTAL JURISPRUDENCE BY INDIAN JUDICIARY

The formulation of certain principles to develop a better regime for protecting the environment is a remarkable achievement of Indian judiciary.[141]

5.12.1. Doctrine of Absolute Liability Principle

This doctrine was evolved by Supreme Court in the case of *Oleum Gas Leak Case*[142]. it was held by the court in the *Bhopal gas case*[143] that where an Enterprise is occupied with inherently dangerous or a hazardous activity and harm result to anybody by virtue of a mishap in the operation of such dangerous of naturally unsafe movement coming about for instance, in escape of poisonous gas the enterprise, irrespective of their negligence, motive or intention is strictly and completely obligated to repay every

[140]NGT, *Mahendra Pandey v. UoI*, Application No. 470/2016, (January 2, 2019).

[141] See Shibani, G., (2019). *Indian Environmental Law: Key Concepts and Principles*. Hyderabad: Orient Black Swan.

[142]*M. C. Mehta* v. *Union of India* AIR 1987 SC 1086.

[143]*Union Carbide Corporation* v. *Union of India* AIR 1990 SC 237.

one of the individuals who are influenced by the accident and such risk is not subject to any exemption. Accordingly, the Supreme Court created another trend of absolute liability without any exemption for harm caused by hazardous and inherently dangerous Industries by interpreting the scope of the power under Article 32 to issue directions or orders which ever may be appropriate in appropriate proceedings.

5.12.2. Polluter Pay Principle

The concept of compensation for environmental degradation has evolved at a snail's pace over a period. It started with the strict liability principle followed by the absolute liability principle and compensation under Article 32 and finally, the polluter pays principle the fundamental basis of this slogan is- 'If one makes a mess, it's your duty to clean it up.' It should be mentioned that in environment law the 'polluter pays principle' does not allude to "fault." Instead, it supports a remedial methodology which is concerned with repairing natural harms. It is a rule in international environmental law where the polluting party pays for harm or damage done to the natural environment. Thus, under this principle, the responsibility of repairing the environment is put on the shoulder of the offending industry. The polluter pays principle means two things: firstly, the polluter should pay for the administration of the pollution control system; secondly, the polluter should pay for the consequences of the pollution.

The Supreme Court in, Indian Council for E*nviro-Legal Action v. Union of India*[144] and *Vellore Citizen Welfare Forum v. Union of India*[145] not only reiterated the same doctrine but moved a step ahead by laying down the 'polluter pays' principle. The concept was elaborated in the Vellore tanneries pollution case, by court by stating that the polluter pay principle as interpreted by this court means that absolute liability for harm to the environment, extended not only to compensate the victims of pollution but also the cost of restoring the environmental degradation. Remediation of the damaged environment is part of the process of sustainable development and as such, the polluter is liable to pay the cost to the individual sufferers as well as the cost for reversing the damaged ecology.

5.12.3. Precautionary Principle

The Supreme Court[146] pointed out that the traditional concept that development and ecology are opposed to each other in is no longer

144 AIR 1996 SC 144.

145AIR 1996 SC 271.

146*M. C Mehta* v. *Union of India* AIR1997 SC 734.

acceptable. "Sustainable development" is the answer. The supreme court after explaining the silent principle of sustainable development expressed the view that the "*precautionary principle*" and the "*polluter pays principle*" are essential features of Sustainable development and that they have been accepted as a part of the law of the land the supreme court explaining the precautionary principle held that the concerns engaged Industries not only to take measures for environmental Protection but also to anticipate prevent and attack the causes of environmental pollution and degradation.

The Supreme Court of India in *Vellore Citizens Forum case*[147], develop the concepts for the precautionary principle or precautionary approach and state, whose action or policy has a suspected risk of causing harm to the public or to the environment then in the absence of scientific consensus that the action or policy is harmful, the burden of proof that it is not harmful false on those taking the action.

5.12.4. Public Trust Doctrine

The Public Trust Doctrine primarily rests on the principle that certain resources like air, water, sea and the forest have such a great importance to people as a whole that is would be wholly and unjustified to make them a subject of private ownership. In *M. C Mehta v. Kamal Nath*[148] the Supreme Court laid down the Public Trust Doctrine. The Supreme Court held that the state is the Trustee of all-natural resources. The natural resources are meant for or public use and enjoyment and it cannot be given into private ownership.

5.12.5. Doctrine of Sustainable Development

"The doctrine sustainable development in the Indian sub-continent's jurisprudence and political theories has long-standing recognition. An ancient hymn from the Upanishads reminds us of this. 'Kautilya's Arthashastra' in 300 B. C. – also refers to preservation of land and its surrounding nature. The concept thus is not new and India has a long cultural tradition of frugality and simple living in harmony with nature. The concept thus is not new and India has a long cultural tradition of frugality and simple living in harmony with nature. All great religions which have traversed in our country have preached the unity of humankind with nature. Presently however, the doctrine of sustainable development is no longer in the realm of a concept to be

---

[147] AIR 1996 SC 271.

[148] 1997 1 SCC 38.

shaped and focused differently in different circumstances but now it stands as an established principle and is placed at par even with the very concept of democracy, human rights and sovereign equality of states as stated by Prof. Lowe in his Article on 'sustainable development and unsustainable arguments.' The doctrine of sustainable development in the present phase though launched initially at the Rio declaration but the same has found its recognition both in major international legal instruments as also in number of judicial decisions since nineties."[149]

The World Commission on Environment and Development (WCED), the mission of Brundtland commission was to unite countries to pursue sustainable development together. The Brundtland Commission official is dissolved in December, 1987 after releasing 'Our Common Future,' also known as Brundtland report in October 1987, a document which coined and defined the meaning of the term "sustainable development." As per Brundtland report sustainable development signifies "development that meets the need of present without compromising the ability of the future generation to meet their own need."

In *Rural Litigation and Entitlement Kendra v. State of UP*[150] the court for the first time dealt with the issue relating to environment and development and held that it is always to be remember that natural resources are the permanent assets of the mankind and are not intended to be exhausted in one generation. Later in *Vellore Citizens Welfare Forum*[151] case the Supreme Court observed that sustainable development has to be accepted as a variable concept to eradicate poverty and improve the quality of human life while living with the carrying capacity of supporting ecosystem.

Climate change and a deteriorating environment is a key challenge to sustainability, bio-diversity, environment and stability across the world. So, it is not exaggeration to say that environment, climate change and sustainable development are related with each other and discussion will be meaningless, if any one of them kept outside. There is a twofold relationship between sustainable development and climate change. On the one hand, climate change influences nature and human living conditions and thereby also the basis for social and economic development, while on the other hand, society's priorities on sustainable development influence both the GHG emissions due

[149] Extract from a Lecture delivered by the Hon'ble Mr. Justice Umesh C. Banerjee to the students of NALSAR on February 22, 2011. Former Judge, Supreme Court of India and Founder Chancellor, NALSAR University of Law, Hyderabad.

[150] AIR 1985 SC 652.

[151] (1996) 5 SCC 647.

to which climate change caused and the vulnerability. The world summit on Sustainable development added a new dimension (social development) in concept of sustainable development. No doubt, development is the today's need but not at the cost of human survival. Development does not mean development of things. Its real meaning is the holistic development of human being. Holistic development is not possible without considering the issue of protection of environment and climate change. Climate change posing threat to the continuance of biodiversity on this earth. It is a herculean task to mitigate and adapt to the climate change. Efforts to cope with the impacts of climate change and attempts to promote sustainable development share common goals and determinants including access to resources, sharing of information and technology, fair and equitable distribution of resources and its benefits, access to risk-sharing mechanisms and abilities of decision-support mechanisms to cope with uncertainty.

5.12.6. Right to Wholesome Environment

In *Charan Lal Sahu v. Union of India*[152] the Supreme Court stated that right to life guaranteed by Article 21 of the Constitution includes the right to a wholesome environment. Later in *Damodar Rao v. S. O. Municipal Corporation, Hyderabad*[153] the court is associated to the constitutional mandate under Article 48A and 51A(g) to support these reasoning and went to the extent of stating that environmental pollution would be a violation of the fundamental right to life and personal liberty as enshrined in Article 21 of the constitution.

5.13. MAINSTREAMING CLIMATE CHANGE IN SUSTAINABLE DEVELOPMENT

Government initiatives for the diffusion of renewable energy and energy efficient technologies, Joint Forest management, water resource management, and environmental education in educational institutions represent a broad spectrum of efforts to integrate climate change concerns in sustainable development.

The National Environment Policy 2006 provides the basis for integration of environmental considerations in the policies of various sector. The policy statement for Abatement of Pollution 1992 stresses on "polluter pay" principle while The Forest Policy 1988, highlights the environmental

152 AIR 1990 SC 1480

153 AIR 1987 AP 171

protection through preservation and restoration of the ecological balance. The policy seeks to augment the forest cover in India[154].

The statutory framework for the environment and energy efficiency includes the Indian Forests Act, 1927, the Water (Prevention and Control of Pollution) Act, 1974, the Air (Prevention and Control of Pollution) Act, 1981, the Forest (Conservation) Act, 1980, and the Environment (Protection) Act, 1986. Other enactments include the Public Liability Insurance Act 1991 the National Environment Tribunal Act, 1995, the National Environment Appellate Authority Act, 1997, the Energy Conservation Act, 2001, and the Electricity Act, 2003. The courts have also elaborated on the concepts relating to sustainable development, and the 'polluter pays' and 'precautionary' principles. In India, matters of public interest, particularly pertaining environment, are too articulated effectively through a vigilant media, an active NGO community, and through the judicial process which has recognized the citizen s right to a clean environment as a component of the right to live and liberty.

"Addressing climate change mitigation and adaptation involves many stakeholders, cuts across short and long timeframes and requires that all development projects be assessed for their sensitivity to climate concerns. This integration of climate concerns in the development process has been mainstreamed in India through high-level multi-stakeholder committees."[155]

5.14. CONCLUSION

India as a country that has continuously taking action on climate change, in the last few years India has taken some substantial steps in undertaking ambitious domestic climate role in catalyzing action. Action at the national level has trickled down to all levels, with involvement from the states, private sector as well as civil society. However, India should remain cautious that its impressive list of climate undertaking is not bogged down by bureaucratic hurdles and is implemented in a timely manner. As India heads into the next few decades of critical economic growth. It will continue to grapple with the competing tensions of tackling economic development and climate change. To ensure that neither priority is compromised, it is essential that the Indian government continues to find that balance where the economy and the environment are not at loggerheads with each other. Instead, by

[154] "Sustainable Development in India a Value Based System- Strategies Necessary for India's Economic Self Sufficiency" *Journal of Institutions of Valuers* (March 2009): 308-15.

[155] India's: Addressing Energy Security And Climate Change, Government of India 9, (Ministry of Environment & Forest, Ministry of Power, Bureau of Energy Efficiency October 2007).

exploring innovative ways to engage the private sector in tackling climate change, continuing a dialog with civil society, and laying the foundation for a low carbon and climate-resilient economy, India can ensure its economic growth is in sync with the environment. While India finalizes the details of the Missions under its National Action Plan on Climate Change (NAPCC), there are several other initiatives being undertaken and facilitated by the Government of India to address the issue **of** climate change.

Recognizing the role of forests as a carbon sink. India is promoting afforestation on an unprecedented scale. India, which has a fifth of its area under forests, is one of the few developing countries in the world where the forest cover is increasing per year, despite the pressures of population growth and rapid economic development. New programs to increase the capacity of frontline forestry personnel, improve forestry infrastructure and control forest fires are also being implemented.

India is also an active participant in CDM under the Kyoto Protocol. With second highest number of projects registered for any country and estimated to offset almost 10% of India's total emission per year by 2012 India has been assessed as the "Best CDM Country" in recent independent study. India has also recently approved its National solar mission setting ambitious solar energy generation targets for the medium term.

The Government of India is also supporting and facilitating major research programs to assess to assess various aspects related to climate change. We have comprehensive scientific climate change program in the place that involved over 220 of the best scientists in the country.

A special focus area is the melting of the Himalayan glaciers and their link to climate change. India is also on the path to launching its own satellite to monitor greenhouse gases in the atmosphere, which will put it in an exclusive club of three countries. It has also released the results of a range of rigorous studies that estimate the GHG emissions profile of India for the next two decades, which show that India's total and per capita emissions over the next two decades will remain modest.

Apart from government initiative the Indian judiciary plays a remarkable role in uplifting the goal of preservation through its various land mark decisions and the Acts and laws which provides a platform so as that one cannot exploit the nature and its gift for his greedy needs. of course, the initiative for the protection of the environment and climate change came from the legislature but the failure of the executive to implement the environmental laws in India created ground for the intervention of judiciary. Where judiciary evolved new doctrine and principles to deal with environment pollution and climate change. The judiciary is also criticized for overstepping the administrative function and for lack of political or administrative will or because of other lacunae. Never the less, the judiciary is actively playing its role in spite of repetitive failure of another organ.

Chapter

6 CLIMATE CHANGE MITIGATION AND PARIS AGREEMENT, 2015: INDIAN INITIATIVES

Contents

6.1. INTRODUCTION

Climate change is the biggest challenge for developing countries like India. Some countries of the world including India are vulnerable to the impacts of climate change. India has the vast population that depends on agriculture and monsoons, as we know that climate change affects the monsoon. India has expensive coastal areas, Himalaya region and Island. Indian population accounts for 17.5% of the world population and the land surface of India is 2.5% of the world surface area. Most of the Indian populations are poor who has no access to electricity[1] and they rely on the solid biomass for cooking and 76 million population of India do not have access to safe drinking water.[2] India 1.77 million people are without house.[3] In a recent social economic survey indicates that economic and social deprivation are much higher in terms of availability of house, access to education, available source of energy and a regular source of income.[4] In coming years most of the population in India will be urbanized and some population will be rural area more industry will be established. People will have the more money. Urbanization industrialization means emission which ultimately result in climate change. The unpleasant twist is that India suffers acutely from the effects of climate change while not contributing substantially to the problem. India is the fourth- largest emitter of greenhouse gases (GHGs) in the world.[5] India being developing country has to meet its development goal along with its obligation under Paris Agreement. India only contributes about 7% to global carbon emission.[6] The quantification of goals in India's NDCs under Paris Agreement is threefold: first, reducing the emission intensity of its gross domestic product (GDP) by 33%–35% (vis-à-vis 2005) by 2030; second, achieving 40% cumulative electric power installed

[1] Kundan, P., (2015). *The Extent of India's Energy Poverty*. Down To Earth. https://www.downtoearth.org.in/news/energy/the-extent-of-india-s-energy-poverty-48966 (accessed on 07 July 2023).

[2] Saptarshi, D., (2017). *76 Million Don't Have Safe Drinking Water: India's Looming Water Crisis*. NDTV, https://swachhindia.ndtv.com/76-million-dont-have-safe-drinking-water-indias-looming-water-crisis-5606/ (accessed on 07 July 2023).

[3]Somesh, J., (2013). 1.*77 Million People Live without Shelter.* Business Standard. https://www.ighhub.org/resource/177-million-people-live-without-shelter (accessed on 07 July 2023).

[4] *India's Intended Nationally Determined Contribution: Working Towards Climate Justice*, https://www4.unfccc.int/sites/ndcstaging/PublishedDocuments/India%20First/INDIA%20INDC%20TO%20UNFCCC.pdf (accessed on 07 July 2023).

[5] The United Nations Environment Program, *Emissions Gap Report 2020*. https://www.unep.org/emissions-gap-report-2020 (accessed on 07 July 2023).

[6] Jacob, K., (2021). *India's Percentage CO_2 Emissions Rose Faster than the World Average*. The Hindu. https://www.thehindu.com/news/national/indias-percentage-co2-emissions-rose-faster-than-the-world-average/article33965283.ece (accessed on 07 July 2023).

capacity from non-fossil fuel-based energy resources by 2030; and third, creating an additional carbon sink of 2.5–3 billion tons of CO_2 equivalent by 2030 through additional forest and tree cover.[7] Even though there would have not been obligation to mitigate but it was incumbent need to mitigate due to India's venerability to climate change.[8]

6.2. INDIA'S NATIONAL CIRCUMSTANCES

India has a diverse geographical condition starting with landscapes varying from snow-capped mountain ranges to deserts, plains, hills, plateaus, coastal regions and islands. The diverse geographical condition of India manifests varied climate regimes ranging from continental to coastal, from extremes of heat to extremes of cold, from extreme aridity and negligible rainfall to excessive humidity and torrential rainfall. India's climate is mainly influenced by the presence of the Himalaya and the Thar Desert, Himalaya acts as an orographic barrier to atmospheric circulation for both the summer monsoon and the winter westerlies and keeps them confined within the subcontinent, thus playing a major role in causing precipitation in India.

The Indian monsoon is one of the most prominent climate systems of the world by providing nearly 75% of the annual rainfall of the country. Rainfall distribution and intensity have a very significant impact over different socio-economic sectors, especially agriculture and hydrology, besides impact on other aspects of various ecosystems.

The rainfall trends for the period from 1989 to 2018 showed a decreasing trend for the annual and monsoon season in the districts of eastern and central India, while the districts in the western part of the country experienced an increase. Overall, the decreasing trend in rainfall is visibly dominant over large part of India. The annual mean temperature during year 1901–2019 showed a significant increasing trend (0.61°C/100 years), with a higher increasing trend in maximum temperature (1.00°C/100 years) and relatively lower increasing trend (0.22°C/100 years) in minimum temperature. The year 2019 was the seventh warmest year on record since 1901.[9] The long coastline of over 8,000 km with flat coastal area, shallow

---

7 *Supra* note 4.

8 See Lakshmanan, Pushpa, K., et al., (2017). Paris agreement on climate change and India. *Journal of Climate Change, 39*(1), 1–10.

9 Jacob, K., (2020). *2019 Was The 7th Warmest Year in India Since 1901, Says IMD*. The Hindu. https://www.thehindu.com/sci-tech/energy-and-environment/2019-was-the-7th-warmest-year-since-1901-says-imd/article30515266.ece (accessed on 07 July 2023).

continental shelf, high population density, particular geographical location and specific physiographic features, makes India vulnerable to cyclones and its associated risks. In the India there are thirteen coastal States and Union Territories (UTs) including 84 coastal districts, are affected by tropical cyclones.[10] In the year 2019, eight cyclonic storms formed over the north Indian Ocean.[11]

Heatwaves normally occur between March and June, and in some cases, it extends till July in India. Heatwaves has become common occurrence over the Indo-Gangetic plains of India. Increasing trends in the heatwaves are being observed over most of the stations across India.[12]

The Himalayan Mountain range is one of the major sources of fresh water due to large concentration of snow and glaciers. Snow and glaciers make possible to support water supply of major Indian rivers such as Indus, Ganga and Brahmaputra. Most of the Himalayan glaciers are retreating and the rates of retreating have been probably accelerated in the past few decades. At present there is threat of rising of sea level along the Indian coast that is estimated to be rising at about 1.7 mm/year and at different rates along various parts of the coast.[13]

6.3. INDIA'S NATIONALLY DETERMINED CONTRIBUTION

India communicated its NDC for the period 2021 to 2030:

1. "To put forward and further propagate a healthy and sustainable way of living based on traditions and values of conservation and moderation.
2. To adopt a climate friendly and a cleaner path than the one followed hitherto by others at corresponding level of economic

[10] See Suma, M., & Balaram, P. S., (2014). Perspective study on coastal disaster management at Andhra Pradesh. *IOSR Journal of Business and Management, 16*(12), 55–60.

[11] Akshit, S., (2019). "*8th cyclone formed in the North Indian Ocean Region: Pawan Joins Kyarr, Fani, Maha, Vayu, Hikaa, Bulbul and Pabuk; to cross Somalia on December 7.* Down To Earth. https://www.downtoearth.org.in/news/climate-change/8th-cyclone-formed-in-the-north-indian-ocean-region-68307 (accessed on 07 July 2023).

[12] Government of India, (2021). *India Third Biennial Update Report to the United Nations Framework Convention on Climate Change*, 54 (Ministry of Environment, Forest and Climate Change), https://unfccc.int/sites/default/files/resource/INDIA_%20BUR-3_20.02.2021_High.pdf (accessed on 07 July 2023).

[13] *Ibid* 69.

development.

3. To reduce the emissions intensity of its GDP by 33 to 35% by 2030 from 2005 level.
4. To achieve about 40% cumulative electric power installed capacity from non-fossil fuel-based energy resources by 2030 with the help of transfer of technology and low-cost international finance including from Green Climate Fund (GCF).
5. To create an additional carbon sink of 2.5 to 3 billion ton of CO_2 equivalent through additional forest and tree cover by 2030.
6. To better adapt to climate change by enhancing investments in development programs in sectors vulnerable to climate change, particularly agriculture, water resources, Himalayan region, coastal regions, health and disaster management.
7. To mobilize domestic and new and additional funds from developed countries to implement the above mitigation and adaptation actions in view of the resource required and the resource gap.
8. To build capacities, create domestic framework and international architecture for quick diffusion of cutting-edge climate technology in India and for joint collaborative R&D for such future technologies."[14]

6.4. INDIA'S OBLIGATION TO MITIGATE CLIMATE CHANGE AND CLIMATE FRIENDLY LIFESTYLE

In COP 21, India decided to bring out a publication on its own climate-friendly traditional practices, which have been part of the psyche of its common man for years together, called 'Parampara.' These efforts eventually succeeded in convincing all the countries in world that the pulse of the problem lies in excessive consumption patterns and lifestyles by developed countries. Indeed, the concern of sustainable lifestyles found an appropriate expression in the preamble to the Paris Agreement. These various communication efforts disseminated the Indian viewpoint and stand on climate change-related negotiating issues in COP 21, which helped in arriving at the 'win-win' Paris Agreement.[15]

---

[14] *Supra* note 4

[15] Ashok, L., (2019). Reaching agreement in Paris: A negotiator's perspectives. In: Navroz, K. D., (ed.), *India in a Warming World: Integrating Climate Change and Development* (p. 180). New Delhi: Oxford University Press.

The environmental consequences of unsustainable lifestyles and way of production and consumption are widely acknowledged. According to the Inter-Governmental Panel of Climate Change (IPCC), anthropogenic GHG, these are mainly caused with the size of population, energy use, lifestyle, economic activities…etc. With the climate change, GHGs the adaptation and mitigation are more effected with by livelihood, culture and behavior. So, this can be reduced eventually by

- Changes in consumption ways
- Ways to reduce usage of power and
- Less wastage of food.

By not utilizing the resources accordingly leads to pressure on environment and nature. Where a part of society is over using resources and a part of society is unable to meet the resources (poor) which is leading to the imbalance in their lifestyle by not meeting the basic facilities like housing, education, food, health.

These can be reduced in ways by

- concentrating on poor to meet their basic needs
- shifting to sustainable lifestyles within nature capacity
- need changes in sustainability in richest sections of society.

India has many traditional practices in reducing the level of carbon footprints and lifestyle. Where these practices should be encouraged rather than the modern equipment's that leads more CO_2 in nature. This is also helpful to other developing countries which are looking for the traditional methods for development and protection of green environment.[16]

6.4.1. Habits and Culture

The Indian culture and life are very famous in world-view in which all its aspects exist according to the human experience and are recognized. In India people have a general rule, to the need-based consumption and fixed responsibility of not wasting the resources and that grows respect in their life. It is estimated that 70% of people live rural areas which are near to nature and not based on high consumption. Many of skills are practiced

[16] On the very first day of COP 21, that is, on 30 November 2015, the Financial Times carried an opinion Article titled, 'Do Not Let Lifestyles of the Rich World Deny the Dreams of the Rest,' by the Prime Minister of India, which we felt set the tone and tenor of the Conference (Modi 2015). The concerns of developing countries regarding climate friendly lifestyles and climate justice stood powerfully articulated by the prime minister in his article.

from one generation to other in master-to-discipline. People are dependent on high energy based, craftsmanship, organic farming that do not let depend more on electricity. Examples Khadi (famous fabric), handlooms, etc., are popular across Indian in gender or generation. Recognizing the bless of nature and eco system, people maintain their devotional worship places surrounded with ponds, gardens, bathing ghats, etc. [17]

6.4.2. Food Wastage

Food production, processing, distribution, consumption are main factors because they include the energy and usage of natural resources. The wastage of food is more in industrialized countries and the loss is mainly in retail and consumer levels. The per capita wastage of food in Europe and North America is 95–115 kg/yr. The wastage of food in developing countries is very less in consumer levels and also trying to reduce waste in pre-harvest levels. In India not wasting food and respect for food considered as important thing in one's life. The per capita wastage of food in southern and African countries is only 6–11 kg/yr.[18]

6.4.3. Local Food

Food transporting is one of the main causes for increase in GHGs in nature world-wide. Food (fruits and vegetables) which are grown in local areas and buying from the local markets helps reduce of GHGs. The ready to eat packaged food are more sold in developed countries than local grown food. Europe is increased with 3% of packages by 2020. In India the local grown food is more preferred than the packed food and large food habits according to the climate. Many practices are being done preserving the food like salt in pickles, neem leaves in rice, sun drying grains, etc.[19]

6.4.4. Meat Consumption

The consumption of meat is more in developed countries (double) compared to developing countries. The production in meat has tripled in last four decades and increased by 20% in ten years. The global meat consumption per capita base in 2011–13 was 33.8kg and expected to meet 36.3kg by 2023 in developed countries.

---

[17] *Climate Friendly Lifestyle Practices In India*. https://archive.nyu.edu/bitstream/2451/42298/2/Climate%20Friendly%20Lifestyle.pdf (accessed on 07 July 2023).

[18] *Ibid.*

[19] *Ibid*

In India 42% of households are vegetarian and remaining are less consumption type. The per capita base in 2011–13 was 3.3kg which is 1/3rd of global average.[20]

6.4.5. Transportation

Globally transport is mainly causing the increase in the 23% of CO_2 in nature. 17–18% is from road transport. The personal vehicles in urban areas have increased which has become a challenge. It is also recognized that urban cities are traveling in public transport for achieving low carbon pathway. In rich societies the cars are mostly used for short distances. In most of the urban areas it is found more release of CO_2 this results in high GHG emissions with transport services.

In India pedal rickshaws, bicycles and walking are more preferred for short distances and use mostly the public vehicles. The NMT is being seen. In most of wealthy cities 40–50% use NMT (cycles, walk) mostly. The public transport is mostly helpful for low-income group and also helps in achieving the NMT goal. [21]

6.4.6. Energy and Water

Global CO_2 emissions are from electricity and heat generation sectors which was of 11% of total emissions in 2012. Many developed countries people especially use most of electricity to heat or cool their homes.

In India simple sustainable consumption is done where using electricity only when it is necessary and is taught to children from young age. Even in very hot summer most of the people prefer fans or coolers rather than A/Cs. Sun drying of clothes and hand washing vessels are preferred rather than electronic machines. This practice is especially followed by south Asian and African countries. This saves 200–300kWh/yr assuming per day of dish washing.

People especially in villages and some in towns and cities prefer bathing with mug and water rather than shower which waste lots of water. Some prefer cold water bath for whole year. With traditional structures such as bore wells, step wells for collecting water in dry areas helps in conservation of natural resources.

---

[20] *Ibid.*

[21] *Ibid.*

- collecting rain water with terrace catchments.
- washing vessels with dry ash and coconut fiber and using soil pots for cool water.
- Drinking water without touching lips to tumbler which makes leftover water can be consumed by others.[22]

6.4.7. Building and Construction

GHG emissions are from building sector which have doubled since 1970. The building sector had approximately 32% of global final energy and 19% of GHG emissions. It says that modern buildings which uses cement for construction, using 1 ton of cement releases 1 ton of CO_2 in atmosphere. It is 9.3% of global emissions in 2013.[23] Traditional building practices like with solar passive direction homes, thermal insulation with mud, jail type windows for natural ventilation will make reduce the use of energy requirements and chemical emissions.[24]

In India the houses are traditionally constructed with bamboo, stone and clay that not only help climatically but also helps in GHG mitigation and reducing cement usage and transportation.[25]

6.4.8. Waste Management (Reduce, Reuse, and Recycle)

A major share of waste generated globally is municipal solid waste (MSW) coming from urban settlements which is 46% of total waste. In western countries the practice of reuse and recycle is limited.

In India the reuse and recycle is a part of their lifestyle. There exists an informal collecting of recycled products (newspapers, metals, plastic.) form door to door linking to recycling industry. In day-to-day life Indian households try to minimize the waste generation and reuse like

- Using packages to store vegetables, grains, storing groceries in kitchen.
- Old clothes are being passed to next generation or to those can get fit.

---

22 *Ibid.*

23 *Ibid.*

24 See Radhika, K., Anna, A., et al., (2021). The what, why, and how of changing cooling energy consumption in India's urban households. *Environmental Research Letters, 16*, 1–12. https://iopscience.iop.org/article/10.1088/1748-9326/abecbc/pdf (accessed on 07 July 2023).

25 *Supra* note 17.

- buying luxury goods in second hand (has a huge market).[26]

6.4.9. Health and Yoga

India is known for traditional medicine systems (ayurvedic, homeopathy, siddha,). This system of medicine existed since ancient times and also scripture in books. 70% of Indians use the traditional system methods to cure health issues.

The United Nations has declared June 21 as international yoga day. Yoga (to unite/join) is an ancient mental, physical, spiritual practice that oriented in India and now is practiced world-wide and gained popularity.

Many people can't afford exercise or fitness centers to make them fit, yoga is a best way to make fit, improve mental and physical health that can be done by anyone irrespective of status and physic. This also reduce the use of resources and maintain sustainability.[27]

6.4.10. Nature Conservation

In India forests are considered with respect and trees worshipped. Forest in India covers 24.01% of area and is dominant. The forests in India are important resource for rural and local people throughout the country providing food, fuel, fodder. They play a main role in soil and water resources and also reducing 12% of GHG emissions.

Forests are important things of Indian culture. These are responsible for flora and fauna diversity and are maintained/protected in well manner.

Planting trees is considered as the good thing ('punya') and celebrate "van Mahotsav" every year. Bishnoi of western Thar Desert, is a group who checks on eco-friendly community, for them harming environment is harming themselves.

In Rajasthan the 'khejri tree' is considered as the water sustaining tree that is not cut even if it interrupts the building of house. Live example for this is Salasar temple in Sikar district.[28]

Celebrating 150 Years of Mahatma Gandhi's Birth Anniversary:

The Government of India and people all over the country and around the world celebrated the 150th birth anniversary of Mahatma Gandhi in 2019.

[26] *Supra* note 17.

[27] *Supra* note 17.

[28] *Ibid.*

In September 2019, during the Climate Action Summit, the United Nations released a commemorative stamp to mark the 150th birth anniversary of Mahatma Gandhi. A simple understanding of Mahatma Gandhi's Cosmo centric approach to human beings is relevant to the challenge of climate change faced by the world today. Following the guidance of Mahatma Gandhi, India drafted its NDCs which was submitted to the UNFCCC in 2015. India's first NDC focuses on propagating a healthy and sustainable way of living based on the traditions and values of conservation and moderation.[29]

6.4.11. Use of Solar Energy and Wind Energy

India has History of using sunlight and wind in their lifestyle and culture Sun was the major source of energy for many things like health for cooking food drying clothes shelter. The festival of ***Uttarayan*** which was specifically devoted for sun used to be celebrated across India. Sunlight first used for doing household things which was environment-friendly such as ventilation and lightening. People were dependent upon the energy or fossil fuels.

Narendra Modi prime minister of India said India has potential to turn into Suryaputra (land of solar power) and use of modern technology to generate green solar energy. India has decided to anchor of global solar eclipse of all countries located between the tropical of cancer and tropic of Capricorn. Other use of sunlight in ancient India was used for saving food for long term like Singh ki which was made by the Tribes of Sikkim which is made of radish taproot another example is *Aam Papad* which is a snack made from the mango pulp recipe which preserves mango which can be eaten in other season.[30]

Science of Surya Namaskar which was used to be done by Indian sent, which is continued even today also helps the person in making good health and making him active. Surya Namaskar is a set of 12 poses, which is done in the morning time. The regular practice of Surya Namaskar improves the blood circulation in the human body and helps in maintaining good health and also helps in making good immunity. There are many benefits of Surya Namaskar to the liver, hurt, stomach, chest, throat, legs. In Jammu Kashmir peoples of Ladakh construct their homes for sustainable living they make two floors. They shift on first floor in winter season and ground floor is

---

[29] *Supra* note 12 at 423.

[30] PARAMPARA *India's Culture of Climate Friendly Sustainable Practices* 5, 6, http://moef.gov.in/wp-content/themes/moef-green/ebook/paramparaebook/Parampara_2016_Ebook.html#page/5 (accessed on 07 July 2023).

occupied by the cattle due to extremely cold weather. In in ancient India Sun was regarded as God and due to this Sun temple was also made in Odisha in 13th century it is one of the UNESCO World Heritage sites.[31]

6.4.12. Use of Sun for Doing Household Things

Andhra and Telangana people dry their green chili for long use which is known as Challa Mirapakayalu. Sunlight was also used for drying clothes in the open and is used to save energy if you look into the modern cloth dryer which accounts for massive 12% of electricity used in a typical household in countries which are developed. India Bing tropical nation enjoys the sunlight and we can see that in whole country drying cloth in sunlight is in use which is also eco-friendly method. Food in drying in Sun like is highly preferred method for preservation in India for example coconut is dried in sunlight which is preserve for the whole year the dried coconut is called Copra.[32]

In Maharashtra in Solapur region there is a tradition of making Sun dried Bajra Bhakri which can be stored for a week without refrigeration which is eaten with groundnut chili chutney.[33]

6.4.13. Wind Energy

Wind energy has been recognized as a very important source of energy for mechanical purpose in India since long time. Wind energy was used for sailing boat. It was also used for the separating grain from the chaff. This method of separation was developed by ancient cultures. In ancient India Mughal architecture used to make air conditioning Mahal. Hawa Mahal is one of the examples of search architecture.[34]

6.5. MITIGATION AND ADAPTATION STRATEGIES FOR COMBATING CLIMATE CHANGE

The problem of climate change poses threat to humanity. India being developing country is committed to addressing the challenge of climate change by adhering to the paramountcy of UNFCCC. India being responsible country in the world and firm in addressing global challenge of climate change is working towards fulfilling its commitment made under

[31] *Ibid* 8.

[32] *Ibid* 9.

[33] *Supra* note 30 at 10.

[34] *Supra* note 30 at 11-12.

Paris agreement. Hereinafter mentioned initiatives are being done by Indian government as mitigation and adaptation strategies for combating climate change.

6.6. CLEAN AND EFFICIENT ENERGY SYSTEM

Energy is a dynamic input for production and growth. Bearing in mind universal energy access and energy security as one of the fundamental development goals of the country, Government of India (GoI) has started a two-pronged approach to cater to the energy demand of its citizens while safeguarding minimum growth in carbon emissions. If we see the power generation in India, the Government is promoting greater use of renewables in the energy mix mainly through solar and wind power and shifting towards supercritical technologies for coal-based power plants. On the demand side, efforts are being made to efficiently use energy through various innovative policy actions under the overall ambit of Energy Conservation Act, 2001.

On August 2021, installed renewable power capacity (excluding hydro above 25 MW) has already crossed 100 GW, contributing to about 24% of the country's installed electricity capacity. India stands at 4th position in the world in terms of installed RE capacity, 5th in solar and 4th in wind in terms of installed capacity in August 2021.[35] Besides, comparison between 2005 and 2020 shows that dependence on other sources like gas and diesel for meeting power demand has declined over the years.

The share of renewable energy continues to be progressively increased in the electricity mix with renewable energy generation in India crossing 100 billion units. The cumulative renewable power installed capacity (excluding hydro above 25 MW) has increased by 2.6 times from 35 GW on 31 March 2014 to 90.39 GW on 30 November 2020 and constitutes over 24% of the country's installed power capacity in India. With the inclusion of large hydro, the total installed capacity would be 146 GW in India and the share of renewable energy in installed capacity would be over 38%. The share of non-fossil sources in installed capacity of electricity generation increased from 32.31% in March 2014 to 37.99% in September 2020.[36]

---

[35] Ministry of New and Renewable Energy, (2021). *India achieves 100 GW Milestone of Installed Renewable Energy Capacity*. Press Information Bureau, https://pib.gov.in/PressReleasePage.aspx?PRID=1745254 (accessed on 07 July 2023).

[36] *Supra* note 12 at 206.

6.7. PROMOTION OF CLEAN ENERGY/ RENEWABLE ENERGY

India is running one of the major renewable capacity expansion programs in the world. This increase in the previous decade in generation of renewable energy is to be significantly scaled up with the aim to achieve 175 GW renewable energy capacity till 2022 which will be enhanced to 450 GW subsequently till 2030.[37] India's RE target is to be made up by 100 GW from solar PV power, 60 GW from wind power, 10 GW from bioenergy, and 5 GW from small hydropower. Installed capacity of renewable energy sources in India has crossed the mile-stone of 100 GW on 12 August 2021.[38] This includes solar power installed capacity of 36.91 GW.[39] The solar and wind energy sectors have seen rapid growth owing to policy and regulatory support, at both the central and state levels in India. State regulators have been periodically issuing policies and regulations for renewable energy with a special focus on solar energy. India has also become global solar alliance, members of International Solar Alliance (ISA) all countries located between the Tropic of Cancer and the Tropic of Capricorn. India now is currently at 5th Global Position for overall installed renewable energy capacity.[40]

6.7.1. Wind Energy

It has been the predominant contributor to the renewable energy growth in India accounting for wind power 38.43 GW of the renewable installed capacity. The cumulative installed capacity of wind power in the country was 38.789 GW (as on 28 February 2021),41 making India fourth highest in the world, after China, USA, and Germany.42 With a potential of more than

---

[37] Prime Minister's Office, (2020). *PM's address on 8th Convocation Ceremony of Pandit Deendayal Petroleum University*. Press Information Bureau, https://pib.gov.in/PressReleseDetail.aspx?PRID=1674707 (accessed on 07 July 2023).

[38] Ministry of New and Renewable Energy, (2021). *India achieves 100 GW Milestone of Installed Renewable Energy Capacity*. Press Information Bureau, https://pib.gov.in/PressReleasePage.aspx?PRID=1745254 (accessed on 07 July 2023).

[39] Ministry of New and Renewable Energy, (2020). *Prime Minister to inaugurate REINVEST – 2020 on 26 November 2020*. Press Information Bureau, https://pib.gov.in/PressReleasePage.aspx?PRID=1674019 (accessed on 07 July 2023).

[40] Ministry of New and Renewable Energy, (2021). *India achieves 100 GW Milestone of Installed Renewable Energy Capacity*. Press Information Bureau, https://pib.gov.in/PressReleasePage.aspx?PRID=1745254 (accessed on 07 July 2023).

[41] https://mnre.gov.in/the-ministry/physical-progress (accessed on 07 July 2023).

[42] *Supra* note 12 at 208.

100 GW, the aim is to achieve a target of 60 GW of wind power installed capacity by 2022.43

6.7.2. Solar Power

Solar power in India is poised to grow significantly with Solar Mission as a major initiative of the Government of India. Installed capacity of renewable energy sources as on 12 August 2021 stands over 100 GW in India.[44] This includes solar power installed capacity of 36.91 GW[45] and now it has increased more than 42 GW as on August 2021.[46] The ambitious solar expansion program seeks to enhance the capacity to 100 GW by 2022, which is expected to be scaled up further thereafter. A scheme for development of 25 *Solar Parks*, *Ultra Mega Solar Power Projects*, canal top solar projects and one hundred thousand solar pumps for farmers is at different stages of implementation. Government of India is also promoting solarization of all the 55,000 petrol pumps across the country out of which about 3,135 petrol pumps have already been solarized.[47] Recent initiative in this field is Bhadla Solar Park, Reva Mega Solar Plant, Kurnool Ultra Mega Solar Park, etc.

6.7.2.1. Recent Initiatives

- **Biggest Floating Solar Power Plant**: The biggest floating solar power plant of India is going to be set up at Ramagundam in Telangana. The capacity of the solar power plant is 100 MW. It is projected to be opened in May 2021. The floating solar power plant is being set up at the Ramagundam Thermal Power Plant reservoir. The area in which solar power plant is to be set up is 450 acres of the reservoir. This project is commissioned by National Thermal Power Corporation (NTPC). NTPC targets to reduce its carbon footprints and increase its green energy production to 30% of its capacity through this solar power plant. The cost of the project has been valued as Rs. 423 crores. The solar power plant

[43] India plans to produce 175 GW of renewable energy by 2022.

[44] Ministry of New and Renewable Energy, (2021). *India achieves 100 GW Milestone of Installed Renewable Energy Capacity.* Press Information Bureau, https://pib.gov.in/PressReleasePage.aspx?PRID=1745254 (accessed on 07 July 2023).

[45] Ministry of New and Renewable Energy, (2020). *Prime Minister to inaugurate REINVEST-2020 on 26 November 2020*. Press Information Bureau. https://pib.gov.in/PressReleasePage.aspx?PRID=1674019 (accessed on 07 July 2023).

[46] Ministry of New & Renewable Energy, Government of India, https://mnre.gov.in/ (accessed on 07 July 2023).

[47] *Supra* note 4.

will have 4.5 lakh photovoltaic panels.[48]

- **World largest Floating Solar Power Plant:** The largest floating solar power plant in the world is being set up by India on Omkar Eshwar Dam on Narmada River in Madhya Pradesh. The cost of this project is likely to be Rs. 3,000 crores. This project is expected to start its power generation by 2022–23.

  The panels of the solar power plant are capable of adjusting their positions when the water level of the dam is low. The financial aid being given to this project by The World Bank, Power Grid and International Finance Corporation The power plant is designed in such a way to withstand strong waves and floods and continue to generate power at the same time.[49]

- **IIT Bombay's Energy Swaraj Yatra Bus:** The professor of Indian Institute of Technology (IIT) Bombay, Dr. Chetan Singh Solanki has built 'Energy Swaraj Yatra' bus on which The Union Education Minister, Ramesh Pokhriyal 'Nishank,' took his ride. The "Energy Swaraj Yatra" bus runs on the solar energy by comprising a complete work-cum-residential unit. The professor of IIT Bombay has been traveling on the bus to raise awareness regarding the use of solar power since the year 2020 and his yatra will continue till 2030. While taking the ride, the Education Minister Ramesh Pokhriyal 'Nishank,' highlighted that, this project is in line with the "National Education Policy" which comprises of an "in-built framework for imparting skills" with respect to the use of solar energy in schools and colleges. The education minister also stated that, this bus has been designed with the objective of creating a public movement for the adoption of 100% solar energy. The bus allows the occupant to perform all the daily activities such as bathing, cooking, sleeping, working, meeting and training. This bus has been fitted with the 3.2 kW solar panels having the capacity of 6 kWh of battery storage.[50]

---

[48] *India's Biggest Floating Solar Power Plant Coming up in Telangana,* (2021). Economic Times, https://energy.economictimes.indiatimes.com/news/renewable/indias-biggest-floating-solar-power-plant-coming-up-in-telangana/81881262 (accessed on 07 July 2023).

[49] *World's largest floating solar project to start in MP by 2023: State Govt,* (2021). Economic Times, https://energy.economictimes.indiatimes.com/news/renewable/worlds-largest-floating-solar-project-to-start-in-mp-by-2023-state-govt/80125623 (accessed on 07 July 2023).

[50] Bhoomika, A., (2021). *Union Education Minister Takes Ride In 'Energy Swaraj Yatra' Bus Built by IIT Bombay Professor*. NDTV. https://www.ndtv.com/education/union-education-minister-takes-ride-in-energy-swaraj-yatra-bus-built-iit-bombay-professor (accessed on 07 July 2023).

- **World's Largest Solar Rooftop:** One of the world's largest solar rooftops was installed on Braborne Stadium at the Cricket Club of India (CCI) in Mumbai, Maharashtra in August 2018.[51] Solar rooftop consists of 2,280 solar panels of 360-watt peak (Wp) rating, each on the roof of the west and north stands of the stadium. The combined power generation capacity of these solar panels commissioned by Tata Power is 820 kW (kilowatt).[52]
- **World's Largest Solar Park:** Bhadla Solar Park, located in Bhadla, in Jodhpur district of Rajasthan is the largest solar park in the world (as on March 2020). The Bhadla Solar Park is spread over 10,000 acres (40 km) and has a capacity of 2,245 MW. This project is a joint effort between the NTPC Limited and the Solar Energy Corporation of India (SECI), both of which are public sector enterprises under the MNRE. Before Bhadla, Solar Park Pavagada, in Karnataka, with a capacity of 2,050 MW was the largest solar park in the world.[53]

NTPC Renewable Energy Ltd (NTPC REL) on 12th July 2021 has been given the go-ahead by MNRE under Mode 8 (Ultra Mega Renewable Energy Power Park) of Solar Park Scheme and it will set up India's largest solar park of 4.75 gigawatt (GW) at Rann of Kutch in Gujarat from where it will also generate green hydrogen on a commercial scale. As a part of its green energy portfolio augmentation, it aims to build 60 GW Renewable Energy Capacity by 2032. Currently, the state owned power major has an installed capacity of 66 GW across 70 power projects with an additional 18 GW under construction.[54] Grid-connected Rooftop Solar (RTS) Program Phase II of the grid-connected rooftop solar program had been approved in February 2019, with a target for achieving a cumulative capacity of 40,000 MW from RTS Projects by 2022.[55]

---

[51] *Brabourne Gets World's Largest Rooftop Solar Plant,* (2018). The Hindu, https://www.thehindu.com/news/cities/mumbai/brabourne-gets-worlds-largest-rooftop-solar-plant/article24795400.ece (accessed on 07 July 2023).

[52] *World's Largest Solar Rooftop Installation on a Cricket Stadium*, https://www.tatapower.com/plants-projects/solar-rooftopprojects/cci.aspx (accessed on 07 July 2023).

[53] Priya, S., (2020). *With 2,245 MW of Commissioned Solar Projects, World's Largest Solar Park is Now at Bhadla.* https://mercomindia.com/world-largest-solar-park-bhadla/ (accessed on 07 July 2023).

[54] *NTPC to Set up India's Single Largest Solar Park at Rann of Kutch,* (2021). Press Releases. https://www.ntpc.co.in/en/media/press-releases/details/ntpc-set-india%E2%80%99s-single-largest-solar-park-rann-kutch (accessed on 07 July 2023).

[55] Cabinet Committee on Economic Affairs (CCEA), (2019). *Cabinet approves Phase-II of Grid Connected Rooftop Solar Program for Achieving Cumulative Capacity of 40,000 MW from Roof-*

- **Rewa Solar Power Project:** The Hon'ble Prime Minister Narendra Modi inaugurated the Rewa solar power project in Madhya Pradesh on 10 July 2020 and dedicated Asia's largest solar power project to the nation. This new solar power project 750 MW solar power plant project will reduce emissions equivalent to approximately, 1.5 million ton of CO_2 per year. It exemplifies India's commitment to attaining the target of 175 GW of installed renewable energy capacity by 2022.[56]
- **Solar Energy in Airports:** Cochin International Airport Limited (CIAL) is the first airport in the world which became fully powered by solar energy in August 2015.[57] Airport Authority of India (AAI) has adopted a similar model for its other airports in India.[58] For example, AAI has also set up solar power plants of 38.787 MWp capacity, including 15 MW solar power plant at Kolkata Airport.[59] Further, AAI has also identified Hubbali, Belgavi and Mangalore airports for operating on solar power. AAI has introduced airport GHG emission management under Airport Council International – Airport Carbon Accreditation (ACI-ACA) program at Kolkata, Varanasi, Bhubaneswar and Trivandrum airports to measure carbon emission saved by clean renewable resources.[60]
- **Solar Dryer cum Space Heating System (SoLDry) for UT of Ladakh:** The UT of Ladakh is among the areas that receive the highest solar radiation in the India. On average, this region gets about 20% more solar energy than other parts of the country. It is estimated that above more than 50% of the fruit and vegetable

---

top Solar Projects by the Year 2022. Press Information Bureau. https://pib.gov.in/Pressreleaseshare.aspx?PRID=1565282 (accessed on 07 July 2023).

[56] *PM Narendra Modi inaugurates Asia's Largest Solar Power Project in MP's Rewa*, (2020). Economic Times. https://energy.economictimes.indiatimes.com/news/renewable/pm-narendra-modi-inaugurates-asias-largest-solar-power-project-in-mps-rewa/76887860 (accessed on 07 July 2023).

[57] *Kochi Airport Becomes World's First to Completely Operate on Solar Power*, https://cial.aero/pressroom/newsdetails.aspx?news_id=360 (accessed on 07 July 2023).

[58] Sasikala, R., (2019). *AAI Adopts Cochin Model at Other Airports*. SAWDUST. https://www.sawdust.online/news/aai-adopts-cochin-model-at-other-airports/ (accessed on 07 July 2023).

[59] Priya, S., (2017). *Kolkata Airport Goes Green, Now Has Fully Functional Solar Power Plant*. The Times of India. https://timesofindia.indiatimes.com/travel/destinations/kolkata-airport-goes-green-now-has-fully-functional-solar-power-plant/as61930628.cms (accessed on 07 July 2023).

[60] Government of India, Ministry of Civil Aviation Lok Sabha, Unstarred Question No.: 3035 (To Be Answered on the 11th July 2019), *Use of Green Energy by Airports*, http://164.100.24.220/loksabhaquestions/annex/171/AU3035.pdf (accessed on 07 July 2023).

produce of Ladakh, including some of the world's best varieties of apricot, is wasted due to lack of access to market. The National Institute of Solar Energy (NISE) has successfully deployed 10 solar dryer units (5 in Leh and 5 in Kargil) on a trial basis in July 2018 to dry apricots and evaluate their performance in actual field conditions.[61] Based on the system's satisfactory performance, the horticulture department of Jammu and Kashmir has requested NISE to supply and install 670 Unit of solar dryer cum space heating system. Presently 300 systems are under installation in phase 1.[62]

- **Solarization of Konark:** Konark which is located in Puri district of the State of Odisha and is home to the 13th century Sun Temple, a UNESCO World Heritage Site. GoI has launched a scheme in May 2020 with the objective to develop the historical Sun Temple town of Konark in Odisha as "Surya Nagri," to convey a message of synergy between the modern use of solar energy and the ancient Sun temple, to give emphasis to the importance of promoting solar energy. This scheme of solarization envisages setting up of a 10 MW grid-connected solar project and various solar off-grid applications like solar trees, solar drinking water kiosks, and off-grid solar power plants with battery storage with 100%. It is anticipated that all the energy requirements of Konark town will be met by solar energy.[63]

6.7.3. Biomass Energy

Biomass energy constitutes about 18% of total primary energy use in the country and more than 70% of the country's population depends on it. However, it is currently used in an inefficient manner with high levels of indoor pollution. A number of programs have been initiated for promotion of cleaner and more efficient use, including biomass-based electricity

[61] Shotopa, (2018). *Apricot Solar Dryer Cum Space Heating Distributed in Kargil. Reach* Ladakh Bulletin. https://www.reachladakh.com/news/social-news/apricot-solar-dryer-cum-space-heating-distributed-in-kargil (accessed on 07 July 2023).

[62] See Al-Busoul, M., (2017). Design of fruits solar energy dryer under climatic condition in Jordan. *Journal of Power and Energy Engineering, 5*(2). https://www.scirp.org/journal/paperinformation.aspx?paperid=74446 (accessed on 07 July 2023).

[63] Ministry of New and Renewable Energy, (2020). *Ministry of New and Renewable Energy. Government of India Launches Scheme for 100% Solarization of Konark Sun Temple and Konark Town.* Press Information Bureau, https://pib.gov.in/PressReleasePage.aspx?PRID=1625340 (accessed on 07 July 2023).

generation. It is envisaged to increase biomass installed capacity to 10 GW by 2022. Bioenergy current capacity is 10.31 GW.[64]

6.7.4. Hydropower

Hydropower contributes about 46.32 GW to current portfolio of installed capacity,[65] of which small hydro 4.74 GW is small hydro. Special programs to promote small and mini hydel projects, new and efficient designs of water mills have been introduced for electrification of remote villages. With a vast potential of more than 100 GW, a number of policy initiatives and actions are being undertaken to aggressively pursue development of country's vast hydro potential.[66]

- **Ladakh to Get India's First Geothermal Power Project:** The first ever geothermal power project of India will be established at Puga village of eastern Ladakh. This decision was taken because, Puga village in Ladakh was identified as the hotspot of geothermal energy by the scientists. This project will be completed in four phases. A tripartite Memorandum of Understanding (MoU) to establish and implement the project the first phase was signed on 6th February, 2021 between Oil and Natural Gas Corporation Limited (ONGC) Energy; LAHDC, Leh and the Power Department of UT Ladakh. This geothermal project will be known as Geothermal Field Development Project and will be commission by the end of 2022.[67]
- **Renewable Energy Country Attractiveness Index:** Ernst and Young Global Limited in its 53rd edition of the Renewable Energy Country Attractiveness Index (RECAI) report 2019 ranked 40 countries on attractiveness based on renewable energy investment and deployment opportunities in which India ranked 4th and in latest report which is published in May 2021 India ranked 3rd. This ranking was based on different parameters such as economic stability, investment climate, security and supply of energy, clean energy gap, affordability of energy, policy enablement through

---

[64] *Supra* note 12 at 207.

[65] *Power Sector at a Glance ALL INDIA*, (2021). https://powermin.gov.in/en/content/power-sector-glance-all-india (accessed on 07 July 2023).

[66] *Supra* note 12 at 207.

[67] Sanjay, D., (2021). *Ladakh to Get India's First Geothermal Power Plant.* Times of India. https://timesofindia.indiatimes.com/business/india-business/ladakh-to-get-indias-first-geothermal-power-plant/articleshow/80755083.cms (accessed on 07 July 2023).

political stability and support for renewables, project delivery in terms of energy market access, infrastructure, finance and technology potential based on natural resource, power off-take attractiveness, potential support, technology maturity and forecast growth and pipeline.[68]

6.7.5. Nuclear Power

India is promoting Nuclear Power as a safe, environmentally benign and economically viable source to meet the increasing electricity needs of the country. At present, the Nuclear Power Corporation India Limited (NPCIL) is operating 22 commercial nuclear power reactors with an installed capacity of 6,780 MW as on July 2021, which is about 2% of the power generation capacity from all sources in the country.[69] NPCIL has eight reactors is under various stages of construction, totaling 6,200 MW capacity. Nuclear projects are given priority and considered as must run projects on account of their inherent advantages and the government has committed nuclear capacity addition of 3,300 MW during 2017–22 and a further capacity addition of 6,800 MW during the period of 2022–27. The current nuclear power capacity of 6,780 MW is proposed to be increased to 22,480 MW by 2031.[70] For ensuring better utilization of nuclear energy, two public sector enterprises of the Department of Atomic Energy (DAE), Nuclear Power Corporation of India Limited (NPCIL) and Bharatiya Nabhikiya Vidyut Nigam Limited (BHAVINI) are already in place for generation of power from nuclear energy.[71]

6.7.6. Clean Coal Policies

Coal based power as of now accounts for about 52.6% (202 GW) of India's installed capacity as on July 2021.[72] In order to secure reliable, adequate and affordable supply of electricity, coal will continue to dominate power generation in future. More than 73% of total electricity generation comes

[68] *Renewable Energy Country Attractiveness Index (RECAI).* https://www.ey.com/en_in/recai (accessed on 07 July 2023).

[69] *Power Sector at a Glance ALL INDIA* (2021). https://powermin.gov.in/en/content/power-sector-glance-all-india (accessed on 07 July 2023).

[70] Department of Atomic Energy, (2020). *Nuclear Power Plants.* Press Information Bureau. https://pib.gov.in/PressReleseDetail.aspx?PRID=1605939 (accessed on 07 July 2023).

[71] *Utilization of Nuclear Energy.* Department of Atomic Energy. Lok Sabha Unstarred Question No. 535. http://www.dae.gov.in/writereaddata/lsusq535(1).pdf (accessed on 07 July 2023).

[72] Central Electricity Authority, Ministry of Power, Government of India, https://powermin.gov.in/en/content/power-sector-glance-all-india (accessed on 07 July 2023).

from coal/lignite-based power plants. Already, 66 supercritical units with a total capacity of 45,550 MW have been installed, and 51,260 MW of supercritical capacity was under construction as of March 2018 in India. The National Electricity Plan of India of 2018 included plans to build 94 GW of new coal fired capacity (mainly supercritical coal units) between 2017/18 and 2026/27. Presently, 22 have been permitted, while 50 are under construction.[73] A total of 4,240 MW in 2017–18, 4,240 MW in 2018–19 and 6,220 MW in 2019–20 of supercritical power units were commissioned.[74]

6.7.8. Coal to Gas

Coal gasification is an alternate use of coal considering the environmental apprehensions of burning coal in thermal power plants. India's effort to gasify coal is affected due to the non-availability of technologies to gasify high ash Indian coal. There is a need to develop efficient low-cost gasification technology and support from technology providers of USA and China to transfer these technologies at reasonable cost. India targets for 100-million-ton (MT) coal gasification by 2030 with investments worth over INR 4 trillion. Encouraging use of clean sources of fuel, government has provided for a concession of 20% on the revenue share of coal used for gasification. This is to boost the production of synthetic natural gas, energy fuel, and urea for fertilizers and production of other chemicals.[75] Government of India has already taken several initiatives to improve the efficiency of coal-based power plants and to reduce its carbon footprint. All new, large coal-based generating stations have been mandated to use the highly efficient supercritical technology. Renovation and Modernization (R&M) and Life Extension (LE) of existing old power stations is being undertaken in a phased manner. About 144 old thermal stations have been assigned mandatory targets for improving energy efficiency. Coal beneficiation has been made mandatory. Introduction of ultra-supercritical technology, as and when commercially available is part of future policy. Besides, stringent emission standards being contemplated for thermal plants would significantly reduce emissions.

---

[73] *Supra note* 12 at 213-214.

[74] *Review of Progress of Thermal Power Projects Under Execution in the Country*. Central Electricity Authority, Ministry of Power, Government of India. https://www.cea.nic.in/reports/quarterly/tpmii_quarterly_review/2020/tpmii_qr06.pdf (accessed on 07 July 2023).

[75] Ministry of Coal, (2020). *India aims to achieve 100 MT coal gasification target by 2030: Pralhad Joshi*. Press Information Bureau. https://pib.gov.in/PressReleaseIframePage.aspx?PRID=1650096 (accessed on 07 July 2023).

Recently The Ministry of Environment, Forest and Climate Change (MoEFCC) recently amended the rules of thermal power plants emission standards. The new rule relating to emission standards has extended the deadline for the thermal power plants within ten kilometers of National Capital Region and in cities with more than ten lakh population to comply with the new emission norms by the end of 2022. The deadline for the thermal power plants in "Non-Attainment Cities" to meet the emission norms has been extended till December 31, 2023. The non-attainment cities are those cities that have consistently failed to meet the National Ambient Air Quality Standards. Around 124 such cities had been identified by the CPCB. The coal fired power plants in the rest of the areas have to comply the new standards by December 2024. In 2015, the Environment Ministry had revised the emission norms for Sulphur Dioxides, Nitrogen Oxides and Particulate Matter released from thermal power plants. According to these norms, the thermal power plants had to install emission control systems by 2017. However, the deadline was pushed to 2022 due to implementation issues. According to a report from the Centre for Research on Energy and Clean Air, the Sulphur Dioxide emissions in India was declined by 6% in 2019 as compared to 2018. Still, India is the largest emitter of Sulphur Dioxide. India has chosen the Flue Gas Desulfurization technique for reducing the emissions of toxic Sulphur Dioxide. The process of removal of Sulphur Dioxide is called Flue Gas Desulfurization. The method removes gaseous pollutants such as Sulphur Dioxide from exhaust flue gases generated from boilers and furnaces.[76]

6.7.9. National Smart Grid Mission

National Smart Grid Mission has been launched with the object to bring efficiency in power supply network and facilitate reduction in losses and outages. 'Green Energy Corridor' projects worth INR (Indian National Rupee) 380 billion (USD 6 billion) also are being unrolled to make sure evacuation of renewable energy. The Government's goal of *Electricity for All* is sought to be achieved by the above programs that might require huge investments, infusion of latest technology, availability of fuel and international support.[77]

---

[76] Jayashree, N., (2021). *New Norms for Thermal Plants May Dent India's Emission Targets: Experts. Hindustan Times,* https://www.hindustantimes.com/india-news/paying-emission-fines-cheaper-for-thermal-plants-than-following-norms-101618167373910.html (accessed on 07 July 2023).

[77] *Supra* note 4.

6.7.10. Enhancing Energy Efficiency

Energy efficiency-related mitigation actions with the goal of reducing energy intensity of the Indian economy, Ministry of Power through Bureau of Energy Efficiency (BEE) has initiated a number of energy efficiency initiatives.

6.7.10.1. The National Mission for Enhanced Energy Efficiency (NMEEE)

NMEEE[78] aims to strengthen the market for energy efficiency by creating a conducive regulatory and policy regime. It seeks to upscale the efforts to unlock the market for energy efficiency and help achieve total avoided capacity addition of 19,598 MW and fuel savings of around 23 million tons per year at its full implementation stage. The programs under this mission have resulted in an avoided generation capacity addition of about 10,000 MW between 2005 and 2012 with government targeting to save 10% of current energy consumption by the year 2018–19. Demand Side Management programs have been launched to replace existing low-efficiency appliances.[79]

6.7.10.2. Efficient Lighting in India

During the last decade, there has been rapid transformation of efficient lighting in India. Efficient lighting in India in the last few years, technological advancements in lighting have led to the growth of energy-efficient lighting systems that consist of one or more components such as low loss ballasts; constant wattage high-intensity electronic ballasts; Energy-efficient luminaires; and better monitoring and control mechanisms. Under the 'Unnat Jyoti by Affordable LED for All' (UJALA) scheme, LED bulbs, LED Tube lights, and Energy efficient fans continue to be provided to domestic consumers for the replacement of conventional and inefficient variants. Under Street Lighting National Program (SLNP), conventional street-lights are being replaced with smart and energy efficient LED street lights across the whole India[80]. The program is being implemented by Energy Efficiency Services Limited (EESL), a Joint Venture (JV) of PSUs under the MoP, GoI. Till November 2020, over 366.85 million LED bulbs, 7.207 million LED tube lights, and 2.340 million energy efficient fans had

---

[78] The National Mission for Enhanced Energy Efficiency (NMEEE) is one of the eight national missions under the National Action Plan on Climate Change (NAPCC).

[79] https://beeindia.gov.in/content/nmeee-1 (accessed on 07 July 2023).

[80] *Supra* note 12 at 222-223.

been distributed by EESL across India). This has resulted in collective emission reduction of 180.08 $MtCO_2$ from 2014–15 to November 2020. Through the scheme, EESL has also been able to engage with the common man at a significant scale. So far, more than 90 million consumers have availed the benefit of using these LED bulbs, and thus making it the largest non-subsidy-based LED lighting program in the world. Under SLNP, till September 2020, EESL has installed over 11.25 million LED street lights in Urban Local Bodies (ULBs) and Gram Panchayats across India. This has caused in cumulative energy savings of 18.071 billion units and emission reduction of 14.82 $MtCO_2$ from 2015–16 to 2019–20. Under SLNP, 1,508 ULBs have been enrolled and out of these ULBs, work has been finished in 972 ULBs. This is the largest installation of LED street lights across the world. UJALA and SLNP are voluntary adoption schemes which run without any budgetary support from the GoI and are based on a sustainable business model where the cost is repaid by consumer/ULBs from savings in energy and maintenance expenditure over some time through savings in electricity bills. During the recent years, MNRE has launched Atal Jyoti Yojana (AJAY) Phase-I and Phase-II Schemes and Off-grid and Decentralized Solar PV Applications Phase-III Scheme. Bearing in mind the success of Phase-I of AJAY, the Phase-II was launched in December 2018 with expansion of Phase-I and Solar LED Street Lights (SSLs) of higher illumination (12 W LED with 75 Wp solar panel). [81]

6.7.10.3. Perform, Achieve, and Trade (PAT)

Perform, Achieve and Trade (PAT) Scheme during the first cycle of PAT (2012–15), an energy saving of 8.67 $MtCO_2$ was achieved against the target of 6.686 MtCO2 assigned for 478 designated consumers (DCs). This translated into avoiding about 31 MtCO2 of emissions. PAT II (2016–17 and 2018–19) had been implemented with the objective of widening the reach by increasing the sectors under the scheme and the number of DCs under it. Three new sectors have been added, making it a total of 11 sectors under the scheme and an additional 89 DCs. This 2nd cycle resulted in the total savings of about 13.28 $MtCO_2$, translating into 61.34 $MtCO_2$.[82]

6.7.10.4. Standards and Labeling Program

Standards and Labeling Program launched by the Government of India

[81] *Supra* note 12 at 222-223.

[82] *Supra* note 12 at 217.

enables consumers to make informed decision by providing information about the energy consumption of an appliance. Currently, 26 equipment and appliances are labeled out of which 10 appliances are under the mandatory regime and the remaining 16 appliances under the voluntary regime. The energy saving achieved under the scheme in the year 2018–19 amounts to 55.69 billion units translating to an abatement of 45.67 MtCO_2 in emission.[83] The program has contributed to an increase of 25% to 30% in the energy efficiency of an average refrigerator or air-conditioner in 2014 compared to those sold in 2007. Super-Efficient Fan (that uses half as much energy as the average fan) program has been launched. Further, two sets of Corporate Average Fuel Consumption standards for cars have been notified, with one coming into force in 2017 and the second set in 2022. The Standards and Labeling (S&L) scheme was initiated with the key objective of providing consumers an informed choice regarding energy savings and thereby the cost-saving potential of various energy-consuming appliances. S&L scheme covers the star labeling program for 26 appliances.[84]

6.7.10.5. Partial Risk Guarantee Fund for Energy Efficiency (PRGFEE)

A risk sharing mechanism to provide financial institutions with a partial coverage of risk involved in extending loans for energy efficiency projects, and Venture Capital Fund for Energy Efficiency (VCFEE), a trust fund to provide "last mile" equity capital to energy efficiency companies, have been established.

6.7.11. Carbon Capture, Storage/Utilization

CO_2 capture is still at an emerging stage of implementation in India, with only a few sectors such as refineries and steel having implemented small-scale carbon capture projects. Some industries which have initiated the process of setting up CCS facilities include are National Aluminum Company

---

[83] Communication received from Ministry of Power on Biomass co-firing at NTPC Dadri, Ultra super critical plants, Standards and labelling scheme, Program on Energy from Urban, Industrial and Agricultural Wastes/Residues, Street Lighting National Program, Building Energy Efficiency Program (BEEP) – retrofitting project, PAT cycles, Promoting Market Transformation for Energy Efficiency in MSMEs, Energy Efficiency Financing Platform (EEFP), Framework for Energy Efficient Economic Development (FEEED), Agriculture Demand Side Management (AgDSM), Municipal Demand Side Management (MuDSM), Capacity building of DISCOMs, Activities & Achievements from FY 2017-2020 under DSM Program, Fuel efficiency standard and Fuel economy norms for light and commercial vehicles, e-mail dated 11 June 2020 and 21 September 2020.

[84] *Supra* note 12 at 218.

(NALCO), ONGC, Bharat Heavy Electrical Limited. (BHEL), and Andhra Pradesh Power Generation Corporation (APGENCO).[85] Dalmia Cement has also announced setting up of a large-scale carbon capture unit (CCU) at one of its plants in Tamil Nadu as a move towards becoming carbon neutral by 2040. From the several technologies available for CO_2 capture, the post-combustion solvent capture and stripping with amine-based pressure swing adsorption system is the most common.[86]

6.7.12. Energy Access and Clean Fuels

Energy access and clean fuels constitute two key areas of concern of the Government, as these are closely related to fundamental aspects of the well-being of the population, especially women.

6.7.13. Pradhan Mantri Ujjwala Yojana (PMUY)

PMUY was launched in 2016, for safeguarding the health of women and children by providing families that fall Below the Poverty Line (BPL) with clean cooking fuel (LPG), to avoid cooking in smoke-filled kitchens using biomass. The primary objective was to reduce and gradually move away from dependence on biomass as the primary cooking fuel. Under this scheme, with a cap of INR 1,600 per LPG connection, cash support for connection is provided to BPL families as a move to empower the female members of these households, especially in rural areas, the LPG connections are issued in their names. The aim for this scheme was raised from 50 million households to 80 million households. The target of providing 80 million LPG connections under the scheme has already been achieved as on 7 September 2019 i.e., 7 months earlier of the scheduled target (March 2020). Till December 2020, a total of 287.4 million households has LPG connections (including PMUY beneficiaries. [87]

6.7.14. Deen Dayal Upadhyaya Gram Jyoti Yojana (DDUGJY)

A scheme of Government of India for rural areas, the Deen Dayal

[85] *Industries Which Have Initiated the Process of Setting up CCS Facilities*. https://www.global-cement.com/news/item/9887-dalmiacement-takes-steps-towards-carbon-capture (accessed on 07 July 2023).

[86] *Supra* note 12 at 227.

[87] LPG Profile, Data on LPG Marketing, Petroleum Planning & Analysis Cell, Ministry of Petroleum & Natural Gas, Government of India. https://www.ppac.gov.in/WriteReadData/Reports/201911220459272617502WebVersion LPGProfile1.10.2019.pdf (accessed on 07 July 2023).

Upadhyaya Gram Jyoti Yojana (DDUGJY) was launched by Ministry of Power on 3 December 2014. Under the scheme of DDUGJY, electrification of inhabited un-electrified census villages taken up in 2018- 19 in 1,515 villages (MoP, 2019). Under the scheme, during the XIth five-year plan, 570 projects of INR 2,88,690 million were sanctioned and a grant of INR 2,28,542 million had been released against these projects as on 31 December 2019. During the XII-year Plan, 560 projects of INR 2,37,356 million had been sanctioned and a Grant of INR 1,59,502 million had been released against these projects as on 31 December 2019.[88]

6.7.15. Saubhagya Scheme Pradhan Mantri Sahaj Bijli Har Ghar Yojana (SAUBHAGYA)

SAUBHAGYA was launched in September 2017 with the objective of universal household electrification by providing last mile connectivity and electricity connections to all remaining un-electrified households in rural and urban areas by 31 March 2019. Electricity connections to 26.284 million households (99.93%) have been provided from 11 October 2017 to November 2020.[89]

6.8. MITIGATION ACTION IN BUILDING SECTOR

Mitigation action by taken under building sector. India has been improving the energy efficiency of buildings through mandatory building energy codes and voluntary rating schemes, as well as through policies and programs for improving the efficiency of appliances and equipment. In order to both recognize energy-efficient buildings, as well as to stimulate their large-scale replication, India has developed its own building- energy rating system Green Rating for Integrated Habitat Assessment (GRIHA), based on 34 criteria like site planning, conservation and efficient utilization of resources etc. A number of buildings including Commonwealth Games Village have been rated using GRIHA system. Indira Paryavaran Bhawan, the headquarters of Central Government's MoEFCC is a model building of Government of India and has received LEED India Platinum and a 5 Star GRIHA rating. It is a 'Net Zero Energy' building with 100% onsite power generation. Till October

---

[88] Deen Dayal Upadhyaya Gram Jyoti Yojana, Lok Sabha, Unstarred Question No. 753, Ministry of Power, Government of India. http://164.100.24.220/loksabhaquestions/annex/173/AU753.pdf (accessed on 07 July 2023).

[89] Pradhan Mantri Sahaj Bijli Har Ghar Yojana – Saubhagya Dashboard. Rural Electrification Corporation. https://saubhagya.gov.in/ (accessed on 07 July 2023).

2020, India has 1,825 GRIHA registered projects with approximately 52.5 million sq. meter of 'green' built-up area. India currently has about 7.61 billion sq. ft. green building footprint, 6,055 registered projects, and 780 certified projects [Indian Green Building Council (IGBC) only] as on 15 October 2020.[90]

6.8.1. Energy Conservation Building Code (ECBC)

The ECBC sets minimum energy standards for new commercial buildings. seventeen states and UTs have already adopted and notified the ECBC, 335 demonstration buildings have been supported with technical assistance for ECBC compliance in the States and UTs, out of which 50 buildings are constructed, with a cumulative built-up area of 0.16 billion sq. m by ensuring an approximate energy saving of 0.17 billion units. GHG emission reduction of 0.14 $MtCO_2$ is estimated annually which will scale up to 3.4 $MtCO_2$ in the 25 years' lifespan of the buildings. The Code would be made more stringent to promote construction of even more (Near-Zero) energy-efficient buildings. "Design Guidelines for Energy Efficient Multi-Story Residential buildings" have also been launched.[91]

Under the Smart City Mission, 100 Smart Cities have been selected through a two-stage competition. A total of 5,151 projects worth INR 20,50,180 million have been included by selected 100 cities in their Smart City proposals.

- **ECO Pact: New Low-Carbon range of Concrete:** ACC Ready Mix Concrete has introduced its new low carbon range of concrete called ECO Pact in February 2021. It is a "Green Concrete" that has been introduced for a high-performance, sustainable and circular construction. In the first phase ECO Pact was launched in Mumbai and Hyderabad. It will be rolled out across the country in a phased manner within a few weeks after its launch. The ECO Pact was rolled out with the objectives of enhancement of the transition towards low-carbon and circular construction. The innovative manufacturing process of the ECO Pact will reduce the CO_2 emissions by up to 100%. The ECO Pact range will help the developers to obtain the Green Points from IGBC. The IGBC is a part of the Confederation of Indian Industry (CII). It was established in 2001. The council seeks "To enable a sustainable

[90] *Supra* note *12* at 229-230.

[91] *Ibid.*

built environment for all" and it also seeks to make India a global leader in sustainable built environment by 2025.[92]

- **The India Cooling Action Plan (ICAP):** ICAP was launched on 8 March 2019.[93] It has a long-term vision with a 20-year time horizon and lists out actions that can inter alia help reduce cooling demand. In addition, refrigerant-based space cooling technologies such as air conditioners, the ICAP provides for the use of non-refrigerant based technologies such as fans and coolers and not-in-kind technologies such as district cooling.[94]

6.9. MITIGATION ACTION IN TRANSPORT SECTOR

The transport sector is rapidly growing sector in India, contributing significantly to the GDP of the country. However, this sector is largely oil-dependent and accounts for 12.1% of the country's total GHG emissions (without LULUCF). While India is working in the same global context as other countries who have adopted an Electric Vehicle (EV) policy, its unique mobility pattern necessitates an EV policy that is tailor-made to India's particular needs. India is one of 16 member countries of the Electric Vehicles Initiative (EVI), a multi-government policy forum dedicated to accelerating the introduction and adoption of electric vehicles globally. One of the most important regulatory announcements was the skipping of Bharat Stage (BS)-V emission norms and the advance introduction of BS-VI from 1 April 2020. An increase in the share of alternative fuels in the overall fuel mix is yet another strategy to reduce emissions from the transport sector. GoI has notified the National Policy on Biofuels-2018, which inter-alia aims to develop alternate fuels and to reduce import dependency as one of the objectives. The policy envisages an indicative target of 26% blending of ethanol in petrol and 5% blending of bio-diesel in diesel till the year 2030.[95]

---

[92] Srishti, (2021). *ECOPact – the Green Concrete to create a Sustainable ImPACT.* Architect Interior India. https://www.architectandinteriorsindia.com/products/16048-ecopact-the-green-concrete-to-create-a-sustainable-impact (accessed on 07 July 2023).

[93] Ministry of Environment, Forest and Climate Change, (2019). *India Cooling Action Plan Launched.* Press Information Bureau. https://pib.gov.in/PressReleaseIframePage.aspx?PRID=1568328 (accessed on 07 July 2023).

[94] *India Cooling Action Plan Launched: All about it Union Minister for Environment, Forest and Climate Change, Dr. Harsh Vardhan released the India Cooling Action Plan (ICAP) on March 8 in New Delhi,* (2019). India Today, https://www.indiatoday.in/education-today/gk-current-affairs/story/india-cooling-action-plan-launched-all-about-it-1474037-2019-03-09 (accessed on 07 July 2023).

[95] *Supra* note *12* at 231.

6.9.1. The Green National Highways Corridor

The Green National Highways Corridor project will help in building safer, greener and more resilient national highway corridors in the Indian states of Rajasthan, Himachal Pradesh and Andhra Pradesh has been approved by World Bank. [96]

6.9.2. The National Electric Mobility Mission Plan (NEMMP)

The National Electric Mobility Mission Plan (NEMMP) aimed to deploy 400,000 passenger battery electric vehicles (BEVs) by 2020 as it was mentioned in second BUR. 3 The sales of electric two-wheelers in India has risen from 54,800 units in 2018 to 1,26,000 units in 2019. The Department of Heavy Industry is administering the Faster Adoption and Manufacturing of (Hybrid) and Electric Vehicles (FAME) India Scheme for promotion of electric/hybrid vehicles in India since 1 April 2015. In the first phase of the scheme 0.28 million hybrid and electric vehicles are supported by way of demand incentive amounting to about INR 3,590 million. Phase-II of FAME India Scheme for a period of three years began in April 2019 with a total budgetary support of INR 100,000 million. From April 2019, the second phase of the FAME II scheme encompasses strict speed, range and energy efficiency requirements. The States of Karnataka, Telangana, Maharashtra, Uttar Pradesh, Kerala, Uttarakhand, Andhra Pradesh, Delhi and a few more have rolled out their EV/Draft EV policies. These states are proposing several fiscal incentives to car and battery manufacturers, charging infrastructure companies and consumers. [97]

6.9.3. Metro Rail

Urban Rail, popularly referred to as Metro Rail, has seen substantial growth in India in the recent years. More cities in India are becoming reliant on the metro rail to meet their day to-day mobility requirements. India has currently eleven operational metro systems in 18 cities, while 900 km Metro/RRTS is under construction in 27 cities.[98]

---

[96] *Supra* note *12* at 236.

[97] *Supra* note *12* at 237.

[98] *Ibid.*

6.9.4. Green Indian Railways (IR)

Indian Railways is the third largest rail network in the world. GoI has considers "IR as the growth engine of nation's Vikas yatra" and is continuously striving towards making IR a greener and cleaner sustainable mass transport system. In the union budget 2020–2021 there was provisions for setting up larger solar power capacity alongside the railway tracks on IR. This is expected to give an immense power boost to the national transporter while cutting down its annual electricity bills and contributing to the efforts of IR to become a "net zero" carbon emitter by 2030.[99] Renewable Energy: Presently, the IR has an installed solar power capacity of 103 MW and wind generation capacity of 104 MW. To boost the usage of solar power, Indian Railways Organization for Alternate Fuels (IROAF) is installing flexible solar panels on coaches.[100]

Some of the key initiatives and targets set by the Indian Railways are as follows:

- Enhancement of the share of railways in the overall land-based freight transport from the present around 35% to 45% by the year 2030
- Swachh Rail, Swachh Bharat Campaign for promoting cleanliness across the entire railway.
- Mission Electrification: Currently, 40,886 route km (RKM) (64%) of broad gauge (BG) routes has been electrified by railway. Balance 22,745 RKM are being planned to be electrified in next 3 – 4 years.
- To utilize at least 10% of its energy requirement from renewable sources, thus contributing to India's solar and wind energy missions.
- No single use plastic material has been utilized from 150th Gandhi Jayanti on 2 October 2019.
- As an initiative has been taken to reduce, recycle and dispose plastic waste generated in stations in an eco-friendly manner. For

---

[99] *Indian Railways gears up to become "Green Railway" by 2030*, (2020). The Economic Times. https://energy.economictimes.indiatimes.com/news/power/indian-railways-gears-up-to-become-green-railway-by-2030/76938990

[100] *Indian Railways Hopes to Become World's First 'Green Railway' by 2030; to Be More Planet-Friendly*, (2020). Times of India (accessed on 07 July 2023). https://timesofindia.indiatimes.com/travel/travel-news/indian-railways-hopes-to-become-worlds-first-green-railway-by-2030-to-be-more-planet-friendly/as76941783.cms (accessed on 07 July 2023).

this comprehensive policy guidelines have been brought out for installation of Plastic Bottle Crushing Machines (PBCMs) by Zonal Railways. At present time, around 500 PBCMs have been installed at 370 stations including several district headquarters railway stations over IR.

- All passenger carrying BG coaches, from April 2020 onwards, has been fitted with bio-toilets and direct discharge of human waste from trains has thus been eliminated.
- IR is committed to improve water use efficiency by 20% by 2030.
- Adoption of good practices for green buildings, industrial units and other establishments for the management of resources and infrastructure to achieve environmental sustainability in the growth of Indian railway.[101]

6.9.5. Dedicated Freight Corridors (DFCs)

DFCs have been introduced across the country. Government established 'Dedicated Freight Corridor Corporation of India Limited (DFCCIL),' to undertake planning and development, mobilization of financial 240 resources and construction, maintenance and operation of the DFCs in the first phase, two corridors viz. 1520 km Mumbai-Delhi (*Western Dedicated Freight Corridor*) and 1856 km Ludhiana-Dankuni (*Eastern Dedicated Freight Corridor*) are being constructed. This project is expected to reduce emissions by about 457-million-ton CO_2 over a 30-year period. Key highlights of the initiatives in 2018–19 are Successful trial run of goods train in Bhadan to Khurja section (194 kms) of EDFC-1 was conducted on in November 2018, Trial run of freight train in Madar-Kishangarh Balawas section (306 km) of WDFC was conducted in December 2018, Track linking with mechanized track laying machine is in progress at 7 locations, Track linking of 774 km has been done in the year taking the cumulative linking to 1,898 km.[102]

- **Delhi to Switch All Its Hired Vehicles to EVs in 6 Months:** Manish Sisodia, The Deputy Chief Minister of Delhi, announced that all departments of Delhi government will now use only electric vehicles. This transition to the electric vehicles will be done within 6 months by the Delhi government. Presently The Delhi government has over 2,000 cars in use. Manish Sisodia,

---

[101] *Supra* note *12* at 238.

[102] *Supra* note *12* at 239–240.

is of the view that this decision will inspire other cities in the country as well as world which is a key decision to tackle the issues like pollution and climate change. This announcement was made subsequently when the government launched 'Switch Delhi' campaign. The transition from the diesel or petrol vehicle to EVs will be monitored by the Transport Department of Delhi. Further, the departments under the government are required to produce monthly action reports on the transition to EVs to the nodal department. Switch Delhi Campaign was launched in order to encourage the private vehicle owners to shift to the electric vehicles and install a charging point in their premises within three years by eight-week mass awareness campaign.[103] India has launched its first-ever diesel tractor that has been converted to Compressed Natural Gas (CNG) on February 12, 2021.

- **Green Tax on Vehicles:** The Road transport and highways minister, Nitin Gadkari has announced to impose a green tax on a certain category of vehicles starting from April 1, 2022. The government has proposed that a green tax could be levied on personal vehicles at the time of renewal of registration certificate and it will be levied after 15 years equivalent to 10–25% of the road tax of a petrol or diesel vehicle. Similar tax would also be levied on transport or commercial vehicles which are older than eight years at the time of renewal of the registration certificate and the fitness certificate.

Under the Green tax norms, the Hybrid vehicles, electric vehicles and those vehicles that runs on cleaner alternative fuels like CNG, ethanol and liquefied petroleum gas (LPG) will get exemption from the green tax. Higher taxes will be implied on the petrol or diesel vehicles. The reason behind this is higher tax on Petrol and diesel cars will force people to shift to CNG.

The Air Quality commission which is twenty-member Commission was formed by Ministry of Environment recently directed all the industries in the National Capital Region to switch over to Piped Natural Gas. The deadline of switch over was January 31, 2021.[104]

---

[103] *All Delhi Govt Vehicles to Switch to EVs within 6 Months. A Panacea for Air Pollution?*, (2021). The Wire. https://thewire.in/government/delhi-govt-electric-vehicles-air-pollution-lithium-ion-batteries-coal (accessed on 07 July 2023).

[104] Ministry of Road Transport & Highways, (2021). *Green Tax to be Imposed on Older Vehicles Soon, to Clean up the Environment by Phasing out Unfit and Polluting Vehicles.* Press Information Bureau. https://pib.gov.in/PressReleasePage.aspx?PRID=1692228 (accessed on 07 July 2023).

- **Flexible Fuel Vehicles (FFVs):** The government of India may soon ask the auto companies in India to manufacture the passenger and commercial vehicles that run on the multiple fuel configuration with the aim of reducing the use of polluting fossil fuels and for cutting down the harmful emissions. The government of India is actively looking up after the use of "FFVs" which will ensure the increased use of the bio-fuels to run the vehicles.

In India, the FFVs will have a different advantage because the vehicles will allow the vehicles to use the different blends of ethanol mixed petrol that are available across the country as opposed to the current regulation. The present regulation allows to mix up to 10% ethanol in petrol. But due to shortage in supply and other transportation challenges, 10% blended petrol is available only across 15 states in India. The FFVs would be allowing the vehicles to use all the blends and would also be running on unblended fuel as well.[105]

6.9.6. Shipping

India has 13 major ports, and 200 notified non-major (minor/intermediate) ports along the coastline and sea-islands of India. Recognizing the fuel efficiency, environmental friendliness, and cost-effectiveness, the Government is promoting the growth of Coastal Shipping and Inland Water Transport. The Ministry of Shipping (MoS) has 242 initiated the Green Port project to make ports across India cleaner and greener. This project includes monitoring environmental pollution, setting up wastewater treatment plants, garbage disposal plant, and energy generation from renewable sources (MoS, 2020).

To enhance the inland waterways transport, Government has announced the implementation of Jal Marg Vikas for capacity augmentation of National Water Way-1. It is also proposed to establish integrated Waterways transportation grid with a view to connecting all existing and proposed National waterways with road, rail and ports connectivity. Another initiative in this direction is the Sagarmala Project with the objective to augment port-led development and promote efficient transportation of goods. Bharatmala Project which envisions constructing about 5,000 km of road network all along the coastal areas will further provide connectivity to these ports.[106]

---

[105] *Govt Explores Introduction of 'Flex-Fuel' Vehicles in India,* (2021). The Times of India. https://auto.timesofindia.com/news/others/govt-explores-introduction-of-flex-fuel-vehicles-in-india/articleshow/81645027.cms (accessed on 07 July 2023).

[106] *Supra* note 12 at 241–243.

- **Fully Solar-Powered Commuter Ferry:** India's first solar-powered ferry, the 'Aditya,' which commenced operations in 2017 in Kerala's Vembanad backwaters, has won the prestigious Gustave Trouvé Award for Excellence in Electric Boats and Boating. The Aditya, from Navalt Boats, is a sun powered commuter ferry of great effort for the future of electric marine propulsion.[107]

On 6 February 2020, the Hon'ble Prime Minister, announced in the lower house of Parliament, the government's resolve to develop Ladakh, the nation's newest UT, as a carbon-neutral region. Carbon neutrality for Ladakh is not a mitigation goal, but a development vision. CESL and Ladakh sign a MoU to make the Union Territory carbon neutral CESL will execute solar mini, micro-grid solutions, energy-efficient lighting, battery storage solutions, energy-efficient cooking stoves, and Electric mobility solutions in the UT of Ladakh.[108]

6.9.7. Civil Aviation

Civil Aviation is regarded as the backbone of the Indian transport industry. The use of air transport has facilitated trade and commerce because of its merits of moving goods and individuals in a safe and less time-consuming way. The industry has witnessed dramatic growth in the past few decades and is expected to continue its positive growth in the coming years. The Indian domestic aviation market is the fastest-growing domestic market in the global market in present time. The aviation is slated to overtake its counterparts in countries such as Germany, Japan, Spain and the United Kingdom in the coming decade. In the same period, India is expected to take the third position (behind China and the United States) in the air passenger market. A big jump from its present position of seventh position in the air passenger market is also expected.[109]

---

[107] *India's First Solar Ferry 'Aditya' Wins Global Honor,* (2020). The Logical Indian. https://thelogicalindian.com/sustainability/indias-first-solar-ferry-global-honour-22638 (accessed on 07 July 2023).

[108] Ministry of Power, (2021). *CESL and Ladakh Sign an MoU to Make the Union Territory Carbon Neutral.* Press Information Bureau. https://pib.gov.in/PressReleaseIframePage.aspx?PRID=1725334 (accessed on 07 July 2023).

[109] Krrishan, S., Sarjana, P., et al., (2019). *India: National Green Aviation Policy: India's Step Towards Emission Cuts.* https://www.mondaq.com/india/aviation/847444/national-green-aviation-policy-india39s-step-towards-emission-cuts (accessed on 07 July 2023).

6.9.8. Green Aviation Policy, 2019

The Ministry of Civil Aviation in India aims to achieve the sustainable and inclusive growth of the aviation industry in the country and remedy the ecological concerns posed by the industry. It is in this regard that the Indian Government has proposed a Green Aviation Policy to create a regulatory framework to remedy the environmental issues created by the civil aviation industry by identifying key policy areas that need guiding principles and regulations.[110]

The Green Aviation Policy has been formulated with a vision to enable, promote and strengthen all inclusive, green and sustainable growth of air transport in India. Its mission states that it seeks to 'provide safe and sustainable air travel to various parts of India and the world by minimizing the adverse environmental impacts of civil aviation activities.[111]

The Green Aviation Policy is crucial for the civil aviation industry, especially at a time when the sector is rapidly growing and is poised to take over as the third-largest air passenger market by the end of 2029. The policy is a major step taken by the Indian Government to reduce the pollutant contributions by this industry. The contribution of Indian civil aviation industry is about one percent of total carbon dioxide (CO_2) emissions, this being considerably less than the global average.

This policy will be helpful in terms of closely regulating the industry and making sure that there is utilization of advanced, less polluting and energy-efficient aircraft. This policy places a responsibility on the Directorate General of Civil Aviation (DGCA) to assess and study emissions, and evaluate their contribution to local air quality.[112]

- **Delhi-Dehradun Flights Using Biofuels:** India's first demonstration civil flight powered by indigenous biofuel was operated on 27 August 2018, marking a new chapter in the fast-growing domestic aviation sector. With this demonstration flight, India has become one of the few countries and probably the first

[110] *White Paper on National Green Aviation Policy*, (2019). (Ministry of Civil Aviation Government of India 2/19/2019), https://www.civilaviation.gov.in/sites/default/files/Whitepaper%20on%20National%20Green%20Aviation%20Policy.pdf (accessed on 07 July 2023).

[111] Krrishan, S., Sarjana, P., et al., (2019). *India: National Green Aviation Policy: India's Step Towards Emission Cuts*. https://www.mondaq.com/india/aviation/847444/national-green-aviation-policy-india39s-step-towards-emission-cuts (accessed on 07 July 2023).

[112] Krrishan, S., Sarjana, P., et al., (2019). *India: National Green Aviation Policy: India's Step Towards Emission Cuts*. https://www.mondaq.com/india/aviation/847444/national-green-aviation-policy-india39s-step-towards-emission-cuts (accessed on 07 July 2023).

among developing nations to use biofuel for planes.[113]

- **Climate Actions by Delhi International Airport Limited:** Delhi Airport Located in India's capital city is, managed and developed by Delhi International Airport Limited (DIAL), a joint venture consortium of GMR Group, AAI, and Fraport AG. Environment Sustainability Management is an integral part of DIAL's business strategy, which help in achieving credibility and business sustainability at Delhi Airport. DIAL has adopted the globally recognized ACA Program of Airports Council International (ACI) for effective GHG management at Delhi Airport. Under this framework, Delhi airport became the first Carbon Neutral Airport in the Asia-Pacific region to receive "Carbon Neutral" Accreditation in 2016.

6.10. ADAPTATION AND MITIGATION ACTION IN AGRICULTURE SECTOR

India's voluntary declaration for cutting emission does not mention the activities in the agriculture sector. However, India has undertaken several climate-friendly initiatives to proactively promote sustainable development of the agriculture sector.[114] These initiatives are discussed below. The agriculture sector in India plays a vital role in ensuring food and nutritional security and engages almost two-third of the workforce in gainful employment. Several industries such as sugar, textiles, jute, food, and milk processing depend on agricultural production for the requirement of its raw materials. This sector contributes to 14.45% of the total GHG emissions in India. The threat of climate change poses a challenge for sustainable agricultural growth and hence this sector must become resilient to increasing climatic variability and change. GoI has taken several initiatives to make the sector resilient to climate change and variability, given the importance of the sector for meeting the growing requirements of the population in the country.[115] The National Mission on Sustainable Agriculture (NMSA) under the National Action Plan on Climate Change (NAPCC), aims at transforming agriculture into

---

[113] Kavita, U., (2018). *Spice Jet Flies India's First Biofuel Flight, from Dehradun to Delhi*. The Indian Express, https://indianexpress.com/article/business/aviation/spicejet-operates-indias-first-biofuel-powered-flight-5326913/ (accessed on 07 July 2023).

[114] *Supra* note 12 at 244.

[115] Government of India, (2019). *Agricultural Statistics at a Glance 2018*. (Ministry of Agriculture & Farmers Welfare, 2019). https://agricoop.gov.in/sites/default/files/agristatglance2018.pdf (accessed on 07 July 2023).

an ecologically sustainable, climate-resilient production system by devising appropriate adaptation and mitigation strategies for ensuring food security, equitable access to food resources, enhancing livelihood opportunities and contributing to economic stability at the national level.[116]

6.10.1. Pradhan Mantri Krishi Sinchayee Yojana (PMKSY)

PMKSY another scheme to achieve convergence of investments in irrigation at the field level, expand cultivable area under assured irrigation, and improve on-farm water use efficiency to reduce wastage of water, has resulted in an emissions reduction of 11.979 $MtCO_2$ during 2017–18 and 2018–19. Crop diversification program has been implemented since 2013–14 to promote alternatives to water-intensive paddy cultivation including crops like pulses, oilseeds, 27 maize, cotton, and agroforestry to tackle the problem of depleting water table in select states.[117]

6.10.2. National Mission for Sustainable Agriculture (NMSA)

NMSA as a programmatic intervention has been made operational from the year 2014–15. It aims at making agriculture more productive, sustainable, and remunerative and climate resilient by promoting location specific integrated/composite farming systems; soil and moisture conservation measures; comprehensive soil health management; efficient water management practices and mainstreaming rain-fed technologies. During the 12th Five-year Plan, these measures have been embedded and mainstreamed onto ongoing/proposed Missions/Progammes/ Scheme of department of Agriculture and Cooperation[118]

6.10.3. Sub-Mission on Agroforestry (SMAF)

SMAF under NMSA is implemented since 2016–17 as part of the recommendation of the National Agroforestry Policy, 2014. The National Agroforestry Policy aims to encourage and increase tree plantation in association with crops and livestock to improve productivity, employment, income and livelihoods of rural households, especially the small and marginal farmers; to protect and stabilize ecosystems and promote resilient cropping

---

[116] *Supra* note 12 at 245.

[117] *Supra* note 12 at 245-246.

[118] National Mission for Sustainable Agriculture (NMSA), https://nmsa.dac.gov.in/ (accessed on 07 July 2023).

and farming systems to minimize the risk during extreme climatic conditions and to increase tree cover to promote ecological stability, especially in the vulnerable regions. Emission reduction of 0.1318 $MtCO_2$ was achieved in years 2017–18 and 2018–19.[119]

6.10.4. National Agro-Forestry Mission (NAM)

Mission for Integrated Development of Horticulture (MIDH) in which National Bamboo Mission (NBM) is being implemented as a sub scheme envisages promoting holistic growth of bamboo sector by adopting area-based, regionally differentiated strategy and to increase the area under bamboo cultivation and marketing. As forestry plants contribute significantly towards carbon sequestration, the scheme is extended for growing other timber species in farmlands under initiatives of NAM. This will not only increase the forest cover and increase carbon sequestration but also fulfill one of the aims of the National agro-forestry Policy, 2014.[120]

6.10.5. National Bamboo Mission (NBM)

The restructured NBM was launched in the year 2018–19. The NBM focuses on the development of complete value chain of the bamboo sector for linking growers with consumers starting from planting material, plantation, and creation of facilities for collection, aggregation, processing, marketing, micro, small and medium enterprises, skill development and brand building initiative in a cluster approach mode. The purposes of the NBM include increasing the area under bamboo plantation in non-forest Government and private lands to supplement farm income as well as fulfill quality raw material requirement of industries. The mission would also help in increasing the green cover and enhance the carbon sequestration potential.[121]

6.10.6. Paramparagat Krishi Vikas Yojana (PKVY)

In 2015, we launched the Paramparagat Krishi Vikas Yojana (PKVY)

---

[119] See Operational Guidelines: Sub-Mission on Agroforestry (SMAF) Under National Mission for Sustainable Agriculture (NMSA), https://nmsa.dac.gov.in/pdfdoc/Agroforestory_Guidelines_English.pdf (accessed on 07 July 2023).

[120] *National Agro-forestry Policy*, (2014). https://agricoop.gov.in/sites/default/files/National_agroforestry_policy_2014.pdf (accessed on 07 July 2023).

[121] Government of India, (2020). *Annual Report 2019-20*. "Department of Agriculture, Cooperation & Farmers' Welfare Ministry of Agriculture & Farmers' Welfare, 2020), https://agricoop.nic.in/sites/default/files/ACFW%20English%20%20Annual%20Report%202019-20.pdf (accessed on 07 July 2023).

scheme that envisages organic production without the use of chemical fertilizers and pesticides. Under PKVY organic farming is promoted through adoption of organic farming is promoted through adoption of organic village by cluster approach and Participatory Guarantee System Certification (PGS Certification). Organic fertilization compared to mineral fertilization increases soil organic carbon and thus, sequesters larger amount of CO_2 from the atmosphere in the soil resulting in reduction of greenhouse gas (GHG) emission coupled with additional benefit of increasing / sustaining soil biodiversity and other environmental services.[122]

6.10.7. Pradhan Mantri Krishi Sinchayee Yojana (PMKSY)

The scheme has been in operation from 2015–16 across the country. Efforts are made to replace flood methods of irrigation with micro-irrigation systems which reduces electricity consumption required for pumping of water. The micro-irrigation system is energy and water-efficient technology that delivers the water directly to the root zone. The adoption of water-efficient irrigation technologies helps in saving water and has the potential to improve the quality and quantity of agriculture produce. Area covered under micro-irrigation under PMKSY-PDMC between 2017–18 and 2019–20 is 3.38 million hectares.[123]

6.10.8. Avoiding Crop Residue Burning

Burning of paddy crop residues in Punjab and Haryana for the timely sowing of wheat is a prevalent practice which poses a more serious problem for local air pollution than for climate change and increase in global warming. Farm equipment that enables timely sowing in the standing paddy residues was made available to the farmers directly and also through the Custom Hiring Centers (CHCs) and Farm Machinery Banks (FMBs) to enable sowing of the wheat crop without burning of paddy residues.[124]

6.10.9. Soil Health Card (SHC)

Central Sponsored Scheme "Soil Health Card (SHC)" operational from the year 2014–15 to improve soil health, balanced/efficient use of fertilizer based on soil tests. Under the scheme SHCs will be provided to all farmers

---

[122] Paramparagat Krishi Vikas Yojana, https://pgsindia-ncof.gov.in/PKVY/index.aspx (accessed on 07 July 2023).

[123] *Supra note 12* at 245, 246.

[124] *Supra note 12* at 248, 249.

in the country in a period of three years. Based on the recommendations on SHC, chemical fertilizer will be used judiciously resulting in improvement in nutrient use efficient, improved soil structure besides reduction in GHG emission and contamination of ground water.[125]

6.10.10. National Food Security Mission (NFSM)

Additional components like location specific crop demonstration on cropping system suitable to agro-climatic zones with Improved agronomic practices including direct seeded rice (DSR), system of rice intensification (SRI), crop diversification etc., have been introduced since 2014–15 to address/ upscale climate change related initiatives.[126]

6.10.11. Mission for Integrated Development of Horticulture (MIDH)

MIDH focuses primarily on increasing both production and productivity through adoption of improved and appropriate technologies for ensuring quality, including genetic up gradation of all horticultural crops and addressing challenges of climate change. Special emphasis is being given for adoption of area-based cluster approach towards developing regionally differentiated crops, which are agro-climatically most suitable for the State/ region and having better adaptability to adverse climatic conditions. The quantum of carbon sequestered is estimated to be 108.96 $MtCO_2$ from 2017–18 to 2018–19.[127]

6.10.12. National Mission on Agriculture Extension and Technologies (NMAE&T)

To synergize extension mechanism NMAE&T has been operational from 2014–15 providing extensive capacity building to farmers, scientists and extension workers at various levels on impact of climate change on agriculture, promotion of locally appropriate adaptation strategies, climate resilient practices etc.[128]

---

[125] Soil Health Card, https://www.soilhealth.dac.gov.in/ (accessed on 07 July 2023).

[126] NFSM: National Food Security Mission, https://nfsm.gov.in/ (accessed on 07 July 2023).

[127] Mission for Integrated Development of Horticulture (MIDH), https://midh.gov.in/ (accessed on 07 July 2023).

[128] *National Mission on Agricultural Extension & Technology*, https://vikaspedia.in/agriculture/policies-and-schemes/crops-related/krishi-unnati-yojana/national-mission-on-agricultural-extension-and-technology (accessed on 07 July 2023).

6.10.13. National Agro-Forestry Policy-2014 (NAP)

NAP has been approved by us to bring about coordination, convergence and |synergy among various elements of agro-forestry scattered in various mission, programs, scheme and agencies. Through horticulture/bund plantation/ Agro-forestry, biomass is retained in the field for a relatively longer period and thus sequesters carbon both in above ground and below ground parts.[129]

6.10.14. Other Initiatives

Solarization of Agriculture, Crop diversification program, SRI, Neem-coated urea produced, balanced ration for livestock. Bypass Proteins for animals are the activities taken by government as mitigation and adaptation strategy.

6.11. FORESTRY SECTOR

Around 80% of the country's terrestrial biodiversity exists in forests, and more than 300 million people in India have a high dependency on the forest for their livelihood. Forests as a carbon sink have a very prominent role in mitigating climate change. Various legislations and acts have been formulated by the government of India for the conservation of forests and their resources. LULUCF sequestered 330.76 Mt of CO2, which is around 15% of India's total CO_2 emissions from all sectors in 2016. The forest and tree cover in India has increased from 8,02,088 km2 in 2017 to 8,07,276 km^2 in 2019. The net change in the carbon stocks of forests was +42.6 Mt or 156.2 $MtCO_2$ equivalent between the assessments of 2017 and 2019.[130] The Forest (Conservation) Act, 1980[131] has laid the foundation for protection and conservation of the country's natural forests. The Compensatory Afforestation Fund Act, 2016[132] has made compensatory afforestation mandatory, in case of diversion of forest land to non-forestry uses. Similarly, the Compensatory Afforestation Fund Management and Planning Authorities (CAMPAs) at

[129] *National Agroforestry Policy*, (2014) https://vikaspedia.in/agriculture/policies-and-schemes/crops-related/national-agroforestry-policy (accessed on 07 July 2023).

[130] Ministry of Environment, Forest and Climate Change, (2019). Total Forest and Tree Cover rises to 24.56 percent of the total geographical area of the Country; In a remarkable feat, forest and tree cover has increased by more than 13 Lakhs hectares in the last four years: Union Environment Minister. Press Information Bureau, https://pib.gov.in/PressReleaseIframePage.aspx?PRID=1597987 (accessed on 07 July 2023).

[131] The Forest (Conservation) Act, 1980 (Act 69 of 1980).

[132] The Compensatory Afforestation Fund Act, 2016 (Act 38 of 2016).

both the Centre and States ensure expeditious and transparent utilization of amounts realized from forest land which is diverted for non-forestry purposes. Progress on various fronts has been made under plantation programs such as the Twenty Point Program, National Afforestation Program, National Green Highways Mission (NGHM), and the NMGI.[133]

It is expected that the afforestation drive under NDC which was launched in July 2016, will help in sequestering approximately 0.156 million ton CO_2 annually. A total of 29 State Forest Department Agency projects are currently operating in India with a total expenditure of INR 38.20 billion to treat an area of 2.19 million ha till 2018–19 under the National Afforestation Program. INR 1.59 billion have been earmarked for the Green India Mission (GIM) which marks an increase of 48.8% over the previous allocation in 2018–19.[134]

6.11.1. National Mission for Clean Ganga (NMCG)

an action plan for 'Forestry Interventions in Ganga' has been drawn out by The Forest Research Institute (FRI), Dehradun, which envisages afforestation in 1,34,106 ha of the Ganga riverbank at an estimated cost of INR 22,937.3 million.[135]

6.11.2. Green India Mission (GIM)

National Mission for a Green India aims to use a combination of adaptation and mitigation measures for enhancing carbon sinks in sustainably managed forests and other ecosystems, adaptation of vulnerable species/ecosystems, and adaptation of forest-dependent communities.[136]

- March 21: International Day of Forests

United Nations on March 21 observed The International Day of Forests. The day commemorate the green cover across the world highlighting its importance. The International Day of Forests Day was observed under the **theme**– "Forest restoration: a path to recovery and well-being."

The Ministry of Water Resources, River Development and Ganga Rejuvenation and Ministry of Drinking Water and Sanitation were merged to form the Ministry of Jal Shakti in May 2019, further strengthening India's efforts

---

[133] *Supra* note 12 at 251–253.

[134] *Supra* note 12 at 254, 255.

[135] *Supra* note 12 at 254 255.

[136] *Green India Mission (GIM) Overview & National Perspective*, http://moef.gov.in/wp-content/uploads/2018/03/Green-India-Mission.pd (accessed on 07 July 2023).

towards resolving the mounting challenges faced by this sector.[137] The Union Defense Minister, Rajnath Singh, inaugurated 57,000 water structures under the "Jalabhishekam Water Conservation Campaign" which are constructed in Madhya Pradesh on February 11, 2021 virtually from New Delhi[138]

6.12. WASTE SECTOR

The waste sector in India contributed 2.65% to India's GHG emissions (without LULUCF) in 2016. Waste sector emissions mostly cover waste management activities such as solid waste disposal and wastewater treatment and discharge.[139] In order to manage waste efficiently, government has significantly invested in Solid Waste Management (SWM) programs in the country. The government is currently focusing on the rapid expansion and modernization of sanitation facilities and waste management infrastructure. These are indispensable for future mitigation in the waste sector. India recognizes the dual benefits that is to be arisen from efficient waste disposal leading to enhanced environmental benefits along with initiatives promoting conversion of waste to energy. As on September 2020, a total of 216 Waste-to-Energy plants with aggregate capacity of 370.45 MW eq have been set up in the India to generate power or biogas or Bio CNG from agricultural, urban, industrial and MSWs. The estimated potential for generating power from MSW is about 500 MW which would be increased to 1,075 MW by 2031. Apart from this, the Ministry of New and Renewable Energy (MNRE) is implementing a program on energy recovery from urban and industrial wastes including wastewater. MNRE has also undertaken a number of other programs on energy from urban, industrial and agricultural wastes/residues. Under Swachh Bharat Mission (SBM), Indian Railway had earlier set a target to fit bio-toilets in the entire fleet of railway coaches by the year 2021–22. This target was achieved earlier than targeted time limit. All passenger carrying BG coaches are fitted with bio-toilets and direct discharge of human waste from trains has thus been eliminated from April 2020 onwards.[140]

[137] *Govt forms 'Jal Shakti' Ministry by merging Water Resources and Drinking Water Ministries*, (2019). The Hindu. https://www.thehindubusinessline.com/news/national/govt-forms-jal-shakti-ministry-by-merging-water-resources-and-drinking-water-ministries/article27378402.ece (accessed on 07 July 2023).

[138] *Rajnath Inaugurates Over 57K water Structures in MP*, (2021). Times of India. https://timesofindia.indiatimes.com/city/bhopal/rajnath-inaugurates-over-57k-water-structures-in-mp/articleshow/80869915.cms (accessed on 07 July 2023).

[139] *Supra* note 12 at 264.

140 *Supra* note 12 at 265.

6.12.1. Waste Management Regulatory Landscape

For efficient waste management, specific laws for different types of wastes have been enacted and amended by the government from time to time to accommodate the changing environmental conditions. These are:

- Manufacture, Storage and Import of Hazardous Chemicals Rules, 1989.
- Hazardous Wastes (Management, Handling & Transboundary Movement) Rules, 2016 (This supersedes 2008 version).
- Fly Ash Notification 1999, amended in 2016.
- E-Waste (Management) Rules, 2016 (This supersedes 2015 version).
- Bio-medical Waste (Management & Handling) Rules, 2016 (This supersedes 1998 version).
- Construction and Demolition Waste Management Rules, 2016.
- Municipal Solid Waste Management Rules, 2016 (This supersedes 2000 notification).
- Plastic Waste (Management and Handling) Amendment Rules, 2018 (This supersedes 2011 and 2016 versions).

Recently, the MoEFCC has released the Draft Battery Waste Management Rules, 2020. This Draft Rules seek to replace the Batteries (Management and Handling) Rules, 2001, which provide details for handling and management of batteries under the Environment (Protection) Act, 1986.[141]

6.12.2. Atal Mission for Rejuvenation and Urban Transformation (AMRUT)

In AMRUT Mission for 500 Class-I cities approved by GoI, one of the eligible components is sewerage and sewage treatment plants which includes recycling and re-use of wastewater. This Mission focuses on a healthy and green environment for citizens and promoting use of non-motorized urban transport with the expected outcome of increased amenity values of cities by developing greenery and well-maintained open spaces which are child and elderly friendly, featured in all Mission cities. Efficient public transport or constructing facilities for NMT.[142]

---

[141] *Supra* note 12 at 264.

[142] Atal Mission for Rejuvenation and Urban Transformation, http://amrut.gov.in/content/ (accessed on 07 July 2023).

6.12.3. Swachh Bharat Mission (SBM)

One of the implementation components of this flagship program is SWM which includes setting up of wet waste processing facilities (municipal waste to compost, waste to biogas), and dry waste recycle and recovery facilities including for plastic waste, with Central support of 35% of project cost. For the first time, awareness generation through structured Information, Education and Communication (IEC) and Behavior Change Communication (BCC) activities on SWM, sanitation under Clean India Mission is being taken up at massive scale at all levels. This will lead to community participation towards effective handing of solid/liquid waste, thereby eventually leading to a clean India and low carbon emissions.[143]

6.12.4. Smart Cities

Smart Cities development is a key area to foster sustainable development solutions. France welcomes our offer to be a partner of the Smart cities development scheme of India and is ready to implement its best technological solutions and share its experience in the fields of integrated sustainable cities, in particular in urban planning, urban girds, water and sanitation, sustainable mobility and digital technologies.[144]

6.12.5. Electricity from Plastic Waste

MoEFCC has notified Plastic Waste Management Rules, 2016 for environmentally sound management of plastic waste and preventing plastic pollution in India. The Ministry has also issued "Standard Guidelines for Single-Use Plastic" on 21 January 2019 to all States and UTs and Ministries. The guidelines necessitate waste management system improvements, legal options for States/UTs to prohibit single-use plastic items through regulatory measures, eco-friendly alternatives, social awareness, and public education.

6.13. EDUCATIONAL INITIATIVE FOR MITIGATION

6.13.1. India Climate Change Knowledge Portal

MoEFCC launched the "India Climate Change Knowledge Portal." On 27 November 2020. It is a "single point Information resource" which provides information on the different climate initiatives taken by various ministries for enabling users to access updated status on these initiatives.

[143] *Supra* note 12 at 265.

[144] https://smartcities.gov.in/ (accessed on 07 July 2023).

This portal captures sector-wise adaptation and mitigation actions taken by the various line ministries in one place, including updated information on their implementation. The knowledge portal will help in disseminating knowledge among citizens about all the major steps Government is taking at both national and international levels to address climate change issues.[145]

6.13.2. Educational and Awareness Initiatives

India recognizes the role of education for environmental protection and its conservation. Numerous efforts have been made to reorient and reorganize the educational process and to raise awareness among the students on environmental issues. Around 100 million students joined the Jal Shakti Abhiyan conversation campaign that is expected to save 36,500 million liters of water each year. Students across the country planted approx. 3.5 million saplings under the "one student one tree" campaign.[146]

- The Green Schools Program led by Centre for Science and Environment (CSE) has honored 172 schools with Climate Change Awards in 2019. These schools participated in a rigorous environmental audit exercise, where students had the opportunity to reduce the usage of resources in the school premises.[147]
- Go Green Kids is an initiative by Green Yatra which objects to sensitize and educate children towards environmental protection and improvement. The initiative reached about 5 million students across the country and inspired them to reduce pollution and contribute towards sustainable development (Green Yatra, 2020).[148]
- Youths across the whole country participated in a series of events, a photography contest and a blog competition at the Youth Climate Conclave. This conclave was jointly organized by EU, GiZ, The Energy and Resources Institute (TERI), CEEW, and MoEFCC, with the aim of increasing awareness on climate change issues.[149]

---

[145] *Climate Change Knowledge Portal*, Ministry of Environment, Forest and Climate Change, Government of India. https://cckpindia.nic.in/ (accessed on 07 July 2023).

[146] *Time for Nature: Education and the Environment*, (2020). The New Indian Express.

[147] *CSE Honors 172 Schools with Climate Change Awards for 2019*, (2020). The Times of India, https://timesofindia.indiatimes.com/articleshow/73983030.cms (accessed on 07 July 2023).

[148] Go Green Kids, http://www.greenyatra.org/gogreen-kids.php (accessed on 07 July 2023).

[149] *Youth Climate Conclave*, http://www.uniindia.com/~/youth-climate-conclave-launched/States/news/1749662.html (accessed on 07 July 2023).

GoI launched another scheme to take forward the Prime Minister's vision of solarization of the sun-temple town of Modhera in Mehsana district in situated in Gujarat. The Scheme envisages setting up of the renewable energy installations which will comprise of 6 MW solar PV power plant, 15 MWh battery storage, rooftop solar PV systems, smart meters, solar electric vehicle charging stations in Modhera.

6.14. INDIA AND INTERNATIONAL SOLAR ALLIANCE

After 2015 Paris Agreement India became a much more tactically critical actor in international climate negotiations and shown not only in its increasing geopolitical significance and rising global presence but also in the efforts it has been taking to control rising temperature.

The ISAs are an UN-sponsored international alliance for countries to invest in solar energy, and cooperate in research and development (R&D) of solar power technologies is actually brainchild of India to restructure the international climate order.

India is having two of the world's largest solar plants, and due to solar energy, which is the bedrock of the country's evolving climate diplomacy. According to India's Intended Nationally Determined Contribution (INDC) goals, solar power generation has to reach 100 gigawatts (GW) by 2022 (Ministry of Environment, Forest and Climate Change, Government of India). Prime Minister Modi is on a path of diligent renewable energy assistance with countries like France, Germany, Japan, the United States (US), Canada and other countries. The ISA was announced by Modi and French President Emmanuel Macron at Paris Agreement. By this India has reached on the position of the important decision-making high table contribution in productive climate diplomacy and global governance. Developing countries like China, India and Brazil are in the race for cleaner energy. By this India has been projected as responsible global climate leader. Thus, ISA is fulfilling two purposes it is aiding India in its quest to rejuvenate its solar energy and 2nd is becoming energy independent, and also translate it into India's geopolitical leverage at the UN.[150]

But there are some domestic problems which need to be overcome. The first problem is that India imports approx. its 90% of solar cells from China

[150] Atmaja, G. B., (2019). *How Can India Become a Global Leader in Solar Power Generation?* (Vol. 54, No. 7). Economic and Political Weekly Engage. https://www.epw.in/engage/article/india-global-leader-solar-energy-g20 (accessed on 07 July 2023).

in 2017–18. Out of 252 billion India imported Rs 224 billion solar cells from China. Export of solar cell from India was only of Rs 9.37 billion. According to the Directorate General of Anti-Dumping and Allied Duties (DGAD), due to over dependency on China for chines solar panel resulted in loss of 2 lakh jobs in India. The solar manufacturers find it low-cost to use the cheaper solar panels than manufacturing it themselves at a higher price. According to the Indian government, however, this is not a feasible plan if it is made in India. It also prevents India from achieving economies of scale in its solar power production. Therefore, the Indian government has imposed 25% safeguard duty on the import of Chinese solar cells till July 2019. This duty was to be decreased to 20% till January 2020, and further to 15% until July 2020.

Second issue is the solar power tariff which is decreasing rapidly in India Rs. 2.44 per unit while the cost of the equipment and the cost of setting-up is still the same. Due to the absence of government incentives, solar stakeholders are not encouraged to generate enough.

Third, the solar power supply chain in the country is not as advanced, reliable and robust as it needs to be if India desires to achieve its goals of obtaining 175 GW of clean energy, where 100 GW is supposed to be derived from solar power. The biggest challenge for India is that India is facing financial crisis at home competing with Chinese solar panels, the government is still eager to offer a $1.4 billion line of credit to African countries, Sri Lanka and Bangladesh for their solar projects at an interest rate cheaper than that offered by China. Thus, it can be observed that India's climate diplomacy today is heavily derived from its assistance and cooperation with external relations. If, a strong domestic base is absent it can make the whole range of development unsustainable. More struggle has to be done to in upgrading the domestic vulnerabilities and expanding the domestic use of solar power.[151]

6.14.1. Scope for India

The ISA could be really very important in advancing technology transfer, finding funding sources and generating better storage technologies. In fact, to reduce dependence on importing solar panels from China, many countries are concentrating on making their own technology and schemes to boost

[151] Atmaja, G. B., (2019). How can India become a global leader in solar power generation? (Vol. 54, No. 7). Economic and Political Weekly Engage. https://www.epw.in/engage/article/india-global-leader-solar-energy-g20 (accessed on 07 July 2023).

renewable energy use. For example, Thailand has already reached 3 GW of solar power installations in 2017, amounting to 50% of their 2036 renewable energy goal. If we look on the progress of five Brazil, Russia, India, China, and South Africa (BRICS) countries, we can trace a lot of progress at different scale. If we look into the situation of Brazil; according to the Electric System Monitoring Report released by the Brazilian Ministry of Mines and Energy 88% of its total energy production is renewable energy. Brazil was also the first country to run the auction-based arrangement to obtain renewables and allocate it among various local and international stakeholders.[152]

Their technology is advanced and has reached a mature level. In fact, according to Brazil's solar association, their solar photovoltaic capacity is expected to attain 2 GW by the end of 2018. South Africa and Russia, which are known to rely heavily on fossil fuel, are also making slow progress. South Africa had increased its renewable capacity and plans to boost it to an ambitious target of 18 GW capacity by 2020 (Renewables 2018 Global Status Report). In Russia, though there has been very little investment in this field (only about an allocation of $1 billion in 2014), the Vladimir Putin government has emphasized on putting more effort in this direction. China and India both want to be the global leaders of renewable energy use, it is important for the two countries to get their policymakers as partners, rather than strategic competitors.[153]

At current time the alliance is offering India a clear chance to prove its global leadership in shaping a potentially game-changing discourse in the environmental governance regime. Partnership of India with European countries has given it the leverage to learn and harness new forms of technology, and secure its financial necessities. India's membership in multilateral forums like G20 and BRICS has allowed it to have a global influence. The developmental aid is being provided by India to the African and Pacific countries improves its soft power, and thus enhances its diplomatic outreach. Lastly, participation in all of these multilateral arenas has shaped the Indian policy-makers' intellect to engage in strategies of formation, procurement and investment. At last, it can be said that to achieve the clean energy target of 175 GW by 2022, India has to either rely on cheaper Chinese imports at the cost of local manufacturing or will have to give China partnership in the ISA and secure its energy goals for a long-term basis.[154]

152 *Ibid.*

153 *Ibid.*

154 *Ibid.*

6.14.2. Solar Energy on the Global Agenda

India's Union Cabinet approved the ratification of the amended Framework Agreement of ISA to open up the ISA membership to all countries that are members of the United Nations and for making ISA a truly global organization in August 2019.[155] The Ministry of External Affairs issued the notification on the universalization of ISA membership on 31 July 2020.[156] A first milestone was reached when the First World Solar Technology Summit was held, on 8 September 2020, with the signing of a tripartite MoU between the ISA, GoI, and the World Bank to implement the One Sun, One World, One Grid (OSOWOG) initiative.[157] The Third General Assembly of the ISA was virtually held due to COVID 19 on 14 October 2020. More than 53 member countries attended the meeting from across the world. Oman, and St. Vincent and the Grenadines have become the newest members of ISA. During the Third General Assembly, the ISA Member countries unanimously elected India as the President and France as the Co-president of the ISA Assembly for another term, i.e., up to 2022.[158]

6.14.3. India's Contribution to ISA

i. India has contributed to the ISA's efforts by overcoming the global resource asymmetry in the solar energy sector, facilitating its organizational work, and providing it with financial and other forms of support. The ISA is now the first intergovernmental organization to be headquartered in India. GoI is providing

---

[155] Ministry of New and Renewable Energy, (2019). *Cabinet Approves Ratification of the Amended Framework Agreement of the ISA for Opening up the ISA Membership to all Countries that are Members of the United Nations*. Press Information Bureau, https://pib.gov.in/PressReleasePage.aspx?PRID=1581387 (accessed on 07 July 2023).

[156] *Universalization of the Membership of the International Solar Alliance (ISA),* https://mea.gov.in/pressreleases.htm?dtl/32866/Universalization_of_the_Membership_of_the_International_Solar_Alliance_ISA (accessed on 07 July 2023).

[157] *First World Solar Technology Summit: Key Things to Know*, (2020). *The Times of India,* https://timesofindia.indiatimes.com/business/india-business/first-world-solar-technology-summit-key-things-to-know/articleshow/77972287.cms (accessed on 07 July 2023).; Ministry of New and Renewable Energy, (2020). *Four partnership agreements to be inked including one between M/o NRE, World Bank & ISA*. Press Information Bureau, https://pib.gov.in/PressReleasePage.aspx?PRID=1651926 (accessed on 07 July 2023).

[158] Ministry of New and Renewable Energy, (2020). *India and France Re-elected as President and Co-President of the International Solar Alliance (ISA) at the Third Assembly of the ISA*. Press Information Bureau, https://pib.gov.in/PressReleasePage.aspx?PRID=1664423 (accessed on 07 July 2023).

financial and material support to the ISA for the construction of its headquarters.

ii. In addition to the financial contribution for creating the ISA corpus fund, GoI is also assisting in the capacity building of ISA Member Countries under its ITEC training program besides establishing a Project Preparation Facility through EXIM Bank for developing bankable Solar Energy projects in the ISA member countries. 426

iii. In 2018, GoI had announced nearly USD 1.4 billion worth of lines of credit covering 27 solar projects in 15 countries, which are currently in various stages of implementation.[159]

The 3rd Global RE-INVEST MNRE organized the 3rd Global RE-INVEST (Renewable Energy Investment Conference and Exhibition) 2020 on a virtual platform from November 26 to 28, 2020. The theme of RE-Invest 2020 was 'Innovations for Sustainable Energy Transition.' The event was inaugurated by Prime Minister of India.[160] The UK, Denmark, Germany, France, Australia and Maldives participated as partner countries. The Hon'ble Prime Minister of India in recent times called for connecting solar energy supply across borders, with the mantra of 'One Sun One World One Grid' (OSOWOG). The idea behind the OSOWOG mantra is "The Sun Never Sets" and is a constant at some geographical location, globally, at any given point of time. MNRE, GoI, has a critical role to play in synergizing over 140 countries, across the Far-East and the Far-West regions. India, through the OSOWOG initiative, is planning to take another leap towards building a global ecosystem of interconnected renewable energy resources that are seamlessly shared for mutual benefits and global sustainability. The initiative is planned across three phases. In Phase I – Indian Grid interconnection with Middle East, South Asia and South East Asian (MESASEA) grids to share solar and other renewable energy resources; in Phase II – MESASEA grid getting interconnected with the African power pools to share solar and other renewable energy power of the countries; Phase III-Global interconnection to achieve the OSOWOG vision attracting investments in renewable energy sources.[161]

---

159 *Supra* note 12 at 425, 426.

160 *PM Modi to Inaugurate Third Global Renewable Energy Meeting and Expo on November 26,* (2020). The New Indian Express, https://www.newindianexpress.com/business/2020/nov/19/pm-modi-to-inaugurate-third-global-renewable-energy-meeting-and-expo-on-november-26-2225472.html (accessed on 07 July 2023).

161 *Supra* note *12* at 426.

6.15. EMISSION REDUCTION DURING COVID-19

Due to COVID 19 pandemic situation all over the world countries were compelled to do nationwide lockdowns and cutbacks in economic activity which resulted in a decline of about 17% in daily global emissions of April 2020 as compared with the average levels of 2019. India alone witnessed a decline of 15% in carbon dioxide (CO_2) emissions in March as lockdown was stared from 22nd March, and a projected decline of about 30% in April 2020. These outcomes, while positive, were transitory. By June 2020 when government allowed to start economic activities with Covid 19 guidelines, emissions had risen to just 5% below the average 2019 levels.[162]

A study was published in Nature Climate Change in July 2020 estimated that "daily global CO_2 emission decreased by –17% by early April 2020 compared with the mean 2019 levels." This decline was attributed to the response of Governments for the containment of COVID-19 across the world. In India, a nation-wide lockdown was declared on 25 March 2020 by prime minister, but a few economic activities, especially in the transport sector, were curtailed on 23 March 2020 in some regions of the country even before the nationwide lockdown came into effect. Another study in this regard was by Carbon Brief. It estimated that India's CO_2 emissions declined by 15% year-on-year during March and are likely to have declined by 30% in the month of April. The nation-wide lockdown continued till the end of May 2020. Throughout this difficult period, India has continued to implement its policies to facilitate the supply of renewable energy. The renewable energy purchase obligations for the distribution companies continue to remain in place despite the severe economic situation in the sector. Other policies such as the "must run" status accorded to renewable energy plants, which ensures that all the energy generated by these plants is purchased by distribution utilities, continue to be implemented during this period. This resulted in very low load factors for India's thermal plants. On an average, only 48.5% of India's total thermal capacity was utilized in the year 2020–21 (up to September 2020). The decision to utilize more renewable energy and back down thermal capacity even during the economic slowdown clearly shows India's commitment to addressing climate change.[163]

---

[162] Lauri, M., & Sunil, D., (2020). *Analysis: India's CO_2 Emissions Fall for First Time in Four Decades amid Coronavirus.* Carbon Brief. https://www.carbonbrief.org/analysis-indias-co2-emissions-fall-for-first-time-in-four-decades-amid-coronavirus (accessed on 07 July 2023).

[163] Le Quéré, C., (2020). Temporary reduction in daily global CO_2 emissions during the COVID-19 forced confinement. *Nature Climate Change,* 647. https://www.nature.com/articles/s41558-020-0797-x (accessed on 07 July 2023).

6.16. DIFFICULTIES FACED BY INDIA IN MEETING ITS PROMISE

In the year 2019 there was an unprecedented increase in the concentration of greenhouse gases globally, particularly CO2, heatwaves, and weather-related damages across the world. It is recorded that 2019 was the seventh warmest year on the record for India (since nation-wide records commenced in 1901). India rank 2nd in most affected country by floods. According to the latest report published by the United Nations Office for Disaster Risk Reduction (UNDRR), India ranks third in the highest number of disaster events recorded for the last 20 years. According to Global Climate Risk Index 2020, India was one of the topmost affected countries due to climate change.[164] The super cyclonic storm Amphan with wind speeds of 190 kmph caused widespread damage in coastal areas of West Bengal and Odisha in May 2020.[165] India is one of the worst-hit countries by the COVID-19 pandemic in the year 2020. The measures which were undertaken to control the pandemic have placed a severe strain on India's economy and resources. The COVID-19 pandemic and the consequent lockdowns have had their impact on both demand and supply-side elements of the economy.[166] This pandemic has adversely impacted the manufacturing and services sector, taken a toll on urban and rural employment and negatively impacted household income and consumption, especially of daily wage earners and those belonging to the unorganized sector.[167] India is a developing country with very limited financial resources that are being overstretched to provide relief and rebuild disaster-hit populations, livelihoods and local and regional economies. Climate-induced events and disasters are growing day by day which is putting an enormous, additional, and unjustified burden on the country's developmental challenges. However, India can only meet its promise made in Paris agreement when climate-specific finance, technology transfer and support will be provided by developed countries.

---

[164] *Global Climate Risk Index 2020.* https://germanwatch.org/en/17307 (accessed on 07 July 2023).

[165] Shiv, S. S., (2020). *Cyclone Amphan Batters West Bengal, Odisha. The Hindu*, https://www.thehindu.com/news/national/other-states/cyclone-amphan-batters-west-bengal-odisha/article31634110.ece (accessed on 07 July 2023).

[166] White paper on Make in India,' (2020). *Post COVID-19 – Provide Local Solutions for Global Challenges,* https://tifac.org.in/images/pdf/2020papers/whitepaper10jul2020.pdf (accessed on 07 July 2023).

[167] Billy, P., (2021). *Officially, India Has the World's Second-Worst COVID-19 Outbreak. Unofficially, It's Almost Certainly the Worst.* TIME, https://time.com/5954416/india-covid-second-wave/ (accessed on 07 July 2023).

India's financial needs to pursue both a low-carbon development pathway, dealing with all development deficits, while simultaneously dealing with the growing impacts of extreme events, other consequences of global warming and Covid 19 impact. Changing over low carbon economy while continuously dealing with weathering climate-induced events and disasters is a really costly affairs which requires new, additional, and climate specific financial resources. India's submission of its INDC had noted, in section 5.1 financial needs for its implementation like India would need around USD 206 billion (at 2014–15 prices) between 2015 and 2030 for implementing adaptation actions in agriculture, forestry, fisheries infrastructure, water resources and ecosystems. Mitigation requirements of India were even more onerous. As Estimated by NITI Aayog indicated that the mitigation activities for moderate low carbon development would cost around USD 834 billion till 2030 at 2011 prices.

India's climate mitigation and adaptation actions are largely financed from domestic sources, including government budgetary support as well as a mix of market mechanisms and fiscal instruments and policy interventions. The eight missions under the National Action Plan for Climate Change (NAPCC), namely solar power, energy efficiency, habitat, water, agriculture, forestry, Himalayan ecosystem, and knowledge management have specific budgetary allocations combined with other sources of funding. Climate Change Action Program (CCAP) is a GoI scheme that was approved in January 2014 at a total cost of INR 2,900 million for five years. The objective was to create and strengthen the scientific and analytical capacity to assess climate change in the country, putting in place an appropriate institutional framework for scientific and policy initiatives and implementation of climate change-related actions.[168] National Adaptation Fund on Climate Change (NAFCC) [169]is another scheme which was launched in 2015 with an initial outlay of INR 3,500 million to fund adaptation actions in the country. The projects under NAFCC are being implemented in 26 Indian States in agriculture, water, forestry and coastal sectors to enhance the adaptive capacity at the national and state level, in terms of availability of improved water and food security, livelihoods, and ecosystem services.[170]

[168] Ministry of Environment, Forest and Climate Change, (2019). *Government Releases Comprehensive publication on Various Initiatives Taken Towards Combating and Adapting to Climate Change*. Press Information Bureau, https://pib.gov.in/Pressreleaseshare.aspx?PRID=1564033 (accessed on 07 July 2023).

[169] *National Adaptation Fund on Climate Change,* https://vikaspedia.in/energy/environment/climate-change/national-adaptation-fund-for-climate-change (accessed on 07 July 2023).

[170] Ministry of Environment, Forest and Climate Change, (2018). *National Adaptation Fund for Climate Change*. Press Information Bureau. https://pib.gov.in/newsite/PrintRelease.aspx?relid=176178 (accessed on 07 July 2023).

6.17. TECHNOLOGY NEEDS AND REQUIREMENTS

Even though India is making a lot of efforts to mitigate climate change, country needs to adopt environmentally friendly technologies. However, under the current climate change regime, none of the required technologies which is required were either transferred, facilitated, or were made available to India[171]

6.18. GREEN CLIMATE FUND

Climate Action Summit which was held in New York City on 23 September, 2019. After completion of summit, the Ministry of Finance released a discussion paper on India's financial perspective with respect to climate change, and called attention to the fact that though India is doing its best to meet promises on climate action, financial problem continues to portray a significant obstacle. This is true for all developing countries of the world. This paper noted that a matter of grave concern is the "trend of developing countries being denied their right to financial resources for climate actions."[172]

The ministry also commented that the commitments made by the developed countries for enhancement and support in relation to climate finance as mandated in the UNFCCC and its Paris Agreement are not clearly translated into reality. The Summary Report of the Standing Committee on Finance of UNFCCC (2018) outlines a clear picture in this regard, where it was clearly indicated that total climate specific finance flowed from Annex II Parties in 2016, amounted to around US$ 38 billion, which was less than 40% of the US$ 100 billion per year target of climate finance. The paper specially emphasized that as of 2019, only $10.3 billion have been pledged to the GCF. Out of which only $7.23 billion has been deposited, $4.60 billion approved and $0.39 billion actually disbursed. The paper concluded that India's capacity to meet its INDC requirements, along with other developing countries, hinges on the availability of adequate climate finance.[173]

---

[171] *Supra* note *12* at 368.

[172] Ministry of Finance, (2019). "*Ministry of Finance Brought out a Discussion Paper titled "Climate Summit for Enhanced Action: A Financial Perspective from India," Examines Various Issues on Climate Finance Comprehensively.*" Press Information Bureau. https://pib.gov.in/PressReleaseIframePage.aspx?PRID=1585332 (accessed on 07 July 2023).

[173] *Climate Summit for Enhanced Action: A Financial Perspective from India*. https://dea.gov.in/sites/default/files/FINAL%2017%20SEPT%20VERSION%20Climate%20Summit%20for%20Enahnce%20Action%20A4%20size.pdf (accessed on 07 July 2023).

6.18.1. India's Accounting of Climate Financing

If we go through, the India has commitment under the Paris Agreement and the UNFCCC, we find dearth of comprehensive climate change policy document. The nebulous legislative outlook requires one to turn to India's "general environmental policy" for clarity on aspects such as transportation, environment clearance, and finance. However, even the National Environmental Policy, 2006 remains blurry on the workability of any of these aspects.

India's domestic climate policy is more complicated by the country's federal structure wherein the legislative domains of the central government and the state government are different. While "climate change" is not mentioned in either list as a distinct head of legislative competence but various related topics are mentioned such as treaty-making, certain specific industries, atomic energy and regulation of interstate waters are in the union list on which parliament can make law. State legislatures, on the other hand, can make laws relating to local government, public health and sanitation, certain roads, bridges, and inland waterways, agriculture, irrigation, canals, water storage, regulation of mines, and certain industries. Some relevant overlapping of subject matter exists under the Concurrent List, for example, in relation to subjects like electricity, forests, and protection of wild animals and birds. Climate change and other environmental concerns, can be traced to a number of areas of legislative competence given under the Seventh Schedule but it does not provide clear responsibility either on the central or the state government.

India in its international proposals to combat climate change has been doing a gross overestimation of co-benefits as compared to cost. The submitted India's INDCs in Paris Agreement, 2015 aimed at reducing the emission intensity by 30%–35% by 2030. It was also stated that the targets could be achieved with low-cost finance and technology. However, the estimation was underestimated the cost of the ambitious targets. Additionally, geographical limitation of renewable energy would incur heavy transmission costs, which would eat up investments in other parts of the Indian economy was not taken into account while doing submission, such as education, healthcare, etc. Particularly in the case of India co-benefits of using renewable sources of energy may not actually compensate for their high price.[174]

---

[174] *Climate Governance is Likely to Fail Without a Discourse on Climate Finance*, (2020). Economic and Political Weekly Engage. https://www.epw.in/engage/article/climate-governance-likely-fail-without-discourse (accessed on 07 July 2023).

It is known that a renewable plant cost more. For example, for the establishment of a solar plant requires twice as much investment per KW as comparison to coal plant.

If we compare solar plant generation with coal-based power plant then we find that 1 KW solar plant will generate 1,600 units of energy, whereas a coal-based plant might generate 6,000 to 7,000 units per year. Thus, if we want to replace a 1 KW coal plant then we will have to invest in a solar plant of around 4 KW, which would require eight times as much investment. Thus, the co-benefits should always be compared with the co-costs.[175]

6.19. CONCLUSION

India has progressively continued decoupling of economic growth from greenhouse gas emission. The emission intensity of GDP (excluding emission from the agriculture sector) has reduced by 24% between 2005 and 2016. Thereby achieved its goal to reduce the emission intensity of GDP by 20 to 25% from 2005 level earlier than the target year of 2020. India is well on its way to meet commitment under NDC to reduce its emissions intensity of GDP by 33 to 35% below 2005 levels will before 2030. The cumulative renewable power installed capacity (excluding hydro above 25 MW) has increased by 2.6 times from 35 GW on 31 March 2014 to 90.39 GW on 30 November 2020 and constitutes over 24% of the country's installed power capacity in India. The generation from renewable energy source has doubled between 2014 to 15 and 2018 to 19 and our Prime Minister has further in now the aspirational target of increasing our renewable energy capacity to 450 GW. India has an age-old tradition of living in harmony with nature. India has stabilized, protected and enhanced its forest tree cover over the years and carbon stock. In India forest has increased. India's total forest and tree cover has increased to 8,07,276 km^2 which is 24.5% of total geographical area of the country. India has launched one of the largest campaigns for cleanliness and our discouraging the uses of single use of plastic.

To further strengthen the transition to clean fuel locally the government has provided LPG connection to total of 80.33 million households in rural area, thus leaving them clean cooking fuel and a healthy domestic and external environment. More than 366 million LED bulbs have been distributed UJALA scheme. In India's efforts towards the use of cleaner automobile fuel, a substantial leap has been made from Bharat stage IV (BS IV) to Bharat stage VI (BS VI) emission norms by 1st April 2020, which was

175 *Ibid.*

earlier scheduled to be adopted only by 2024. The country's also working to considerable increase the promotion proportion of the biofuels blend in petrol diesel. India's commitment to rapidly increasing her renewable energy capacity reflects its strong commitment to limiting the rise in global temperature. Honorable prime minister believes in convenient action to combat climate change, is optimistic and committed to use technology and innovation to ensure that our future generation inherit a better world. Apart from championing and safeguarding the tenets of multilateralism, India is hosting, supporting and nurturing the ISA and Coalition for Disaster Resilient Infrastructure (CDRI) to help others. With Sweden India is leading the world on industry track to reduce emission from energy e intensive and hard to a b sector by promoting international collaboration between countries and industry group.

India follows a "dualist" system, due to which international agreements must be translated into domestic law in order to be enforceable in the country. But it can be seen that there is the absence of a time-bound obligation to enforce decisions, such as the 2015 Paris Agreement, has allowed the Indian government to dither in its promise to mitigating climate change. India's only official recognition of climate change initiative is the National Action Plan on Climate Change (NAPCC), 2008, but it lacks specifics on mitigation targets. It has been pointed out that the NAPCC is also ill-suited institutionally to effectively pursue a synergy between climate and development concerns in India. The institutional, systemic and process barriers, together with financial constraints, inter-ministerial coordination, lack of technical expertise and project clearance delays, stand as main challenges in the efficient implementation of the mitigation goal. INDC submitted by India in 2015, also noted that a substantial scaling up of the climate action plans would require greater resources and that a detailed and full-scale assessment of international climate finance needs would be finalized at a later stage. It would depend on the gap between the actual cost of implementing India's plans and what could be made available from domestic sources. India's financial needs to pursue both a low-carbon development pathway, dealing with all development deficits, while simultaneously dealing with the growing impacts of extreme events, other consequences of global warming and COVID 19 impact. Changing over low carbon economy while continuously dealing with weathering climate-induced events and disasters is a really costly affairs which requires new, additional, and climate specific financial resources. India's submission of its INDC had noted, in section 5.1 financial needs for its implementation like

India would need around USD 206 billion (at 2014–15 prices) between 2015 and 2030 for implementing adaptation actions in the field of agriculture, forestry, fisheries infrastructure, water resources and ecosystems. Mitigation requirements of India were even more onerous. An Estimate estimated by NITI Aayog, indicated that the mitigation activities for moderate low carbon development would cost around USD 834 billion till 2030 at 2011 prices. India's climate mitigation and adaptation actions are largely financed from domestic sources, including government budgetary support as well as a mix of market mechanisms and fiscal instruments and policy interventions. Climate-induced events and disasters are growing day by day which is putting an enormous, additional, and unjustified burden on the country's developmental challenges. However, India can only meet its promise made in Paris agreement when climate-specific finance, technology transfer and support will be provided by developed countries.

Chapter

7 CLIMATE CHANGE MITIGATION: PROSPECTS

Climate change is a change in the statistical distribution of weather over period of time that range from decades to millions of years. It can be defined as a change in the average weather condition or a change in the distribution of weather events. Climate change is perhaps the biggest challenge being faced by the world today, and the very existence of human being depends on how effectively this change is tackled. We all know that today the threat of climate change has become imminent. This is a fact that that climate change is the most devastating environmental problem not only for our country but for the whole world. Global temperature is rising day by day, glaciers are melting, sea levels are rising and established climate patterns are dramatically changing and also threatening many species of both flora and fauna. We have often been knocked over by dreadful reports of glaciers vanishing and sea coasts submerging in floods and the news of hurricanes unleashing casualties world-wide. The first ever report on climate change assessment has been prepared by Union Ministry of Earth Sciences (MoES) named as "Assessment of Climate Change over Indian Region." Report states that India's average temperature has already increased by around 0.7°C during the Period of 1901–2018 due to greenhouse gas emission and by the end of 2100 it is expected to rise approximately 4.4°C. The report states that rapid change in the temperature means adverse impact on India's natural ecosystem, agriculture, fresh water resources, while also causing escalating damage to infrastructure. This ultimately leads to adverse impact on countries bio-diversity, food water, energy security and public health due to change in the climate. Some other indicators of climate change are also there, that range from earlier spring arrival, coral reef bleaching, spreading diseases, heavy downpours, disproportionate snow falling, etc., to extinction of various species and plant. Now question arises whether it was man who started the climate change? Answer of this question is negative. Earth's climate has been changing continuously over its long years of history, as it metamorphosed from a hot ball of fire to the present-day world. But there is difference between natural climate change taking its own time and climate change greatly caused by the act of human. For today's climate change, it was the industrial revolution that changed the human activity altogether. As human being started moving fast, he started consuming more and pumped more greenhouse gases to the atmosphere. Apart from industrial revolution, population growth, deforestation, chemical revolution, environmental pollution, increased fossil fuel consumption, etc., are also responsible for the climate change.

The basic principle of climate change can be understood easily. As we know that several types of gas used to reflect or trap heat, and thus cause the world to warm up as if a blanket has been wrapped around it. Greenhouse gases play a vital role in causing the change in climate. These gases include carbon dioxide, methane, CFC gases, ozone, water vapors and nitrous oxide. Carbon dioxide is the main contributor to the greenhouse effect. However, to an extent, the greenhouse effect is desirable because the earth would be too cold without these gases. But the problem arises because human activity has increased the quantity of greenhouse gases. About 80% of this extra man-made carbon dioxide comes from the burning of oil, coal and gas, whilst about 20% comes from deforestation and other land changes that have prevented atmospheric carbon dioxide from being naturally absorbed. The likely physical and social consequences of climate change have made it the single biggest global environmental concern. Climate change has become one of the most controversial and politicized environmental issues. Compared with other environmental issues, it is high on the international political agenda.

We the human being are actually responsible for climate change. Climate change is not seen but already happening. Disregarding the fact of climate change does not make it less true and it is going to get extreme worse. Climate change will make food system more unstable. It will also exacerbate health problems and will result into displacement of people. Due to unexpected climate change infrastructure may be weaken and may lead to fuel conflict. The entire ecosystem will be forced to move. It will be touching every area of life. The repercussion of climate change could be worse than which we are predicting. So, there is incumbent need of mitigation and adaptation. Mitigation and adaptation both are complementary and supplementary of each other. Climate change mitigation refers to the actions which are taken to reduce or limit the magnitude of long-term climate change. Climate change mitigation generally comprises of reductions in anthropogenic emissions of greenhouse gases (GHGs) by increasing the capacity of carbon sinks, e.g., through reforestation. The option through which climate change can be mitigated is removing carbon sequestration in agriculture soil, reduction in carbon emission decarbonization of energy input carbon offset through bio-organic farming switching to renewable energy, bio electricity, individual emission cut, educating people about global warming, creating carbon sink etc. Through mitigation future effects of climate change can be reduced. So, there is need of adaption also; adaption alone is not sufficient because it is not a solution but it ensures in survival ship on climate change and it requires heavy amount of money. There is incumbent need of mitigation policy and laws and rules.

The adverse impacts of climate change on the developmental prospects of the country are amplified enormously because of the existence of widespread poverty and heavy dependence of a large proportion of the population on climate sensitive sectors for livelihood like agriculture, water, etc. Hence for India adaptation becomes inevitable and an imperative for the development process. Adaptation for India is of immediate importance and requires action right now because of vulnerabilities in India that differ among states, among regions and among different groups of people within the same region due to substantial variations in topography, climatic conditions, ecosystems as well as diversity in its social structures, economic conditions and needs of different communities. A wide range of actions have been introduced to address it. Out of the eight National Missions on Climate Change five mission focus on adaptation in sectors like agriculture, water, Himalayan ecosystems, forestry, Capacity building and Knowledge management.

If we know the solutions to the problem of climate change then what is the issues due to which this problem is not being solved. The reason is the world know the problem and solutions both but no one is ready to take responsibility to mitigate it. The world is divided in two parts developing and developed developing nations blames developed nations for climate change so they should take responsibility to mitigate it. Developed nations holds the opinion that each one is contributing to climate change so each country should be responsible for mitigation these contentions are the problem for the countries like India which is developing. The cruel twist is that India really suffers acutely from the effects of climate change while not contributing substantially to the problem. India is the fourth- largest emitter of greenhouse gases (GHGs) today. But this ranking alone is a misleading measure. India only contributes about 7% to global carbon emission. The more meaningful question is, even India were to act miraculously wave and shut down all her old and new coal plants today, would this heroic (probably suicidal) action shift the needle enough on global emissions? May be not. Parts of the developed world, who are mainly responsible for major share in cumulative emissions, have been benefitted from global warming and are unwilling to cut emission drastically. Another important uncomfortable fact is that greenhouse gases, especially carbon dioxide (CO_2), 'hang around' in the atmosphere for a long time. This means a fraction of any CO2 which has been emitted today will remain in atmosphere for a long time. The atmosphere, continuing to warm this planet for centuries to come. Even if the whole world were to drastically cut emissions today, some amount of warming would continue?

Developing countries like India and China has been developing through industrialization and this rapid growth results in emission of greenhouse gases and due to this, these countries are becoming major source of greenhouse gases at international level. The major reason for emission in developing countries are use of fossil fuels in energy production and transportation. And for the development of any country these are basic thing, without which a country cannot be developed. Development is the right of every country but it can't be at the cost of environmental degradation. The main challenge which is being faced by developing country regarding climate change mitigation is the low potential for financing projects focused on a low carbon. The other challenge is the developing country have contributed least towards global climate change so it is inequitable for them to make effort and funding into mitigation technologies in the first place.

There are many issues in international climate regime like issue of differentiation in mitigating has been one of the most controversial in the context of mitigation. The principle of common but differentiated responsibilities and respective capabilities (CBDRRC), first articulated in UNFCCC, represents a departure from the traditional approach of international agreements, namely, to define a common set of obligations for all parties it gives expression to the profound equity concerned raised by climate change challenges, by providing that the climate change commitments of party should be differentiated, Based on their different responsibilities and capability. Although there is a Universal support on the principle of CBDRRC, there is very little agreement on its rational, Core content, and application in particular situations. With respect to the rational for differentiation, developing countries have tended to focus on the term 'responsibilities,' which they understand to be the function of 'historical emissions,' where like US in particular have focused on term 'capabilities,' if the different historical contributions of countries to the climate change problem provide the basis for differentiation as developing countries content, then differentiation will change relatively slowly. In contrast, if capabilities provide the basis for differentiation, then countries obligation could evolve more rapidly, as it develops and gains greater financial technological and administrative capabilities. The principal of CBDRRC preserves the position of both sides in this debate by including both responsibilities and their respective capabilities as base for differentiation.

Another Difficulty in Mitigating Climate Change is that we the human have known that our action has resulted in unnatural warming of the earth. Even after knowing this fact no concrete action has been taken from our side and due to our inaction, it is increasing day by day. Our inaction towards mitigation has accelerated climate change. As Robert Gifford puts it, we too face 'dragons of inaction'- Mental barriers that stymie action. These are in our mind, but are no less real for that. Indeed, they are harder to fight. Because they are part of our self. These barriers are ignorance bias and doubt. Ignorance comes in two part the first part begins with not knowing whether problem exists. In 2012 survey 41% of Indian were ignorant of global warming saying that they had never heard of it. A much larger section of the population is ignorant about what causes the global warming and what to do about. To overcome this problem the communicator must be simpler. Biasness is another form of hindrance at personal level until or unless man is not affected personally and has personal damage, he will not take action immediately. Most Indian thinks that global warming affects future generations more than present one, and that they and their immediate surroundings would not be affected as rest. This result in discounting that makes them less willing to act. Understanding climate change is an exercise in probabilistic thinking- how to respond in the face of uncertain facts; but most of us are notoriously poor at this. Most of us ignore climate change until the next flood, drought, or heatwaves hits us. Another barrier is doubt in under which we think that we are contributing towards climate change mitigation but in fact we don't. We think that our policy maker and industries would be acting in good faith for mitigating climate change but in reality, they are not taking any action for this. The debate on climate change has clearly shifted in recent years, and the current focus is on mitigation and adaptation. To stop worsening and speeding-up of climate change. And to cut global GHG emissions by at least half by 2050. Prevention, mitigation, and especially adaptation is crucial and need to be done on a global basis on public and private\corporate level. Reduction measures must be implemented immediately and sweeping actions have to be prepared to profoundly transform the world's production and consumption patterns and systems.

As the time passed climate change became the most critical issue to resolve. International effort to protect environment has started in early 20th century. The first conference which made known the international nature of environment and introduce the idea of relationship between environment and development international nature of environment was Stockholm conference. The name of conference was United Nation Conference on Human and Environment. It took place on Stockholm from 5 June to 16 June 1972.

Climate change for the first time was gained importance in 1992 at united nation convention on economic development, where United Nations framework convention on climate change was signed by many countries. It came into force at 21 march 1994. The final objective was to stabilize atmospheric concentration of greenhouse gases. At a level that would prevent harm to the climate system. The parties to UNFCCC had agreed to number of commitments for dealing with climate change issues such as resubmission of periodical report by member country about steps taken for green house emission and implementation of UNFCCC, financial help to developing countries to meet their commitments, development and use of technologies that are climate friendly. But it had some drawbacks also due to which further step was needed to control greenhouse gas emission. The major drawback of UNFCCC, was its failure to set targets and timetable for stabilizing the emission of CO_2. This was the reason of adoption of Kyoto protocol to the UNFCCC at the third session of conference of parties (COP) to the UNFCC at Kyoto, Japan on 11 December 1997. This protocol made mandatory target for reducing greenhouse gas emission. The main difference between Kyoto and UNFCC was that UNFCCC only encourages industrialized countries to stabilize greenhouse gas emissions. Whereas Kyoto commits them to do so. Even Kyoto was binding but there were many problems associated with Kyoto. The first and foremost problem was that climate change is global problem and all the countries of the world should contribute in reducing emission. But in Kyoto all the members of UN were not the legally bounded by Kyoto. Some members such as Russia and USA refused to sign initially later on USA refused to ratify. Along with USA Afghanistan, and South Sudan did not ratify. Canada ratified but it withdrew from protocol in 2011. This protocol set target for developed country and developing country had no obligation to reduce emission and they were free. Financial and technological problem with poor and developing country was one of the problems. Even though Kyoto has provision of financial and technological assistance it was not being enforced by developed country. Kyoto had set the target but it had no stipulated defined sanction against non-obliging country. Many times, conferences of parties could not come to concrete decision or solution because of lack of consensus. Most times action or proposal are frustrated by a group of developing countries. Climate change is a consequence of large-scale emission in past so it needs immediate solution but Kyoto itself came late 1997 and it took 8 years to get enforced. Due to which it had been extended to 2020.

Due to the above problems, parties agreed at Durban that a universal legal agreement on climate change shall be adopted as soon as possible but not later than that. And the Paris Agreement came in 2015. Where all the countries of the world except Nicaragua and Syria participated. All the member states of UN were not member of Kyoto protocol and was not actively involved in mitigation efforts. After the failure of Kyoto Conferences of Party (COP) realized that it won't be possible for world to mitigate climate change and no one will be exempted from the adverse impact of climate change. Climate change is global problem and each and every country has responsibility to mitigate it. And for drafting such agreement where all country be accountable and participate in mitigating climate change, they learned from Kyoto protocol. In the history of international environmental law, the Paris agreement may be treated as triumph for achieving universal consensus. All the countries of the world except two countries accepted the agreement. The most important thing was that all the countries understood the urgency of mitigating climate change need of addressing it and there was no dissent from agreed objective of limiting global average temperature to 'well below 2 degree Centigrade above pre-industrial levels' and 'to pursue efforts to limit the temperature increase to 1.5 degree centigrade'; or the aim to reach a balance between emissions and removals by the second half of the century. In other words, no net increase in greenhouse gases after 2050. The agreement required all countries (not just developed nations) to put forth self-defined or nationally determined contributions (INDC or NDC) that outlined their climate actions up to 2030. This was a philosophical break from the past of differentiated responsibilities between developed and developing nations. The agreement also requires countries to prepare and submit successively more stringent INDC with supporting information (including emission inventories) to peer review process, along with a global stock-taking of progress every five years. This is key and the enshrined hope is that the peer review process, the tightening targets and the periodic stock-taking should eventually get the world to an emission regime that is consistent with a warming of less than 2°C.

The agreement also reaffirms developed countries' financial support, by requiring developed countries to jointly provide at least $100 billion a year to help developing countries adapt to climate change and reduce their emissions. The Paris Agreement, as such, is a powerful market signal for Private Parties to financially invest in renewable technologies. As it was assumed that all the country of the world will make effort to mitigate climate change and will help developing countries by providing financial assistance and technology support. When Paris was signed it was thought that for the first time in the world all the nations of the world were together. But in June 2017 USA who is the largest emitter of greenhouse gas withdrew. It has hit the other country. In December 2015, the then-president Barack Obama envisioned we would leave today's children "A world that is safer and more secure, more prosperous, and more free." when he announced that the United States, along with nearly 200 other countries, had committed to the Paris Agreement, an ambitious global action plan to combat climate change. But even less than two years later, the then-president Donald Trump put that future in jeopardy by announcing his plan from withdrawing the United States from the Paris Agreement on 1 June 2017. Trump also indicated his willingness to renegotiate the agreement or negotiate a new one. But other countries, reaffirmed their strong support for the Paris Agreement and said that they were not open to a new negotiation.

The U.S. formally initiated withdrawal from the Paris Agreement on November 4, 2019. A step in this regard that became official on November 4, 2020 as part of a larger effort to dismantle decades of U.S. environmental policy. The delay in the withdrawal was due the complex rules that were built into the Paris agreement to cope with the possibility that a future US president might decide to withdraw the country from the deal. President Obama's negotiators wanted to ensure that if there was a change in leadership then it would take time for the US to get out if. While the US is one of the biggest global greenhouse gas emitters and it is world's biggest and most powerful economy. So, when it became the only country to withdraw from a global solution to a global problem it raised questions of trust. The withdrawal from Paris Agreement became official on 4 November 2020 – by chance, the day after Mr Trump lost the presidential election. After becoming Joe Biden as president of US; The United States has officially re-joined the Paris climate accord on February 19, 2020. The US has re-joined the deal after 107 days it left. With the re-joining agreement by of US, the world leaders are expecting that now the country will prove its seriousness as it was mostly absent for four years. Sweden is the first country in the world which has passed the law to switch on green energy till 2045 in phase-out manner. France has passed the law for banning of exploration of gas and oil in country till 2040. China is taking a lot of steps in this regard but one another truth is that China being the second biggest emitter is shutting down coal plants on its soil, but building more elsewhere in the world.

Recently Initial Nationally Determined Contributions (NDC) Synthesis Report has been recently released by The United Nations Framework Convention on Climate Change (UNFCCC). While commenting on the report; the Secretary-General of the United Nations, Antonio Guterres, stated that the year 2021 is a "make or break" year so as to confront the global climate emergency. This report covered the submissions from countries up to December 31, 2020. As per the submission, 75 countries which are Parties to the Framework Convention communicated a new or updated NDC which represents 30% of global greenhouse gas emissions.

Coming to the national perspective India's domestic approach to climate change largely mirrored its multilateral stance. Domestic policies per se did not exist and climate change was address only in a piecemeal manner, mainly through energy and forestry policies in which climate was never the central focus. The main thrust of these policies for social and economic development. Climate change was mentioned, if at all, only as a "co benefit" of policy. As a non-annexe one party to the UNFCCC and the Kyoto protocol, India had no legal obligation to reduce it GHG emissions. Instead, India has long argued that the historical responsibilities to deal with climate change lies with the developed countries as they were the creators of the problem and that they must come first to take on a leadership role in addressing climate change. The overriding priority of India (and other developing countries) is to address development and poverty eradicates and any action to reduce emissions should be voluntary. This principle, commonly known as the principle of *Common but Differentiated Responsibilities*, is one that India has long been an advocate of. The firewall created the principle between developed and developing countries shielding India from undertaking any meaningful action to reduce emissions.

The 21st century saw the arrival of India on the world stage as a global power. With the Indian economy growing at a phenomenal rate of 8% a year, India was no longer viewed solely as a "developing country." In fact, during the 2007 financial crisis both India and China were considered to be critical partners in the G20 talks. This acknowledgment of India's status as a major emerging economy brought with it more responsibility for it. To be recognized as a legitimate global player the world needed to see India as a country willing to shoulder responsibilities. This expectation was reflected within the multilateral climate negotiations where India began to feel international pressure to be more proactive and take action to reduce its own GHG emissions. The shift in India's stance on taking action domestically and launching the National Action Plan on Climate Change (NAPCC) can be attributed to a large extent to this growing international pressure. Facing the fear of isolation and continued international pressure, prior to the much-hyped Copenhagen Summit in 2009, India further announced that it would voluntarily reduce its emissions intensity between 20 and 25% below 2005 levels by 2020.

While international diplomatic pressure played a critical role that led to India as more proactive stance, one cannot displace the importance that domestic priorities played in this context as well. A realization by the Indian government that it is extremely vulnerable to climate change impacts meant that it was in India's own interest to take action both domestically and internationally. With its accelerating energy demands and India importing more than 70% of its oil, it made sense for India to privatize energy security and focus on Energy efficiency and renewable. This is reflected in specific missions of the NAPCC. Energy access, another domestic concern, could put India's development at risk. The Indian government's focus on developing renewable energy helps it to achieve this goal.

Another driven worth noting is the immense opportunity that the clean Energy technology sector provides for the 21st century. Which the race for fossil fuels becoming increasingly competitive many countries is looking at building their own domestic capacities to shore up energy supplies. Potential growth in this sector and the economic opportunities that it opens up in the global marketplace has driven great interest in this sector.

If we see legal framework relating to climate change in India then we find that the constitution of India under fundamental rights provides a regime of protection for privileged set of rights including right to life and liberty, the right to enjoyment of pollution free water and air. However, no climate change related principles have been evolved by the apex courts. But the Indian courts Cherished out the environmental right by integrating certain *Principles Like Polluter Pays Principles, Precautionary Principle, Intergenerational Equity Principles, Principle of Sustainable Development* and notion of the state as a Trustee of earth and its natural resources. Hence, the issue of Climate Change could well be related to the duty of a state as a Trustee to protect the atmosphere from greenhouse gases emission and to follow these principles in making the law and policies for mitigating the climate change. Besides, the citizen of this India has been assigned fundamental duty to protect environment wildlife and natural resources which could also help in mitigating climate change. Climate change and a deteriorating environment is a key challenge to sustainability, bio-diversity, environment and stability across the world. So, it is not exaggeration to say that environment, climate change and sustainable development are related with each other and discussion will be meaningless, if any one of them kept outside. There is a twofold relationship between sustainable development and climate change. On the one hand, climate change influences nature and human living conditions and thereby also the basis for social and economic development, while on the other hand, society's priorities on sustainable development influence both the GHG emissions due to which climate change caused and the vulnerability. The world summit on Sustainable development added a new dimension (social development) in concept of sustainable development. No doubt, development is the today's need but not at the cost of human survival. Development does not mean development of things. Its real meaning is the holistic development of human being. Holistic development is not possible without considering the issue of protection of environment and climate change. Climate change posing threat to the continuance of biodiversity on this earth. It is a herculean task to mitigate and adapt to the climate change. Efforts to cope with the impacts of climate change and attempts to promote sustainable development share common goals and determinants including access to resources, sharing of information and technology, fair and equitable distribution of resources and its benefits, access to risk-sharing mechanisms and abilities of decision-support mechanisms to cope with uncertainty.

In addition, India has wide array of laws, rules, regulations, plans and policies with administrative authorities to address the environmental pollution and energy conservation, but there are no specific laws on the climate change yet. Then, all such norms and principles could well utilize to address the reduction of greenhouse gases emission and other mitigation measures. Like EPA, 1986 empowers the central government to take all necessary measures for protecting and improving environment and preventing controlling and abetting environmental pollution. The Air (Prevention and Control of Pollution) Act, 1981 which defines 'air pollutant' any solid, liquid or gaseous substance including noise present in the atmosphere in such concentration as may be or tend to be injurious to human being or other living creatures or plants for environment. Such definition could also arguably cover for greenhouse gases for regulation and mitigation of climate change. The Forest Conservation Act, 1980 restrict the conservation of forest land to non-forest use, which could be useful for reforestation. The Wildlife Protection Act, 1972 specifically provides the provision for protection and conservation of wildlife, birds and plants. The Biological Diversity Act, 2002 and its rule addresses the issues concerning conservation, sustainable use and the access to genetic resources and associated traditional knowledge and equitable sharing of benefits arising out of the use of these resources and associated knowledge which support the purpose of mitigation of climate change. The National Green Tribunal Act, 2010 has also formed and empowered is specified Tribunal to hear the cases on environment degradation and climate change as well. The National Environment Policy 2006 provides the basis for integration of environmental considerations in the policies of various sector. The policy statement for Abatement of Pollution, 1992 stresses on "polluter pay" principle while The Forest Policy, 1988, highlights the environmental protection through preservation and restoration of the ecological balance. The policy seeks to augment the forest cover in India. The courts have also elaborated on the concepts relating to sustainable development, and the 'polluter pays' and 'precautionary' principles. In India, matters of public interest, particularly pertaining environment, are to be articulated effectively through a vigilant media, an active NGO community, and through the judicial process which has recognized the citizen s right to a clean environment as a component of the right to live and liberty.

India has also record of plans and policies addressing the mitigation mechanism for climate change. The National Climate Change Council issued India's comprehensive National Action Plan in June 2008 called as National Action Plan on Climate Change. There have been 8 missions announced under this action plan to combat the climate change. National Solar Mission seek to deploy 20,000 megawatt of solar electricity capacity in the country by 2020. National Mission for Enhanced Energy and Efficiency creates new institutional mechanism to enable the development and strengthening of Energy Efficiency markets. National Mission on Sustainable Habitat promotes the introduction of sustainable transport, Energy efficient building and sustainable waste management in cities. National Water Mission promotes the integrated management of water resources. NMGI seek to afforestation in additional 10 million hectare of forest land, water land and community lands. National Mission on Sustainable Agriculture focuses on enhancing productivity and resistance of agriculture so as to reduce vulnerability to extremes of weather, long day spells, flooding and variable moisture availability. National Mission on Strategic Knowledge for Climate Change identify challenges arising from climate change promotes the development and diffusion of knowledge on responses to the challenges in the area of health, demography, migration and livelihood of coastal communities. Thus, these different missions, in addition with 'Seven Policy Thrust Areas' together encompasses action for greenhouse gases mitigation as well as adaptation as well.

Government initiatives for the diffusion of renewable energy and energy efficient technologies, Joint Forest management, water resource management, and environmental education in educational institutions represent a broad spectrum of efforts to integrate climate change concerns in sustainable development.

Historically, India, in keeping with its non-aligned stance, has emphasized equity and development over emission cuts, and rightly called out developed nations for their large role in historical emissions. This stance gave shape to the common-but-differentiated responsibilities that were a key principle underlying climate change negotiations until 2009. India then began to dilute its stance on differentiated responsibilities. It surprised many experts by agreeing to the 1.5°C stretch target in the Paris Agreement, even though this concession could make its future development tricky. India's non-insistence that developed countries first fulfil their pre 2020 commitments before ratifying the Paris Agreement was also perplexing. Experts believe this change in India's stance has to do with its desire to be part of the rule-writing process and its closer ties to the US. Coming back to international climate negotiations, the Paris Agreement was believed to be successful, or was it? Our best understanding today is that even after adding them up, the INDCs do not keep the Earth below the 2°C target. Indeed, experts give only a 5% chance to a warming of less than 2°C, with a median predicted temperature rise of 3.2°C from pre-industrial times." What is more certain is, as scientists have consistently warned, that the climate might throw up unpleasant Surprises if we continue emit as usual.

The problem of Climate Change has been precipitated primary because of excessive consumption pattern and profligate lifestyle of developed nation of the world. Recognizing the importance of lifestyle issue in climate change India's first NDC goal promotes a healthy and sustainable way of living based on our tradition and value of conservation and moderation India's NDCs under Paris Agreement mainly rely on the traditional practices in reducing the level of carbon footprints and sustainable lifestyle. Unlike developed country India will not have to focus on individual emission cut as we know that The Indian culture and life are very famous in world-view in which all its aspects exist according to the human experience are recognized. In India people have a general rule, to the need-based consumption and fixed responsibility of not wasting the resources and that grows respect in their life. The quantification of goals in India's NDCs under Paris Agreement is threefold: first, reducing the emission intensity of its gross domestic product (GDP) by 3 3%–35% (vis-à-vis 2005) by 2030; second, achieving 40% cumulative electric power installed capacity from non-fossil fuel-based energy resources by 2030; and third, creating an additional carbon sink of 2.5–3 billion tons of CO_2 equivalent by 2030 through additional forest and tree cover. Even though there would have not been obligation to mitigate but it was incumbent need to mitigate due to India's venerability to climate change.

It is estimated that 70% of people live rural areas which are near to nature and not based on high consumption. Many of skills are practiced from one generation to other in master-to-discipline. People are dependent on high energy based, craftsmanship, organic farming that do not let depend more on electricity. Examples Khadi (famous fabric), handlooms, etc., are popular across Indian in gender or generation. Recognizing, the bless of nature and eco system people maintained their devotional worship places surrounded with ponds, gardens, bathing Ghats, etc. People especially in villages and some in towns and cities prefer bathing with mug and water rather than shower which waste lots of water. Some prefer cold water bath for whole year. With traditional structures such as bore wells, step wells for collecting water in dry areas helps in conservation of natural resources. Using packages to store vegetables, grains, storing groceries in kitchen, old clothes are being passed to next generation or to those can get fit, buying luxury goods in second hand (has a huge market). In India the houses are traditionally constructed with bamboo, stone and clay that not only help climatically but also helps in GHG mitigation and reducing cement usage and transportation. Forests are important things of Indian culture. These are responsible for flora and fauna diversity and are maintained/protected in well manner. Planting trees is considered as the good thing ('punya') and celebrate "van Mahotsav" every year. Bishnoi of western Thar Desert, is a group who checks on eco-friendly community, for them harming environment is harming themselves. In Rajasthan the 'khejri tree' is considered as the water sustaining tree that is not cut even if it interrupts the building of house. Live example for this is Salasar temple in Sikar district. India is really lucky to have person like Jadav "Molai" Payeng, an environmental activist and forestry worker from Majuli. He is also popularly known as the Forest Man of India for contributing his major portion of life in planting trees on a sandbar of the river Brahmaputra and turning it into a forest reserve. The forest is called Molai forest and is located near Kokilamukh of Jorhat, Assam, and India and encompasses an area of about 1,360 acres / 550 hectares. In 2015, he was honored with Padma Shri, the fourth highest civilian award in India.

The problem of climate change poses an existential challenge for humanity. For India this challenge is compounded by immediate concern of eradicating poverty and accelerating development even though it had a relatively very limited role in causing the problem. However, India is committed to address the challenge of climate change. India has proactively contributed to multilateral efforts to combat climate change and continues to do so while and taking her own independent enhanced initiative in climate mitigation and adaptation besides meeting on its commitment under the UNFCCC, Kyoto protocol and Paris agreement.

India has progressively continued decoupling of economic growth from greenhouse gas emission. The emission intensity of GDP (excluding emission from the agriculture sector) has reduced by 24% between 2005 and 2016. Thereby achieved its goal to reduce the emission intensity of GDP by 20 to 25% from 2005 level earlier than the target year of 2020. India is well on its way to meet commitment under NDC to reduce its emissions intensity of GDP by 33 to 35% below 2005 levels will before 2030. The cumulative renewable power installed capacity (excluding hydro above 25 MW) has increased by 2.6 times from 35 GW on 31 March 2014 to 100 GW on 12 August 2021 and constitutes over 24% of the country's installed power capacity in India. The generation from renewable energy source has doubled between 2014 to 15 and 2018 to 19 and our Prime Minister has further in now the aspirational target of increasing our renewable energy capacity to 450 GW. India has an age-old tradition of living in harmony with nature. India has stabilized, protected and enhanced its forest tree cover over the years and carbon stock. In India forest has increased. India's total forest and tree cover has increased to 8,07,276 km^2 which is 24.5% of total geographical area of the country. India has launched one of the largest campaigns for cleanliness and our discouraging the uses of single use of plastic.

To further strengthen the transition to clean fuel locally the government has provided LPG connection to total of 80.33 million households in rural area, thus leaving them clean cooking fuel and a healthy domestic and external environment. More than 366 million LED bulbs have been distributed under UJALA scheme. In India's efforts towards the use of cleaner automobile fuel, a substantial leap has been made from Bharat stage IV (BS IV) to Bharat stage VI (BS VI) emission norms by 1st April 2020, which was earlier scheduled to be adopted only by 2024. The country is also working to considerable increase the promotion proportion of the biofuels blend in petrol diesel. India's commitment to rapidly increasing her renewable energy capacity reflects its strong commitment to limiting the rise in global temperature. Honourable Prime Minister believes in convenient action to combat climate change, is optimistic and committed to use technology and innovation to ensure that our future generation inherit a better world. Apart from championing and safeguarding the tenets of multilateralism, India is hosting, supporting and nurturing the International Solar Alliance (ISA) and Coalition for Disaster Resilient Infrastructure (CDRI) to help others.

Due to Covid 19 pandemic situation all over the world, countries were compelled to do nationwide lockdowns and cutbacks in economic activity which resulted in a decline of about 17% in daily global emissions of April 2020 as compared with the average levels of 2019. India alone witnessed a decline of 15% in carbon dioxide (CO_2) emissions in March as lockdown was stared from 22nd march, and a projected decline of about 30% in April 2020. These outcomes, while positive, were transitory. By June 2020 when government allowed to start economic activities with COVID 19 guidelines, emissions had risen to just 5% below the average 2019 levels. Throughout this difficult period, India has continued to implement its policies to facilitate the supply of renewable energy. The renewable energy purchase obligations for the distribution companies continue to remain in place despite the severe economic situation in the sector. Other policies such as the "must run" status accorded to renewable energy plants, which ensures that all the energy generated by these plants is purchased by distribution utilities, continue to be implemented during this period. This resulted in very low load factors for India's thermal plants. On an average, only 48.5% of India's total thermal capacity was utilized in the year 2020–21 (up to September 2020). The decision to utilize more renewable energy and back down thermal capacity even during the economic slowdown clearly shows India's commitment to addressing climate change.

The authors see many difficulties faced by India in keeping its promise. According to Global Climate Risk Index 2020, India was one of the topmost affected countries due to climate change. India is one of the worst-hit countries by the COVID-19 pandemic in the year 2020. The measures which were undertaken to control the pandemic have placed a severe strain on India's economy and resources. The COVID-19 pandemic and the consequent lockdowns have had their impact on both demand and supply-side elements of the economy. This pandemic has adversely impacted the manufacturing and services sector, taken a toll on urban and rural employment and negatively impacted household income and consumption, especially of daily wage earners and those belonging to the unorganized sector. India is a developing country with very limited financial resources that are being overstretched to provide relief and rebuild disaster-hit populations, livelihoods and local and regional economies. Climate-induced events and disasters are growing day by day which is putting an enormous, additional, and unjustified burden on the country's developmental challenges. However, India can only meet its promise made in Paris agreement when climate-specific finance, technology transfer and support will be provided by developed countries.

The growth-starved world looking to developing nations to be the new engines of growth but climate change severely diminishes the likelihood of this happening. This is an important point worth emphasizing, as money that could have been used for development in hot countries like India, must at least partially be earmarked for dealing with the after-effects of droughts, floods and heat waves.

India's financial needs to pursue both a low-carbon development pathway, dealing with all development deficits, while simultaneously dealing with the growing impacts of extreme events, other consequences of global warming and COVID 19 impact. Changing over low carbon economy while continuously dealing with weathering climate-induced events and disasters is a really costly affairs which requires new, additional, and climate specific financial resources. India's submission of its INDC had noted, in section 5.1 financial needs for its implementation like India would need around USD

206 billion (at 2014–15 prices) between 2015 and 2030 for implementing adaptation actions in agriculture, forestry, fisheries infrastructure, water resources and ecosystems. Mitigation requirements of India were even more onerous. As Estimated by NITI Aayog indicated that the mitigation activities for moderate low carbon development would cost around USD 834 billion till 2030 at 2011 prices. India's climate mitigation and adaptation actions are largely financed from domestic sources, including government budgetary support as well as a mix of market mechanisms and fiscal instruments and policy interventions.

India's domestic climate policy is more complicated by the country's federal structure wherein the legislative domains of the central government and the state government are distinct. In India international agreements must be translated into domestic law in order to be enforceable in the country. But it can be seen that there is the absence of a time-bound obligation to enforce decisions, such as the 2015 Paris Agreement, has allowed the Indian government to dither in its promise to mitigating climate change. While "climate change" is not mentioned in either list as a distinct head of legislative competence but various related topics are mentioned such as treaty-making, certain specific industries, atomic energy and regulation of interstate waters are in the union list on which parliament can make law. State legislatures, on the other hand, can make laws relating to local government, public health and sanitation, certain roads, bridges, and inland waterways, agriculture, irrigation, canals, water storage, regulation of mines, and certain industries. Some relevant overlapping of subject matter exists under the Concurrent List, for example, in relation to subjects like electricity, forests, and protection of wild animals and birds. Climate change and other environmental concerns, can be traced to a number of areas of legislative competence given under the Seventh Schedule but it does not provide clear responsibility either on the central or the state government. India's only official recognition of climate change is the National Action Plan on Climate Change (NAPCC), 2008, but it lacks specifics on mitigation targets. It has been pointed out that the NAPCC is also ill-suited institutionally to effectively pursue a synergy between climate and development concerns in India.

India in its international proposals to combat climate change has been doing a gross overestimation of co-benefits as compared to cost. As part of the Paris Agreement, 2015, India submitted its INDCs to the UNFCCC. The submitted INDCs aimed at reducing the India's emission intensity by 30%–35% by 2030. It was stated that the targets could be achieved with low-cost finance and technology. However, the estimation was underestimated the cost of the ambitious targets. Additionally, geographical limitation of renewable energy would incur transmission costs, which would eat up investments in other parts of the Indian economy was not taken into account, such as education, healthcare, etc. Particularly in the case of India co-benefits of using renewable sources of energy may not actually compensate for their high price. As we know that a renewable plant cost more.

United Nations (UN) has conferred upon Narendra Modi a "Champion of the Earth" tag for leading the International Solar Alliance (ISA), an organization set up to facilitate a shift to solar energy, there is no clarity on in what way the government is planning on funding the $21 million which is required for the alliance, or on how the aim of producing 100 gigawatts of solar energy will be met by 2021.

We know that Climate is changing since the evolution of the Earth, but recently it was observed that it has taken up speed which has become threat to mankind and very existence of the planet. Rapid population growth and there energy demand is the major reason for GHGs emission. Environmental concerns are associated with all forms of energy sources including fossil fuels, nuclear energy and even renewable energy throughout the energy chain from exploration/mining, transportation and generation/manufacture to the end-use/ decommissioning. International law does not provide for right to "climate refugees" to enter into any other country. Climate change also poses high risk to female population.

Climate change is inevitable and Mitigation is only remedy available against this anthropogenic problem simultaneously there is urgent need to focus on adaptation for ensuring survival ship. The authors find that mitigation as a solution but the problem is that who will take the responsibility of mitigation. The best method suggested by IPCC recent report published in 2021 is shutting down old thermal power plant.

The problems associated with tackling climate change are essentially global in both their causes and effects, respecting no national boundaries. As such, they require a significant element of global cooperation if they are to be tackled appropriately. But The Paris Agreement was world's greatest diplomatic successes and universally accepted agreement. The Paris Agreement sets an ambitious direction for the climate change, and complements this direction with a set of common core obligations for all countries, including legally binding obligations of conduct in relation nationally determined mitigation contributions, and an expectation of progression over time. The obligation which has been imposed on members under Paris agreement is of obligation of conduct not of result and mitigating efforts are based on national circumstances and are self-binding obligation. The Paris Agreement operationalizes the CBDRRC principle not through the FCCC and Kyoto Protocol's annex structure, but in a tailored way and operationalizes the CBDRRC principle through self-differentiation. The recent report of IPCC envisage that even though the world restrict to the temperature to 1.5 degree centigrade glaring effects of the climate change would be witnessed.

The environmental law existed even before Stockholm declaration on human and environment indirectly in IPC and law of Tort. After 1972 direct legislation came for environment protection is Water Act, Air Act Environment Protection Act etc. NGT was established because it was realized that Supreme Court was feeling many problem technically. They were doing with constituting technical committee. Highly technical issues are involved in environmental matter. Due to overcoming this problem national green tribunal act was brought. Till now there has been no comprehensive legislation to address climate change mitigation or adaptation. However, efforts have been made by introducing Climate Change Bill, 2015, 'to fill this gap, but it failed to pass. There is also as yet no litigation in which climate concerns have been at issue except one case, i.e., *Ridhima Pandey v. Union of India & others* but unfortunately it was dismissed by NGT. The Supreme Court, High Courts and NGT have acknowledged and even endorsed the relevance of climate concerns in the context of environment development trade-offs and decision-making, however, a climate-centric rights based or other claim is even yet too brought to the portals of the Indian Courts. For India climate change challenge is compounded by immediate concern of eradicating poverty and accelerating development even though it had a relatively very limited role in causing the problem. However, India is committed to address the challenge of climate change. India has proactively contributed to multilateral efforts to combat climate change and continues to do so while and taking her own independent enhanced initiative in climate mitigation and adaptation besides meeting on its commitment under the UNFCCC, Kyoto protocol and Paris agreement. India has progressively continued decoupling of economic growth from greenhouse gas emission. However, India can only meet its promise made in Paris agreement when climate-specific finance, technology transfer and support will be provided by developed countries. Climate change is opulent; assuming and conceding that climate change mitigation is only remedy available against this anthropogenic problem, believing that UNFCCC and Paris Convention,

2015 have made a vigorous attempt to have a direct and concerted action of the World community to mitigate climate change, understanding and believing that India being a concerned state and a champion of developing nations have initiated actions with regard to mitigation of climate change and sustainable environment without losing sights of its development goal and energy needs; the authors in this work have examined and investigated the legal issues of climate change mitigation with special reference to Indian perspective to suggest for a better and holistic climate change mitigation without compromising development. Climate change mitigation is only remedy available against this anthropogenic problem simultaneously there is urgent need to focus on adaptation for ensuring survival ship. For testing UNFCCC and Paris Convention, 2015 have made a vigorous attempt to have a direct and concerted action of the World community to mitigate climate change. International effort to mitigate climate change culminated in Paris agreement which is really vigorous attempt to mitigate climate change at international level, the Paris agreement may be treated as victory for achieving universal consensus for the first time. The most important thing is that all the countries of the world understood the urgency of mitigating climate change need of addressing it. For testing that India being a concerned state and a champion of developing nations have initiated actions with regard to mitigation of climate change and sustainable environment without losing sights of its development goal and energy needs.

India being developing country and non-annex 1 country (under UNFCCC) had no responsibility to mitigate under protocol but India did a lot of actions in this regard like launch of National Action Plan on climate change and state action plan on climate change (SAPCC). India being developing country compounded by immediate concern of eradicating poverty and accelerating development even though it had a relatively very limited role in causing the problem. Even though it is doing a lot for mitigating climate change in each and every sector starting from promotion of clean and renewable energy, enhancement of forest area, making Indian railway green, enhancing coal based power plant efficiency, establishment of world biggest solar plant in Reva, using solar energy for metro projects, generating electricity from waste, shutting down the old and low efficient thermal power plants, distributing the LPG and LED lights free of cost. Even during the difficult period of COVID 19 pandemic, India has continued to implement its policies to facilitate the supply of renewable energy. The renewable energy purchase obligations for the distribution companies continue to remain in place despite the severe economic situation in the sector. According to third biennial report submitted by India to UNFCCC, India has achieved its goal to reduce the emission intensity of GDP by 20 to 25% from 2005 level earlier than the target year of 2020 and is in the line of meeting its commitment under Paris agreement.

After this elaborate and critical discussion, following submissions may be made relating to climate change mitigation;

i. For mitigating climate change everyone should use energy-efficient appliances, including LED light and energy-efficient fans. Maintaining air conditioner temperatures at 24–25°C and cleaning the filters regularly will help in lower electricity consumption. By painting roof white and using solar powered DC micro grids where it is appropriate. Compact Fluorescent Lamps (CFL) should be used instead of traditional electric bulbs as it consumes very less power, it does not radiate heat. Government should take some initiatives in the form of providing subsidy, incentive on the purchase of energy-efficient appliances. Wearing Indian style of dress like dhotis etc., and discarding western dresses in offices (not in factories for safety reasons) cuts down on energy bills.

ii. People should walk for going to near places. If there are problems with pavements (or lack thereof) waste and dogs, they should talk to Municipal Corporation regarding the issue. Carpooling must be tried for going to work place, Pooling of cars by business executives in companies will help in saving petrol. Everyone should try to take public transport for longer distances rather than personal connivances. Everyone should Check and maintain optimum tire pressure and service your vehicle regularly and maintain the appropriate gear for a given speed while driving. We should also develop habit of riding cycle at least once in a week or in a month. By doing this, the concentration of greenhouse gas emission will be decreased. Blending of bio-fuels and ethanol with petrol and diesel to be encouraged to cut down oil use.

iii. The population in India is increasing day by day, unless this population growth is controlled; government would not be able to supply enough energy to its citizen. Population can be controlled by making the people aware about the disadvantages of population explosion. As it is not easy in a complicated society like India, so government should take some hard steps in controlling the population in the form of providing some incentives to those who have only one or two children.

iv. Afforestation should be promoted and deforestation should be avoided because plants and trees act as carbon sink and help on decreasing the concentration of carbon dioxide in the atmosphere.

v. Every person should measure their water consumption. Installation of meters should be at different points of use, it will help in finding leaks. Everyone must use bucket to bath and turn the tap off when they brush. Lowering water pressure, treating the sewage water and reusing gray water to flush, or in gardens will help in saving water. We should rely upon the rain water harvesting system. All the new building should be constructed with an open tank on the top most roof to collect the rain water which can be used for various useful purposes.

vi. Human being should eat less meat; eat fewer dairy products and should avoid food waste. Storing provisions at eye-level in transparent containers and do not stack items behind one another, and should be conscious of food waste. Everyone should measure, monitor, reduce wastage and Segregate their waste. Wet waste should be managed locally.

vii. Herbal beauty cosmetics should be used by everyone instead of other traditional cosmetic items which are made of harmful chemicals. These herbal organic cosmetics are eco-friendly and do not affect the climate.

viii. For meeting energy requirement the dependency upon renewable energy such as solar energy, wind energy, etc., should be enhanced. Renewable energy is one of the strongest devices to combat climate change. Government should also concentrate on better research and technological advancement regarding using the solar energy. There is also need of providing subsidy to the public at large so that the people could purchase the solar plate along with other devices. This is best alternate of electricity which will help in combating the climate change.

ix. Everyone should rely upon the new ecological principle “think globally, act locally, consume less and share better” in place of only “think globally but act locally.”

x. As we know manufacturing process of papers consumes excessive power; so, habit of not wasting the papers would help in adapting and mitigating the climate change. Everyone should use papers as minimum as possible. We should develop habit of doing online transactions for depositing telephone bills, electricity bills, insurance premium, etc. As it decreases power consumption as well as complication of works. The habit of taking print out of every material should be avoided and if there is need of hard copy, print should be taken on both sides of paper. Recycled paper should be used as maximum as possible because it consumes 60% less power as the normal paper in the manufacturing process. Typing on both sides of paper in offices will also help in saving paper use and consequently cuts down energy consumption.

xi. Bio-degradable and eco-friendly carrying bags, pouches, packets, toys, baskets, etc., should be used in place of plastic items. Everyone should develop habit of carrying carry bag when they are going outside. The dependency on plastics and fibers has become an essential part of life right from computer chips to the aero planes, so it is not possible for us to remove the plastics completely from our requirements. The solution of this problem that research should be done on manufacturing the bio-degradable plastics.

xii. For implementing any action towards the mitigation of climate change, first of all people should be made aware through the campaigns, seminars, advertisements in newspapers, broadcasting from the radio transistors about the different aspects of climate change and need of taking actions for mitigating the same. The help of booklets, pamphlets, posters, WhatsApp, Facebook etc., should be taken for communicating the message of such initiatives. Seminars, symposiums, panel discussions and workshops on climate change mitigation must be organized for spreading awareness in society. Environment as a subject is thought at school level. Student should also be thought also for energy conservation since school times because children are more sensitive towards anything.

xiii. Building should be constructed with modern eco-friendly technology. It is said that modern buildings for which cement is used for construction, using 1 ton of cement releases 1 ton of CO_2 in atmosphere. Traditional building practices like with solar passive direction homes, thermal insulation with mud, jail type windows for natural ventilation will help in less use of energy requirements.

xiv. Marriage ceremonies, Birth day parties, Get-together parties with the friends and juniors, Fare well parties, engagement ceremony, etc. Should be organized on day time instead of night time. By doing this Power consumption could be less by following this habit.

xv. Traditional fuels such as coals, charcoals, woods, bio-fuels (dried dung), etc., should be avoided as it causes air pollution which ultimately lead to global warming. Liquefied petroleum gas (LPG) should be used for cooking purposes for domestic as well as commercial use, it is also good for human health.

xvi. It is necessary that people should change the mindset towards the protecting and preserving the environment as all these things cannot be left on the government. People at large at the local level and non-governmental organizations (NGOs) should come forward actively in implementation of National action Plan and other governmental programs for combating climate change.

xvii. India should enact law for renewable energy and other energy law should be amended with concern of environment protection. All projects dealing with Renewable energy requires heavy investment and for sustainable energy, government should provide financial incentives for renewable energy projects. Indian government should provide subsidy on solar panels. Metro projects should use renewable energy. The recent example is Nagpur metro which is using solar energy and along with this they are planting on pillars to create carbon sink to absorb emissions. Solar energy should be used for non-productive purposes such as water heating in place of using energy produce from conventional source. Scientific and technological research should be encouraged for renewable energy production.

xviii. Indian power plants are mostly coal-based. Super-Critical-pulverized fuel fired boiler technologies should to be introduced in the power plant, which will enhance their efficiency.

xix. Government should open up urban waste management to the privet sector. Mandate bulk generators of waste like apartment, complexes, factories, and hospitals, academic institutions to go Zero Waste in house. Government should also give high priorities to Schemes like 'Swachh Survekshan' that emphasizes process, progress and foster healthy competition between cities. While there is already a regulation that requires manufacturers to be responsible for their packaging waste, including post-consumer waste, there needs to be an emphasis on uniform enforcement of this rule.

xx. There should be Institution of a congestion tax for central business districts and receipts should be used for funding public transport, and/or develop walking and cycling infrastructure. Parking fees should also be used tactically and strategically to reflect the preference for public and no motorized transport. Government should adopt a universal carbon tax. India already has this in several pockets including excise duty on petrol and diesel, and cess on coal. The tax-net and method, either implicitly or explicitly, needs to cover other forms of carbon emissions.

xxi. Drip irrigation should be encouraged, it will reduce water consumption. As we know agriculturists get free electricity, they tend to use inefficient pump sets to siphon out ground water. Nobody can compel them to use efficient pumps; so at least if water consumption is cut down by drip irrigation, electricity can be saved.

xxii. CDM projects must be guided by meaningful public participation and local benefit-sharing policies for ensuring that they promote sustainable development as well as climate change mitigation. Clear guidelines are needed so that the many competing uses of land areas supporting biomass projects are considered, including the livelihoods of indigenous and marginalized communities, different ethnic groups, and women. In addition, in order to establish a critical knowledge base for sound and profitable bioenergy management. There is needs of collaborative partnership among researchers, governments and industries in developing as well as industrialized countries.

xxiii. New ideas, ever new concepts have arisen recently in the environment and energy arena. The most modern integrated Steel Plant at Visakhapatnam recently developed the concept of SHE – Safety, Health and Environment. Meticulous planning was done to break the evil nexus between environmental degradation and industrial growth with emphasis on waste heat recovery, waste minimization/recycling and water conservation. This concept, if popularized in industries, will go a long way in improving the energy efficiency of Indian industries without sacrificing environmental and safety concerns.

xxiv. Expansion of nuclear energy should only proceed if the problems of disposal of nuclear wastes are addressed. Some relevant provision In Indian Law: Nuclear power flatters to deceive "The Atomic Energy Regulatory Board (AERB) is envisaged with the regulatory and safety functions in Sections 16,17 and 23 of Atomic Energy Act 1962. "Even though there are the Atomic Energy (Disposal of Radio Active waste) Rules 1987, there is no comprehensive SWM plan on the part of AERB"

xxv. Video Conferencing should be carried out for doing official work instead of traveling to and fro by cars, trains and planes. This will reduce the consumption of fuels. Due to COVID 19 the culture of work from home has been develop it can be continued for a work which can be done from home.

xxvi. The National Action Plan should be implemented in synchronized manner by every Ministries of the Government of India. All the private and public organizations should cooperate in implementing the NAPCC. *India's domestic climate policy is outdated and there is a pressing need to revisit the National Action Plan on Climate Change, 2008, and to reformulate domestic climate policy in India.* Government should take help of academicians, public spirited and expert individuals in the concerned fields and NGOs in making the policies and plans concerning the climate change. Simultaneously, the Parliament, state legislatures, and courts must carefully re-examine the existing policy framework through the lens of climate change, and should revise the framework, where it is necessary. This is a prospect for future research and deliberation in this field. For example, the Parliament can make laws on atomic energy (List I), the state legislatures can make laws on agriculture (List II), and both the Parliament and state legislatures can make laws on forests (List III). Each of these areas given in seventh schedule is related to climate change and to the environment. As we know that 'climate change' and 'environment' are not expressly mentioned in the Seventh Schedule, it is unclear whether the Parliament or the state legislatures have the overall authority and responsibility for making laws relating to climate change and to environmental concerns.

BIBLIOGRAPHY

International Documents:

1. Agenda 21
2. Bali Action Plan2007
3. Basel Convention on The Control of Trans-boundary Movements of Hazardous Wastes and Their Disposal, 1989
4. Conference of the Parties 22
5. Conference of the Parties 23
6. Conference of the Parties 24
7. Conference of the Parties 25
8. Convention Concerning Protection of Word Culture and Natural Heritage, 1972
9. Convention on Biological Diversity, 1992
10. Convention on International Trade in Endangered Species of Wild Fauna and Flora, 1973
11. Convention on the Conservation of Antarctic Marine Living Resources, 1980
12. Convention on the Conservation of Migratory Species of Wild Animals, 1983
13. Convention on Wetlands of International Importance Especially as Waterfowl Habitat, 1971

14. Earth Summit, 1992
15. International Convention for the Prevention of Pollution from Ships, 1973
16. Kyoto Protocol, 1997
17. Montreal Protocol on the Substances That Deplete Ozone Layers, 1987
18. Paris Agreement under the United Nations Framework Convention on Climate Change, 2015
19. Rio+20
20. The Antarctic Treaty, 1959
21. The Convention on Climate Change, 1992
22. The United Nations Convention on the Law of the Sea, 1982.
23. United Nation Conference on Human and Environment, 1972
24. United Nations Conference on Environment and Development, 1992
25. United Nations' World Summit on Sustainable Development, 2002
26. Vienna Convention for the Protection of the Ozone Layer, 1985

Commission and Reports:

1. Emissions Gap Report 2020
2. India First Biennial Update Report (BUR) to the United Nations Framework Convention on Climate Change, 2015
3. India Second BUR to the United Nations Framework Convention on Climate Change, 2018
4. India Third BUR to the United Nations Framework Convention on Climate Change, 2021
5. India's Intended Nationally Determined Contribution: Working Towards Climate Justice, 2015
6. IPCC Fifth Assessment Report, 2013
7. IPCC First Assessment Report, 1990
8. IPCC Fourth Assessment Report, 2007
9. IPCC Second Assessment Report, 1995
10. IPCC Sixth Assessment Report, 2021

11. IPCC Third Assessment Report, 2001
12. Report of the UN Conference on the Human Environment,1972
13. Report on Assessment of Climate Change over Indian Region, 2020
14. United Nation Environment Program Reports
15. World Commission on Environment and Development, 1987

Indian Documents:

1. Energy Conservation Rules, 2012
2. Hazarduous Wastes (Management and Handling) Rules, 1989.
3. Indian Forest Act 1927
4. The Air (Prevention and Control of Pollution) Act, 1981.
5. The Atomic Energy Act, 1962.
6. The Bengal Smoke Nuisance Act, 1905
7. The Biological Diversity Act 2002.
8. The Bombay Smoke Nuisance Act, 1912
9. the Cattle Trespass Act, 1871
10. The Climate Change Bill, 2015
11. The Compensatory Afforestation Fund Act, 2016
12. The Constitution of India
13. The Electricity Act, 2003
14. The Employees State Insurance Act, 1948.
15. The Energy Conservation Act, 2001
16. The Environment (Protection) Act, 1986.
17. The Forest (Conservation) Act, 1980.
18. The Indian Easement Act, 1882
19. The Indian Forest Act, 1927.
20. The Major Port Trusts Act, 1963.
21. The Motor Vehicles Act, 1988
22. The Mysore Destructive Insects & Pests Act, 1917.
23. The National Environment Appellate Authority Act, 1997.
24. The National Environment Tribunal Act, 1995.
25. The National Green Tribunal Act, 2010.

26. The National Rural Employment Guarantee Act (NAREGA), 2005
27. The Oriental Gas Company Act, 1857
28. The Public Liability Insurance Act, 1991.
29. The Shore Nuisance (Bombay and Colaba) Act, 1853.
30. The Water (Prevention & Control of Pollution) Act, 1974.
31. The Water (Prevention and Control of Pollution) Amendment Act, 1978.
32. The Water (Prevention and Control of Pollution) Cess Act, 1977.
33. The Wildlife (Protection) Act, 1972.
34. The Wildlife (Protection) Rules, 1995.
35. Wild Life (Protection) Amendment Act, 2002.

Indian Policy Documents:

1. Integrated Energy Policy, 2006
2. National Action Plan for Climate Change, 2008
3. National Agroforestry Policy 2014
4. National environment policy 2006
5. National Forest Policy, 1952
6. Policy statement for Abatement of Pollution, 1992
7. The Forest Policy 1988

BOOKS:

1. Ackerman, F., & Stanton, E. A., (2014). *Climate Change and Global Equity.* New York: Anthem Press.
2. Aldy, J. E., & Stavins, R. N., (2010). *Post-Kyoto International Climate Policy: Implementing Architectures for Agreement.* New York: Cambridge University Press Research from Harvard Project on International Climate Agreements.
3. Anbumozhi, V., Breiling, M., et al., (2012). *Climate Change in Asia and the Pacific How Can Countries Adapt.* New Delhi: Sage Publications India Pvt Ltd.
4. Anderson, S., et al., (2002). *Leaders, Pushers and Laggards in the Working of the Climate Regime.* Columbia: Global Environmental Change.

5. Bodansky, D., Brunnee, J., et al., (2017). *International Climate Change Law.* London: Oxford University Press.

6. Boyle, G., (2012). *Renewable Energy Power for Sustainable Future.* London: Oxford University Press.

7. Bradbrook, A. J., et al., (2005). *Law of Energy for Sustainable Development.* London: Cambridge University Press.

8. Chatterjee, B., (2011). *Climate Change, Trade and Natural Disasters.* New Delhi: Deep & Deep Publication Pvt Ltd.

9. Churchil, R., Freestone, D., et al., 1991). *International Law and Global Climate Change.* London: Graham & Trotman.

10. Clarissefrass-Ehrfeld, (2009). Renewable *Energy Sources A Chance to Combat Climate Change.* The Netherlands: Wolters Kluwer Law & Business, Kluwer Law International.

11. Cullet, P., (2003). *Differential Treatment in International Environmental Law*. New York: Ashgate Publishing Company.

12. Dash, S. K., (2007). *Climate Change an Indian Perspective.* Ahmedabad: Centre for Environment Education.

13. Dash, S. K., (2007). *Climate Change an Indian Perspective.* Ahmedabad: Centre for Environment Education.

14. Dhall, Lt. Col. Vivek, (2006). *India's Energy Security.* New Delhi: Vij Book Pvt Ltd.

15. Dhawan, T. K., (2010). *Global Warming & Climate Change.* New Delhi: Mohini Publishers & Distributors.

16. Dubash, N. K., &Kale, S. S., et al., (2018). *Mapping Power: The Political Economy of Electricity in India's States* New Delhi: Oxford University Press.

17. Dubash, N. K., (2011). *Handbook of Climate Change and India: Development, Politics and Governance*. New York: Routledge.

18. Dubash, N. K., (2019). *India in a Warming World: Integrating Climate Change and Development.* New Delhi: Oxford University Press.

19. Faria, J. A., & Ruiz, M. P., (2016). *Climate Change Mitigation: Greenhouse Gas Reduction and Biochemical.* New York: CRC Press Taylor &Francis.

20. Filho, W. L., et al., (2013). *Climate-Smart Technologies: Integrating Renewable Energy and Energy Efficiency in*

Mitigation and Adaptation Responses. Verlag Berlin Heidelberg: Springer.

21. Gardiner, S. M., et al., (2010). *Ethics And Climate Change: An Introduction.* New Jersey: John Wiley & Sons, Ltd.
22. Ghosh, S., (2019). *Indian Environmental Law: Key Concepts and Principles*. Hyderabad: Orient Blackswan.
23. Gillespie, A., (2000). *International Environmental Law, Policy, and Ethics.* London: Oxford University Press.
24. Griffths, J., Rao, M., et al., (2011). *The Health Practitioner's Guide to Climate Change; Diagnosis and Cure.* London: Earthscan.
25. Gupta, J., (1997). *Climate Change Convention and Developing Countries: From Conflict to Consensus.* The Netherlands: Kluwer Academic Publishers.
26. Hollo, E. J., et al., (2013). *Climate Change and The Law*. Dordrecht: Springer.
27. Hulme, M., (2009). *Why We Disagree About Climate Change: Understanding Controversy, Inaction and Opportunity.* New York: Cambridge University Press.
28. Jaswal, P. S., & Jaswal, N., (1999). *Environmental Law Environment Protection, Sustainable Development and the Law.* Allahabad, India: Allahabad Law Agency.
29. Jaswal, P. S., (1996). *Directive Principles Jurisprudence and Socio Economic Justice in India.* New Delhi: APH Publishing Corporation.
30. Jodoin, S., & Marie-Claire, C. S., (2013). *Sustainable Development, International Criminal Justice, and Treaty Implementation.* New York: Cambridge University Press.
31. Kaur, G., (2006). *Environmental Crime.* New Delhi: Shree Publishers & Distributors.
32. Keyhani, A., Marwali, M. N., et al., (2010). *Integration of Green and Renewable Energy in Electric Power System.* New Jersey: Wiley.
33. Kohli, K., & Menon, M., (2016). *Business Interests and The Environmental Crisis.* New York: Sage.
34. Kunich, J. C., (2010). *Betting the Earth How We Can Still Win*

the Biggest Gamble of All Time. Marion: Parkhurst Brothers Publishers Inc.

35. Leelakrishnan, P., (2016). *Environmental Law in India.* New Delhi: LexisNexis.
36. Leichenko, R. M., & Obrien, K. L., (2008). *Environmental Change and Globalization: Double Exposures.* London: Oxford University Press.
37. Lever-Trracy, C., (2011). *Confronting Climate Change.* New York: Routledge.
38. Lord, R., Goldberg, S., et al., (2012). *Climate Change Liability: Transnational Law and Practice.* New York: Cambridge University Press.
39. May, J. R., & Daly, E., (2014). *Global Environmental Constitutionalism.* London: Cambridge University Press.
40. Mccaffrey, S. C., & Salcido, R. E., (2009). *Global Issues in Environmental Law.* New York: American Case Book Series.
41. Mittal, V., (2012). *Crop Stress Management and Global Climate Change.* New Delhi: SBS Publishers and Distributors Pvt. Ltd.
42. Modi, N., (2015). *Convenient Action Continuity for Change*. New Delhi: LexisNexis.
43. Naseem, M., (2021). *International Energy Law*. The Netherlands: Wolters Kluwer.
44. Paola, M. D., & Pellegrino, G., 2014). *Canned Heat: Ethics and Politics of Global Climate Change.* New York: Routledge Taylor & Francis Group.
45. Peak, S., & Smith, J., (2009). *Climate Change from Science to Sustainability.* London: Oxford University Press.
46. Pitmaurice, M., Ong, D., & Merkouris, P., (2010). *Research Book on International Environmental Law.* United Kingdom: Edward Elgar Publishers.
47. Rajamani, L., (2006). *Differential Treatment under International Environmental Law.* London: Oxford University Press.
48. Rajimwale, A., (2006). *Environment Versus Development: A Dialectical Dilemma of Earth' Existence.* New Delhi: Sunrise Publication.
49. Ramaswamy, S., & Kumar, G. S., (2009). *Energy Environment*

and Sustainable Development Issues and Policies. New Delhi: Regal Publication.

50. Ramesh, M., (2018). *The Climate Solution: India's Climate-Change Crisis and What We Can Do About It.* New Delhi: Hachette India.
51. Salpekar, A., & Sharma, K., (2010). *Green House Effect and Global Climate Change Management.* Meerut India: Raj Publication.
52. Sands, P., (2003). *Principles of International Environmental Law.* London: Cambridge University Press.
53. Sarkar, A. N., (2010). *Global Climate Change: Beyond Copenhagen.* New Delhi: Pentagon Earth.
54. Segger, Marie-Claire, C., & Khalfan, A., (2004). *Sustainable Development Law Principle Practices and Prospects*. London: Oxford University Press.
55. Sen, S., (2012). *Public Interest Litigation in India: Implications for Law and Development.* Kolkata: Mahanirban Calcutta Research Group.
56. Shanthakumar's, S., (2009). *Introduction to Environmental Law.* Nagpur: Wadhwa and Company.
57. Sherman, D., & Smith, J. W., (2007). *The Climate Change Challenge and the Failure of Democracy.* Westport: Praeger Publishers.
58. Shue, H., (2014). *Climate Justice Vulnerability and Protection.* London: Oxford University Press.
59. Singh, G., & Bahri, A., (2016). *Environmental Law*. Lucknow, India: Eastern Book Company.
60. Singh, S. K., (2010). *Environment Law and Climate Change.* New Delhi: SBS Publishers & Distributors Pvt. Ltd.
61. Sinha, A., & Mitra, S., (2006). *Economic Development, Climate Change, and the Environment.* UK: Routledge Taylor & Francis Group.
62. Soltau, F., (2009). *Fairness in International Climate Change Law and Policy*. New York: Cambridge University Press.
63. Sovacool, B. K., & Dworkin, M. H., (2014). *Global Energy Justice Problem, Principles, and Practices.* UK: Cambridge

University Press.

64. Sprinz, D., & Luterbacher, U., (2018). *International Relations and Global Climate Change: New Perspectives.* Massachusetts, MIT Press.

65. Sprinz, D. F., et al., (2018). Global *Climate Policy: Actors, Concepts, and Enduring Challenges*. Massachusetts: MIT Press.

66. Srinivasan, A., (2008). *Climate Regime Beyond 2012: Reconciling Asian Development Principle and Global Climate Interest.* Hayama, Japan: Institute for Global Environmental Strategies.

67. Stone, C. D., (2010). *Should Trees Have Standing? Law, Morality, and the Environment*. London: Oxford University Press.

68. Syed, M. H., (2009). *Preventing Efforts for Global Warming.* New Delhi: Surabhi Publications.

69. Tandon, U., (2018). *Energy Law and Policy*. New Delhi: Oxford University Press.

70. Tandon, U., (2007). *Environment and Climate Change Policy and Law.* Hyderabad: The ICFAI University Press.

71. Tondon, U., (2016). *Climate Change Law Policy and Governance*. Lucknow, India: Eastern Book Company.

72. Warren, A., & French, J. R., (2001). *Habitat Conservation: Managing the Physical Environment.* New Jersey: Wiley.

73. Weiss, E. B., (1989). *In Fairness to Future Generations: International Law, Common Patrimony, and Intergenerational Equity.* United Nations University Press.

74. Zahar, A., (2008). *International Climate Change Law and State Compliance.* New York: Routledge.

75. Zedillo, E., (2008). *Global Warming Looking Beyond Kyoto.* Washington D.C.: Brookings Institution Press.

Articles:

1. Aggarwal, P. K., (2003). Impact of climate change on Indian agriculture. *Journal of Plant Biology, 30*, 189–198.

2. Al-Busoul, M., (2017). Design of fruits solar energy dryer under climatic condition in Jordan. *Journal of Power and Energy Engineering, 5*(2), 123–137.

3. Atmaja, G. B., (2019). How can India become a global leader in

solar power generation? *Economic and Political Weekly Engage, 54*(7).

4. Balasudarsun, N. L., & Panda, S., (2020). Review on emergence of electric mobility and feasibility of electric mobility adoption in India. *Aegaeum Journal, 8*(9), 1407–1414.
5. Balat, M., (2005). Usage of energy sources and environmental problems. *Energy Exploration & Exploitation, 23*(2), 141–168.
6. Baruah, A. G., (2019). How can India become a global leader in solar power generation? *Economic & Political Weekly Engage, 54*(7).
7. Bauer, N., & Brecha, R. J., (2012). Economics of nuclear power and climate change mitigation policies. *Proceedings of the National Academy of Sciences of the United States of America, 109*(42), 16805–16810.
8. Bhattacharya, S., Sharma, C., et al., (2006). Climate change and malaria in India. *Current Science, 90*(3), 369–375.
9. Bodansky, D., (2016). The legal character of the Paris agreement. *Review of European, Comparative and International Law, 25*(2), 142–150.
10. Bodansky, D., (1993). The United Nations framework convention on climate change: A commentary. *Yale Journal of International Law, 18*, 453–554.
11. Chakravarty, B. K., (2006). Environmentalism: Indian Constitution and Judiciary. *Journal of Indian Law Institute*, *48*, 100.
12. Chaturvedi, R. K., Tiwari, R., & Ravindranath, N. H., (2008). Climate change and forests in India. *The International Forestry Review, 10*(2), 256–268.
13. Coghlan, M., (2002). Prospects and pitfalls of Kyoto protocol to the United Nations framework convention on climate change. *Melbourne Journal of International Law, 3*(1) 165.
14. Dankelman, I., (2002). Climate change: Learning from gender analysis and women's experiences of organizing for sustainable development. *Gender and Development*, *10*(2), 21–29.
15. Das, P., & Kushwaha, H. S., (2011). CBM: A clean energy option for India to address climate change issue. *Asia-Pacific Journal of Social Science, SI*(2), 11–16.

16. Davies, M., Guenther, B., et al., (2008). Adaptive social protection: Synergies for poverty reduction. *Institute of Development Studies, 39*(4), 105–111.

17. Dechezlepretre, A., et al., (2007). *The Clean Development Mechanism and the International Diffusion of Technologies: An Empirical Study* (Vol. 105). Nota Di Lavoro.

18. Dhara, V. R., Schramm, L., Paul, J., & George, (2013). Climate Change and infectious diseases in India: Implications for health care providers. *Indian J. Med. Res., 138*, 847–852.

19. Dutta, S., (2021). *Ladakh to Get India's First Geothermal Power Plant.* Times of India.

20. Ghosh, P. K., (2013). Judicial activism and public interest litigation in India. *Galgotias Journal of Legal Studies, 1*, 77–96.

21. Ghosh, S., (2020). Litigating climate claims in India. *American Journal of International Law,* 114, 45–49.

22. Gupta, N., Mishra, A., & Agrawal, N. K., (2018). Interstate cooperation for climate change adaptation in Indian Himalayan Region. *Economic & Political Weekly, 53*(12), 35–38.

23. Harish, S., Singh, N., & Tongia, R., (2020). Impact of temperature on electricity demand: Evidence from Delhi and Indian States. *Energy Policy, 140*, 111445. https://doi.org/10.1016/j.enpol.2020.111445.

24. Hiller, A., (2011). Climate change and individual responsibility. *The Monist*, *94*(3), 349–368.

25. Holtsmark, B., et al., (1998). Emission trading under Kyoto protocol. *Centre For International Climate and Environmental Research, 11*.

26. Jakob, M., Steckel, J. C., et al., (2020). The future of coal in a carbon-constrained climate. *Nature Climate Change, 10*, 704–707.

27. Jayaraman, T., & Kanitkar, T., (2016). The Paris agreement. *economic & Political Weekly, 51*(3), 10–13.

28. Jotzo, F., (2005). Developing countries and future of the Kyoto protocol. *Global Change, Peace & Security, 17*(1), 77–86.

29. Khosla, R., Agarwal, A., et al., (2021). The what, why, and how of changing cooling energy consumption in India's urban households. *Environmental Research Letters, 16*, 1–12.

30. Klein, R. J. T., & Nicholls, R. J., (2001). Technological options for adaptation to climate change in coastal zones. *Journal of Coastal Research, 17*(3), 531–543.

31. Kooten, G. C. V., Shaikh, S. L., et al., (2002). Mitigating climate change by planting trees: The transaction costs trap. *Land Economics, 78*(4), 559–572.

32. Kumar, K. S. K., Shyamsundar, P., et al., (2010). Economics of climate change adaptation in India. *Economic and Political Weekly, 45*(18), 25–29.

33. Kumar, K. S. K., (2007). Climate change studies in Indian agriculture. *Economic and Political Weekly, 42*(45, 46), 13–14.

34. Kumar, S. K., (2007). Climate change studies in Indian agriculture. *Economic and Political Weekly, 42*(45, 46), 13–18.

35. Lakshmanan, A. R., Pushpa, K., et al., (2017). Paris agreement on climate change and India. *Journal of Climate Change, 39*(1), 1–10.

36. Martin, S., (2010). Climate change, migration, and governance. *Global Governance, 16*(3), 397–414.

37. McInerney-Lankford, S., Darrow, M., et al., (2011). *Human Rights and Climate Change: A Review of the International Legal Dimensions* (pp. 8–10). World Bank Studies the World Bank.

38. Mehdi, A., (2010). Climate change and biodiversity: India's perspective and legal framework. *Journal of the Indian Law Institute, 52*(3, 4), 343–365.

39. Mishra, V. S., (2010). National green tribunal: Alternative environment dispute resolution mechanism. *Journal of the Indian Law Institute, 52*(3, 4), 522–552.

40. Myllyvirta, L., & Dahiya, S., (2020). *Analysis: India's CO_2 Emissions Fall for First Time in Four Decades amid Coronavirus.* Carbon Brief.

41. Batjes, N. H., (1998). Mitigation of atmospheric CO_2 concentrations by increased carbon sequestration in the soil. *Biology and Fertility of Soils, 27*, 230–235.

42. Nandi, J., (2021). *New Norms for Thermal Plants May Dent India's Emission Targets: Experts.* Hindustan Times.

43. Nayak, A., (2020). The promise of environmental action for

economic growth. *Economic and Political Weekly Engage, 55*(1).

44. Nibleus, K., & Lundin, R., (2010). Climate change and mitigation. *AMBIO, Supplement 1. Special Report: Energy 2050, 39*, 11–17.
45. Ninan, K. N., (2019). Climate change and rural poverty levels in India. *Economic And Political Weekly, 54*(2), 36–43.
46. Nomani, Md. Z. M., (2000). Enviro-constitutional ethos in right duty discourse: Towards the creation of an equitable and sustainable socio-legal order. *International Journal of English and Literature, 1*, 61.
47. Perrigo, B., (2021). *Officially, India Has the World's Second-Worst COVID-19 Outbreak. Unofficially, It's Almost Certainly the Worst*. TIME.
48. Petroni, S., (2009). Policy review: Thoughts on addressing population and climate change in a just and ethical manner. *Population and Environment, 30*(6), 275–289.
49. Pillai, A. V., & Dubash, N. K., (2021). The limits of opportunism: The uneven emergence of climate institutions in India. *Environmental Politics*.
50. Quéré, C. L., (2020). Temporary reduction in daily global CO_2 emissions during the COVID-19 forced confinement. *Nature Climate Change*, 647.
51. Rajamani, L., (2013). Rights based climate litigation in the Indian courts: Potential, prospects & potential problems. *Centre For Policy Research Climate Initiative (Working Paper 2013/1 May).*
52. Rajamani, L., (2012). The Durban platform for enhanced action and the future of the climate regime. *International and Comparative Law Quarterly, 61*(2), 501–518.
53. Rajamani, L., (2014). The Warsaw climate negotiations: Emerging understandings and battle lines on the road to the 2015 climate agreement. *International and Comparative Law Quarterly*, *63*(3), 721–740.
54. Rajamani, L., (2016). Ambition and differentiation in the 2015 Paris agreement: Interpretative possibilities and underlying politics. *International and Comparative Law Quarterly, 65*(2), 493–514.
55. Randers, J., Goluke, U., (2020). An earth system model shows self-sustained thawing of permafrost even if all man-made GHG

emissions stop in 2020. *Scientific Reports, 10*.

56. Rawat, V. R. S., & Kishwan, J., (2008). Forest conservation-based, climate change-mitigation approach for India. *The International Forestry Review, 10*(2), 269–280.
57. Sahu, A., Yadav, N., et al., (2016). Floating photovoltaic power plant: A review. *Renewable and Sustainable Energy Reviews, 66*, 815–824.
58. Sanwal, M., (2016). Paris agreement on climate change challenges of urban transition remain. *Economic & Political Weekly, 2*(13), 21, 22.
59. Sharma, A., (2010). *Right Based Legal Guarantee as a Social Protection Framework: The Mahatma Gandhi National Rural Employment Guarantee Act* (p. 5). Asian Development Bank.
60. Singh, P. K., & Dhiman, R. C., (2012). Climate change and human health: Indian Context. *Journal of Vector Borne Diseases, 49*(2), 55.
61. Singh, S. S., (2020). *Cyclone Amphan Batters West Bengal, Odisha.* The Hindu.
62. Somanathan, E., & Somanathan, R., (2009). Climate Change: Challenges Facing India's Poor. *Economic and Political Weekly, 44*(31), 51–58.
63. Suma, M., & Balaram, P. S., (2014). Perspective study on coastal disaster management at Andhra Pradesh. *IOSR Journal of Business and Management, 16*(12), 55–60.
64. Szulczewski, M. L., & MacMinn, C. W., (2012). Lifetime of carbon capture and storage as a climate-change mitigation technology. *Proceedings of the National Academy of Sciences of the United States of America, 109*(14), 5185–5189.
65. Upadhyay, K., (2018). *Spice Jet flies India's First Biofuel Flight, from Dehradun to Delhi*. The Indian Express.
66. Wei, T., Yang, S., et al., (2012). Developed and developing world responsibilities for historical climate change and CO_2 mitigation. *Proceedings of the National Academy of Sciences of the United States of America, 109*(32), 12911–12915.
67. Wesemael, B. V., & Lambin, E. F., (2001). Carbon sinks and conserving biodiversity. *Science, New Series, 294*(5549), 2094–2095.

Periodicals:

1. *Indian Railways Hopes to Become World's First 'Green Railway' By 2030; To Be More Planet-Friendly,* (2020). Times of India.
2. Climate change and gendered vulnerabilities: Accounting for women and patriarchal systems in climate governance policy. *Economic and Political Weekly Engage,* (2020).
3. Climate governance is likely to fail without a discourse on climate finance. *Economic and Political Weekly Engage,* (2020).
4. *PM Narendra Modi Inaugurates Asia's Largest Solar Power Project in MP's Rewa,* (2020). Economic Times.
5. Sustainable development in India a value based system – strategies necessary for India's economic self-sufficiency, (2009). *Journal of Institutions of Valuers,* 308–315.

INDEX